Student Solution

to accompany

Intermediate Algebra with P.O.W.E.R. Learning

Sherri Messersmith
College of DuPage

Lawrence Perez
Saddleback College

Robert S. Feldman
University of Massachusetts—Amherst

Prepared by
diacriTech, Inc.

McGraw Hill Education

STUDENT SOLUTIONS MANUAL TO ACCOMPANY
INTERMEDIATE ALGEBRA WITH P.O.W.E.R. LEARNING

1 2 3 4 5 6 7 8 9 0 DOH/DOH 1 0 9 8 7 6 5 4 3

ISBN: 978-0-07-748384-5
MHID: 0-07-748384-7

www.mhhe.com

Contents

Chapter 4   Linear Equations in Two Variables and Functions

Chapter 5   Solving Systems of Linear Equations

Chapter 6   Polynomials and Polynomial Functions

Chapter 11 Exponential and Logarithmic Functions

Chapter 12 Nonlinear Functions, Conic Sections, and Nonlinear Systems

Chapter 1: Real Numbers and Algebraic Expressions

Section 1.1

1) Given the set of numbers
$$\left\{-14,6,\frac{2}{5},\sqrt{19},0,3.\overline{28},-1\frac{3}{7},0.95\right\}$$

   a) Whole: $6,0$

   b) Integers: $-14,6,0$

   c) Irrational: $\sqrt{19}$

   d) Natural: 6

   e) Rational: $-14,6,\frac{2}{5},0,3.\overline{28},-1\frac{3}{7},0.95$

   f) Real: $-14,6,\frac{2}{5},\sqrt{19},0,3.\overline{28},-1\frac{3}{7},0.95$

3) True

5) False

7) True

9)

11)

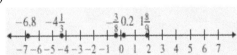

13)

15) 13

17) $\dfrac{3}{2}$

19) $-|10|=-(10)=-10$

21) $-|-19|=-(19)=-19$

23) -11

25) 7

27) 4.2

29) $-10,-2,0,\dfrac{9}{10},3.8,7$

31) $-9,-4\dfrac{1}{2},-0.3,\dfrac{1}{4},\dfrac{5}{8},1$

33) True

35) True

37) False

39) False

41) True

43) -27

45) -0.5%

47) -4371

49) -6333

Section 1.2

1) Add the absolute values of the number and keep the sign.

3) Change addition to subtraction. Find the additive inverse of the second number. Subtract the absolute values of the numbers and keep the sign of the larger number.

5) $9+(-13)=-4$

7) $-2-12=-2+(-12)=-14$

9) $-25+38=13$

Section 1.2: Operations on Real Numbers

11) $-1-(-19)=-1+19=18$

13) $-794-657=-794+(-657)=-1451$

15) $-\dfrac{3}{10}+\dfrac{4}{5}=-\dfrac{3}{10}+\dfrac{8}{10}=\dfrac{5}{10}$

17) $-\dfrac{5}{8}-\dfrac{2}{3}=-\dfrac{5}{8}+\left(-\dfrac{2}{3}\right)$

$\qquad =-\dfrac{15}{24}+\left(-\dfrac{16}{24}\right)$

$\qquad =-\dfrac{31}{24}$ or $-1\dfrac{7}{24}$

19)

$-\dfrac{11}{12}-\left(-\dfrac{5}{9}\right)=-\dfrac{11}{12}+\dfrac{5}{9}=-\dfrac{33}{36}+\dfrac{20}{36}=-\dfrac{13}{36}$

21) $7.3-11.2=7.3+(-11.2)=-3.9$

23) $-5.09-(-12.4)=-5.09+12.4=7.31$

25) $-1-4.2=-1+(-4.2)=-5.2$

27) $18-|-12|=18-(12)=18+(-12)=6$

29) $|13|+9=13+9=22$

31)

$|-5.2|-4.8=5.2-4.8=5.2+(-4.8)=0.4$

33) True

35) False; for the same reason as #34.

37) True, by the additive inverse property of addition.

39)

Difference = High Elevation − Low Elevation

$\qquad =29,028-(-36,201)$

$\qquad =29,028+36,201$

$\qquad =65,229$

There is a 65,229-ft difference between Mt. Everest and the Mariana Trench.

41) Difference = 2005 income − 2004 income

$\qquad =51,700-51,081$

$\qquad =51,700+(-51,081)$

$\qquad =619$

The median income for a male with a bachelor's degree rose by $619 from 2004 to 2005.

43)

High temperature = Low temperature + Difference

$\qquad =-79.8+213.8$

$\qquad =134$

The highest temperature on record in the United States is 134°F.

45) Net yardage $=7+4+1+6-10$

$\qquad =18+(-10)$

$\qquad =8$

The Patriots' yardage on this offensive drive was 8 yards.

2

47) a) $173,000 - 178,000 = -5,000$
   The decrease is represented by $-5,000$.

   b) $183,000 - 173,000 = 10,000$
   The increase is represented by $10,000$.

   c) $201,000 - 183,000 = 18,000$
   The increase is represented by $18,000$.

   d) $180,000 - 201,000 = -21,000$
   The decrease is represented by $-21,000$.

49) positive

51) negative

53) $-5 \cdot 9 = -45$

55) $-14 \cdot (-3) = 42$

57) $-2 \cdot 5 \cdot (-3) = 30$

59) $\dfrac{7}{9}\left(-\dfrac{6}{5}\right) = \dfrac{7}{\cancel{9}_{3}}\left(-\dfrac{\cancel{6}^{2}}{5}\right) = -\dfrac{14}{15}$

61) $(-0.25)(1.2) = -0.3$

63) $8 \cdot (-2) \cdot (-4) \cdot (-1) = -64$

65) $(-8) \cdot (-9) \cdot 0 \cdot \left(-\dfrac{1}{4}\right) \cdot (-2) = 0$

67) $-42 \div (-6) = 7$

69) $\dfrac{56}{-7} = -8$

71) $\dfrac{-3.6}{0.9} = -4$

73) $-\dfrac{12}{13} \div \left(-\dfrac{6}{5}\right) = -\dfrac{12}{13} \cdot \left(-\dfrac{5}{6}\right)$

   $= -\dfrac{\cancel{12}^{2}}{13}\left(-\dfrac{5}{\cancel{6}_{1}}\right)$

   $= \dfrac{10}{13}$

75) $\dfrac{0}{-4} = 0$

77) $\dfrac{360}{-280} = -\dfrac{9}{7}\ or\ -1\dfrac{2}{7}$

79) $\dfrac{\dfrac{20}{21}}{\dfrac{5}{7}} = \dfrac{20}{21} \div \dfrac{5}{7} = \dfrac{20}{21} \cdot \dfrac{7}{5} = \dfrac{\cancel{20}^{4}}{\cancel{21}_{3}} \cdot \dfrac{\cancel{7}^{1}}{\cancel{5}_{1}} = \dfrac{4}{3}\ or\ 1\dfrac{1}{3}$

81) $\dfrac{-0.5}{10} = -0.05$

Section 1.3

1) $9 \cdot 9 \cdot 9 \cdot 9 = 9^4$

3) negative

5) positive

7) $4 \cdot 4 \cdot 4 = 4^3$

9) $0.2 \cdot 0.2 \cdot 0.2 \cdot 0.2 \cdot 0.2 = 0.2^5$

11) $\dfrac{3}{4} \cdot \dfrac{3}{4} = \left(\dfrac{3}{4}\right)^2$

13)

$$(-7)\cdot(-7)\cdot(-7)\cdot(-7)\cdot(-7)\cdot(-7)=(-7)^6$$

15) $2^5 = 2\cdot2\cdot2\cdot2\cdot2 = 32$

17) $(11)^2 = 11\cdot11 = 121$

19) $(-2)^4 = (-2)\cdot(-2)\cdot(-2)\cdot(-2) = 16$

21) $-7^2 = -1\cdot7\cdot7 = -49$

23) $-2^3 = -1\cdot2\cdot2\cdot2 = -8$

25) $\left(\dfrac{1}{5}\right)^3 = \dfrac{1}{5}\cdot\dfrac{1}{5}\cdot\dfrac{1}{5} = \dfrac{1}{125}$

27) $(0.5)^2 = 0.25\,\text{or}\,\dfrac{1}{4}$

29)   False; the $\sqrt{}$ symbol means to find only the positive square root.

31)   True

33) $8\,\text{and}-8$

35) $20\,\text{and}-20$

37) $\dfrac{5}{4}\,\text{and}-\dfrac{5}{4}$

39) $\sqrt{36}=6$

41) $-\sqrt{1}=-1$

43)   not real

45) $\sqrt{\dfrac{100}{121}}=\dfrac{10}{11}$

47) $-\sqrt{\dfrac{1}{64}}=-1\cdot\sqrt{\dfrac{1}{64}}=-1\cdot\left(\dfrac{1}{8}\right)=-\dfrac{1}{8}$

49)   Parentheses, Exponents, Multiplication & Division from left to right, Addition & Subtraction from left to right

51) $23+4\cdot5 = 23+20 = 43$

53) $4\big[9-(-3)\big] = 4(9+3) = 4(12) = 48$

55)

$$3-(-4)6+12\sqrt{100} = 3-(-24)+12(10) = 3+24+120 = 147$$

57) $-50\div10+15 = -5+15 = 10$

59) $20-3\cdot2+9 = 20-6+9 = 14+9 = 23$

61) $\dfrac{1}{2}\cdot\dfrac{4}{5}-\dfrac{2}{5}\cdot\dfrac{3}{10} = \dfrac{4}{10}-\dfrac{6}{50} = \dfrac{4\cdot5}{10\cdot5}-\dfrac{6}{50}$

$$= \dfrac{20}{50}-\dfrac{6}{50} = \dfrac{14}{50} = \dfrac{7}{25}$$

63) $15-3(6-4)^2 = 15-3(2)^2$
$$= 15-3(4)$$
$$= 15-12 = 3$$

65) $-6\big[21\div(3+4)\big]-9 = -6\big[21\div7\big]-9$
$$= -6\big[3\big]-9$$
$$= -18-9 = -27$$

67) $4+3\Big[(3-7)^3\div(10-2)\Big] = 4+3\Big[(-4)^3\div(8)\Big]$
$$= 4+3(-64\div8)$$
$$= 4+3(-8)$$
$$= 4+(-24) = -20$$

69) $\dfrac{12(5+1)}{2\cdot5-1}=\dfrac{12(6)}{10-1}=\dfrac{72}{9}=8$

71) $\dfrac{4(7-2)^2}{(12)^2-8\cdot3}=\dfrac{4(5)^2}{144-24}$

$=\dfrac{4(25)}{120}$

$=\dfrac{100}{120}$

$=\dfrac{10}{12}=\dfrac{5}{6}$

Section 1.4

1) Evaluate $2j^2+3j-7$ when

a) $j=4$ b) $j=-5$

a) $2(4)^2+3(4)-7=2(16)+12-7$

$=32+12-7$

$=37$

b) $2(-5)^2+3(-5)-7=2(25)+(-15)-7$

$=50+(-15)-7$

$=50+(-22)$

$=28$

3) $8(-2)+7=-16+7=-9$

5) $(-2)^2+(-2)(7)+10=4+(-14)+10$

$=-10+10=0$

7) $\dfrac{2(-2)}{7+(-3)}=\dfrac{-4}{4}=-1$

9) $\dfrac{(-2)^2-(7)^2}{2(-3)^2+7}=\dfrac{4-49}{2(9)+7}$

$=\dfrac{-45}{18+7}$

$=\dfrac{-45}{25}$

$=-\dfrac{9}{5}$

11) 0

13) $\dfrac{1}{6}$

15) associative

17) commutative

19) associative

21) distributive

23) identity

25) distributive

27) $7(u+v)=7u+7v$

29) $k+4=4+k$

31) $-4z+0=-4z$

33) No, subtraction is not commutative.

35) $5(4+3)=5\cdot4+5\cdot3=20+15=35$

37)

$-2(5+7)=-2\cdot5+(-2)\cdot7=-10-14=-24$

39)

$-7(2-6)=-7\cdot2+(-7)(-6)=-14+42=28$

41)

$$-(6+1) = -1 \cdot 6 + (-1)(1) = -6 + (-1) = -7$$

43)

$$(-10+3)5 = -10 \cdot 5 + 3 \cdot 5 = -50 + 15 = -35$$

45) $\quad 9(g+6) = 9 \cdot g + 9 \cdot 6 = 9g + 54$

47) $\quad -5(z+3) = -5 \cdot z + (-5) \cdot 3 = -5z - 15$

49)

$$-8(u-4) = -8 \cdot u + (-8)(-4) = -8u + 32$$

51) $\quad -(v-6) = -1 \cdot v + (-1)(-6) = -v + 6$

53)

$$10(m+5n-3) = 10 \cdot m + 10 \cdot 5n + 10 \cdot (-3)$$
$$= 10m + 50n - 30$$

55)

$$-(-8c + 9d - 14) = -1 \cdot (-8c) + (-1) \cdot 9d + (-1)(-14)$$
$$= 8c - 9d + 14$$

Chapter 1 Review

1) Given the set of numbers

$$\left\{ \sqrt{23}, -6, 14.38, \frac{3}{11}, 2, 5.\overline{7}, 0, 9.21743819... \right\}$$

a) Whole: $0, 2$
b) Natural: 2
c) Integers: $-6, 0, 2$

d) Rational: $-6, 14.38, \dfrac{3}{11}, 2, 5.\overline{7}, 0$

e) Irrational: $\sqrt{23}, 9.21743819...$

3) $\quad |-10| = 10$

5) $\quad 60 - (-15) = 60 + 15 = 75$

7) $\quad 0.8 - 5.9 = -5.1$

9) $\quad (-10)(-7) = 70$

11) $\quad (3.7)(-2.1) = -7.77$

13) $\quad (-1)(6)(-4)\left(-\dfrac{1}{2}\right)(-5) = 60$

15) $\quad \dfrac{-24}{-12} = 2$

17) $\quad -\dfrac{20}{27} \div \dfrac{8}{15} = -\dfrac{20}{27} \cdot \dfrac{15}{8}$

$$= -\dfrac{\cancel{20}^{5}}{\cancel{27}_{9}} \cdot \dfrac{\cancel{15}^{5}}{\cancel{8}_{2}}$$

$$= -\dfrac{25}{18} \text{ or } -1\dfrac{7}{18}$$

19) $\quad -5^2 = -1 \cdot 5 \cdot 5 = -25$

21) $\quad (-3)^4 = (-3)(-3)(-3)(-3) = 81$

23) $\quad -2^6 = -1 \cdot 2 \cdot 2 \cdot 2 \cdot 2 \cdot 2 \cdot 2 = -64$

25) $\quad \sqrt{49} = 7$

27) $\quad -\sqrt{36} = -6$

29) $\quad 64 \div (-8) + 6 = -8 + 6 = -2$

31) $\quad -11 - 3 \cdot 9 + (-2)^1 = -11 - 3 \cdot 9 + (-2)$

$$= -11 - 27 + (-2)$$
$$= -38 + (-2) = -40$$

33) $\dfrac{3}{4} \cdot \left|\dfrac{5}{7}\right| = \dfrac{3}{4} \cdot \dfrac{5}{7} = \dfrac{15}{28}$

35) $\dfrac{4^2 - (3 \cdot 5)}{|-4-2|} = \dfrac{16-15}{|-6|} = \dfrac{1}{6}$

37) $12 + \sqrt{16} - 2^3 + 1 = 12 + 4 - 8 + 1$
$$= 16 - 8 + 1$$
$$= 8 + 1 = 9$$

39) Terms: $c^4, 12c^3, -c^2, -3.8c, 11$
Coefficients: $1, 12, -1, -3.8$

41) $\dfrac{5 - 6(-4)}{(-4)^2 - (5)^2} = \dfrac{5+24}{16-25} = \dfrac{29}{-9}$ or $-3\dfrac{2}{9}$

43) associative

45) commutative

47)
$3(10-6) = 3 \cdot 10 + 3 \cdot (-6) = 30 + (-18) = 12$

49) $-(12+5) = (-1)(12) + (-1)(5)$
$$= -12 + (-5) = -17$$

Chapter 1 Test

1) Given the set of numbers
$$\left\{ 41, -8, 0, 2.\overline{83}, \sqrt{75}, 6.5, 4\dfrac{5}{8}, 6.37528861... \right\}$$

   a) Integers: $41, -8, 0$
   b) Irrational: $\sqrt{75}, 6.37528861...$
   c) Natural: 41
   d) Rational: $41, -8, 0, 2.\overline{83}, 6.5, 4\dfrac{5}{8}$
   e) Whole: $41, 0$

3) $\dfrac{9}{14} \cdot \dfrac{7}{24} = \dfrac{\cancel{9}^{3}}{\cancel{14}_{2}} \cdot \dfrac{\cancel{7}^{1}}{\cancel{24}_{8}} = \dfrac{3}{16}$

5) $5\dfrac{1}{4} - 2\dfrac{1}{6} = \dfrac{21}{4} - \dfrac{13}{6}$
$$= \dfrac{21(3)}{4(3)} - \dfrac{13(2)}{6(2)}$$
$$= \dfrac{63}{12} - \dfrac{26}{12}$$
$$= \dfrac{37}{12} \text{ or } 3\dfrac{1}{12}$$

7) $\dfrac{4}{7} - \dfrac{5}{6} = \dfrac{4(6)}{7(6)} - \dfrac{5(7)}{6(7)} = \dfrac{24}{42} - \dfrac{35}{42} = -\dfrac{11}{42}$

9) $25 + 15 \div 5 = 25 + 3 = 28$

11) $-8 \cdot (-6) = 48$

13)
$$30 - 5\left[-10 + (2-6)^2 \right] = 30 - 5\left[-10 + (-4)^2 \right]$$
$$= 30 - 5(-10 + 16)$$
$$= 30 - 5(6)$$
$$= 30 - 30 = 0$$

15) $2^5 = 2 \cdot 2 \cdot 2 \cdot 2 \cdot 2 = 32$

17) $|-92| = 92$

19) $-2(7 + \sqrt{25}) = -2(7+5)$
$$= -2(12) = -24$$

Chapter 1: Test

21) $\dfrac{(-\sqrt{36} \div 2)^3 + 7}{-|-11+3| \cdot (-4)} = \dfrac{(-6 \div 2)^3 + 7}{-|-8| \cdot (-4)}$

$= \dfrac{(-3)^3 + 7}{-8 \cdot (-4)}$

$= \dfrac{-27 + 7}{32}$

$= \dfrac{-\overset{5}{\cancel{20}}}{\underset{8}{\cancel{32}}} = -\dfrac{5}{8}$

23) False

25) False

27) $9(-1)^2 + 3(-1) - 6 = 9 \cdot 1 + (-3) - 6$

$= 9 + (-3) - 6$

$= 6 - 6 = 0$

29) a) $-2(5+3) = -2 \cdot 5 + (-2) \cdot 3$

$= -10 + (-6) = -16$

b) $5(t + 9u + 1) = 5 \cdot t + 5 \cdot 9u + 5 \cdot 1$

$= 5t + 45u + 5$

Section 2.1

1) equation

3) expression

5) No, it is not an equation.

7) Both b) and d) are linear equations in one variable. Item a) is not because the first term has an exponent of 2 and item c) is not because it is not an equation.

9) $-8p = 12; p = -\dfrac{3}{2}$

$$-\overset{4}{\cancel{8}}\left(-\dfrac{3}{\underset{1}{\cancel{2}}}\right) = 12$$

$$12 = 12$$

Yes, $-\dfrac{3}{2}$ is a solution.

11) $2(t-5)+7 = 3(2t-9)-2; t = 4$

$2(4-5)+7 = 3(2\cdot 4-9)-2$

$2(-1)+7 = 3(8-9)-2$

$-2+7 = 3(-1)-2$

$5 = -3-2$

$5 \neq -5$

No, 4 is not a solution.

13) $r - 6 = 11$

$r - 6 + 6 = 11 + 6$

$r = 17$

15) $-16 = k - 12$

$-16 + 12 = k - 12 + 12$

$-4 = k$

$k = -4$

17) $a + \dfrac{5}{8} = \dfrac{1}{2}$

$a + \dfrac{5}{8} - \dfrac{5}{8} = \dfrac{1}{2} - \dfrac{5}{8}$

$a = \dfrac{4}{8} - \dfrac{5}{8}$

$a = -\dfrac{1}{8}$

19) $3y = 30$

$\dfrac{3y}{3} = \dfrac{30}{3}$

$y = 10$

21) $-6 = \dfrac{k}{8}$

$8(-6) = \dfrac{k}{8}\cdot 8$

$-48 = k$

$k = -48$

23) $\dfrac{2}{3}g = -10$

$\dfrac{3}{2}\cdot\dfrac{2}{3}g = -\overset{5}{\cancel{10}}\cdot\dfrac{3}{\underset{1}{\cancel{2}}}$

$1g = -15, \ g = -15$

25) $-\dfrac{5}{3}d = -30$

$\left(-\dfrac{3}{5}\right)\left(-\dfrac{5}{3}\right)d = -\overset{6}{\cancel{30}}\left(-\dfrac{3}{\underset{1}{\cancel{5}}}\right)$

$1d = 18, \ d = 18$

Section 2.1: Linear Equations in One Variable

27) $0.5q = 6$

$$\frac{0.5}{0.5}q = \frac{6}{0.5}$$

$$q = 12$$

29) $3x - 7 = 17$

$$3x - 7 + 7 = 17 + 7$$

$$3x = 24$$

$$\frac{3}{3}x = \frac{24}{3}$$

$$x = 8$$

31) $8d - 15 = -15$

$$8d - 15 + 15 = -15 + 15$$

$$8d = 0$$

$$\frac{8d}{8} = \frac{0}{8}$$

$$d = 0$$

33) $\frac{4}{9}w - 11 = 1$

$$\frac{4}{9}w - 11 + 11 = 1 + 11$$

$$\frac{4}{9}w = 12$$

$$\frac{9}{4} \cdot \frac{4}{9}w = \overset{3}{\cancel{12}} \cdot \frac{9}{\underset{1}{\cancel{4}}}$$

$$1w = 27, \; w = 27$$

35) $\frac{10}{7}m + 3 = 1$

$$\frac{10}{7}m + 3 - 3 = 1 - 3$$

$$\frac{10}{7}m = -2$$

$$\frac{7}{10} \cdot \frac{10}{7}m = -\overset{1}{\cancel{2}} \cdot \frac{7}{\underset{5}{\cancel{10}}}$$

$$1m = -\frac{7}{5}, \; m = -\frac{7}{5}$$

37) $5 - 0.4p = 2.6$

$$5 - 5 - 0.4p = 2.6 - 5$$

$$-0.4p = -2.4$$

$$\frac{-0.4p}{-0.4} = \frac{-2.4}{-0.4}$$

$$p = 6$$

39) $10v + 9 - 2v + 16 = 1$

$$8v + 25 = 1$$

$$8v + 25 - 25 = 1 - 25$$

$$8v = -24$$

$$\frac{8v}{8} = \frac{-24}{8}$$

$$v = -3$$

41) $5 = -12p + 7 + 4p - 12$

$$5 = -8p - 5$$

$$5 + 5 = -8p - 5 + 5$$

$$10 = -8p$$

$$-8p = 10$$

$$\frac{-8p}{-8} = \frac{10}{-8}$$

$$p = -\frac{5}{4}$$

43) $-12 = 7(2a - 3) - (8a - 9)$

$$-12 = 14a - 21 - 8a + 9$$

$$-12 = 6a - 12$$

$$-12 + 12 = 6a - 12 + 12$$

$$0 = 6a$$

$$6a = 0$$

$$\frac{6a}{6} = \frac{0}{6}$$

$$a = 0$$

45)
$$2y+7=5y-2$$
$$2y+7-7=5y-2-7$$
$$2y=5y-9$$
$$2y-5y=5y-5y-9$$
$$-3y=-9$$
$$\frac{-3y}{-3}=\frac{-9}{-3}$$
$$y=3$$

47)
$$6-7p=2p+33$$
$$6-6-7p=2p+33-6$$
$$-7p=2p+27$$
$$-7p-2p=2p-2p+27$$
$$-9p=27$$
$$\frac{-9p}{-9}=\frac{27}{-9}$$
$$p=-3$$

49)
$$-8x+6-2x+11=3+3x-7x$$
$$-10x+17=3-4x$$
$$-10x+17-17=3-4x-17$$
$$-10x=-14-4x$$
$$-10x+4x=-14-4x+4x$$
$$-6x=-14$$
$$\frac{-6x}{-6}=\frac{-14}{-6}$$
$$x=\frac{7}{3}$$

51)
$$4(2t+5)-7=5(t+5)$$
$$8t+20-7=5t+25$$
$$8t+13=5t+25$$
$$8t+13-13=5t+25-13$$
$$8t=5t+12$$
$$8t-5t=5t-5t+12$$
$$3t=12$$
$$\frac{3t}{3}=\frac{12}{3}$$
$$t=4$$

53) $-9r+4r-11+2=3r+7-8r+9$
$$-5r-9=-5r+16$$
$$-5r+5r-9=-5r+5r+16$$
$$-9\neq16$$

The solution set is $\varnothing$.

55)
$$j-15j+8=-3(4j-3)-2j-1$$
$$-14j+8=-12j+9-2j-1$$
$$-14j+8=-14j+8$$
$$-14j+14j+8=-14j+14j+8$$
$$8=8$$

The solution is all real numbers.

57)
$$8(3t+4)=10t-3+7(2t+5)$$
$$24t+32=10t-3+14t+35$$
$$24t+32=24t+32$$
$$24t-24t+32=24t-24t+32$$
$$32=32$$

The solution is all real numbers.

59) $8-7(2-3w)-9w=4(5w-1)-3w-2$
$$8-14+21w-9w=20w-4-3w-2$$
$$-6+12w=17w-6$$
$$-6+6+12w=17w-6+6$$
$$12w=17w+0$$
$$12w-17w=17w-17w+0$$
$$-5w=0$$
$$\frac{-5w}{-5}=\frac{0}{-5}$$
$$w=0$$

61) $7y+2(1-4y)=8y-5(y+4)$
$$7y+2-8y=8y-5y-20$$
$$-y+2=3y-20$$
$$-y+2-2=3y-20-2$$
$$-y=3y-22$$
$$-y-3y=3y-3y-22$$
$$-4y=-22$$
$$\frac{-4y}{-4}=\frac{-22}{-4}$$
$$y=\frac{11}{2}$$

63)
$$\frac{1}{6}+\frac{5}{4}=\frac{1}{2}x-\frac{5}{12}$$
$$12\left(\frac{1}{6}x+\frac{5}{4}\right)=12\left(\frac{1}{2}x-\frac{5}{12}\right)$$
$$12\cdot\frac{1}{6}x+12\cdot\frac{5}{4}=12\cdot\frac{1}{2}x-12\cdot\frac{5}{12}$$
$$2x+15=6x-5$$
$$2x-6x+15=6x-6x-5$$
$$-4x+15=-5$$
$$-4x+15-15=-5-15$$
$$-4x=-20$$
$$\frac{-4x}{-4}=\frac{-20}{-4}$$
$$x=5$$

65)
$$\frac{2}{3}d-1=\frac{1}{5}d+\frac{2}{5}$$
$$15\left(\frac{2}{3}d-1\right)=15\left(\frac{1}{5}d+\frac{2}{5}\right)$$
$$15\cdot\frac{2}{3}d-15\cdot1=15\cdot\frac{1}{5}d+15\cdot\frac{2}{5}$$
$$10d-15=3d+6$$
$$10d-15+15=3d+6+15$$
$$10d=3d+21$$
$$10d-3d=3d-3d+21$$
$$7d=21$$
$$\frac{7d}{7}=\frac{21}{7}$$
$$d=3$$

67)
$$\frac{m}{3}+\frac{1}{2}=\frac{2m}{3}+3$$

$$6\left(\frac{m}{3}+\frac{1}{2}\right)=6\left(\frac{2m}{3}+3\right)$$

$$6\cdot\frac{m}{3}+6\cdot\frac{1}{2}=6\cdot\frac{2m}{3}+6\cdot3$$

$$2m+3=4m+18$$

$$2m+3-3=4m+18-3$$

$$2m=4m+15$$

$$2m-4m=4m-4m+15$$

$$-2m=15$$

$$\frac{-2m}{-2}=\frac{15}{-2}$$

$$m=-\frac{15}{2}$$

69)
$$\frac{1}{3}+\frac{1}{9}(k+5)-\frac{k}{4}=2$$

$$36\left[\frac{1}{3}+\frac{1}{9}(k+5)-\frac{k}{4}\right]=36\cdot2$$

$$36\cdot\frac{1}{3}+36\cdot\frac{1}{9}(k+5)-36\cdot\frac{k}{4}=36\cdot2$$

$$12+4(k+5)-9k=72$$

$$12+4k+20-9k=72$$

$$32-5k=72$$

$$32-32-5k=72-32$$

$$-5k=40$$

$$\frac{-5k}{-5}=\frac{40}{-5}$$

$$k=-8$$

71)
$$0.05(t+8)-0.01t=0.6$$

$$100\left[0.05(t+8)-0.01t\right]=100(0.6)$$

$$5(t+8)-1t=60$$

$$5t+40-1t=60$$

$$4t+40=60$$

$$4t+40-40=60-40$$

$$4t=20$$

$$\frac{4t}{4}=\frac{20}{4}$$

$$t=5$$

73)
$$0.1x+0.15(8-x)=0.125(8)$$

$$1000\left[0.1x+0.15(8-x)\right]=1000\left[0.125(8)\right]$$

$$100x+150(8-x)=125(8)$$

$$100x+1200-150x=1000$$

$$-50x+1200=1000$$

$$-50x+1200-1200=1000-1200$$

$$-50x=-200$$

$$\frac{-50x}{-50}=\frac{-200}{-50}$$

$$x=4$$

75)
$$0.04s+0.03(s+200)=27$$

$$100\left[0.04s+0.03(s+200)\right]=100(27)$$

$$4s+3(s+200)=2700$$

$$4s+3s+600=2700$$

$$7s+600=2700$$

$$7s+600-600=2700-600$$

$$7s=2100$$

$$\frac{7s}{7}=\frac{2100}{7}$$

$$s=300$$

77) Let x = the number
$$x+4=15$$
$$x+4-4=15-4$$
$$x=11$$
Check: $11+4=15$

79) Let x = the number
$$x-7=22$$
$$x-7+7=22+7$$
$$x=29$$
Check: $29-7=22$

81) Let x = the number
$$2x=-16$$
$$\frac{2x}{2}=\frac{-16}{2}$$
$$x=-8$$
Check: $2(-8)=-16$

83) Let x = the number
$$2x+7=35$$
$$2x+7-7=35-7$$
$$2x=28$$
$$\frac{2x}{2}=\frac{28}{2}$$
$$x=14$$
Check: $2(14)+7=35$

85) Let x = the number
$$3x-8=40$$
$$3x-8+8=40+8$$
$$3x=48$$
$$\frac{3x}{3}=\frac{48}{3}$$
$$x=16$$
Check: $3(16)-8=40$

87) Let x = the number
$$\frac{1}{2}x+10=3$$
$$\frac{1}{2}x+10-10=3-10$$
$$\frac{1}{2}x=-7$$
$$2\cdot\frac{1}{2}x=-7\cdot2$$
$$x=-14$$
Check: $\frac{1}{2}\cdot(-14)=-7$

89) Let x = the number
$$2x-3=x+8$$
$$2x-x-3=x-x+8$$
$$x-3=8$$
$$x-3+3=8+3$$
$$x=11$$
Check: $2\cdot11-3=11+8$

91) Let x = the number
$$\frac{1}{3}x+10=x-2$$
$$\frac{1}{3}x-x+10=x-x-2$$
$$\frac{1}{3}x-\frac{3}{3}x+10=-2$$
$$-\frac{2}{3}x+10=-2$$
$$-\frac{2}{3}x+10-10=-2-10$$
$$-\frac{2}{3}x=-12$$
$$-\frac{3}{2}\left(-\frac{2}{3}\right)x=-\overset{6}{\cancel{12}}\left(-\frac{3}{\underset{1}{\cancel{2}}}\right)$$
$$1x=18,\ x=18$$
Check: $\frac{1}{3}\cdot18+10=18-2$

93) Let x = the number

$$2(x+5)=16$$
$$2x+10=16$$
$$2x+10-10=16-10$$
$$2x=6$$
$$\frac{2x}{2}=\frac{6}{2}$$
$$x=3$$

Check: $2(3+5)=16$

95) Let x = the number

$$3x=\frac{1}{2}x+15$$
$$3x-\frac{1}{2}x=\frac{1}{2}x-\frac{1}{2}x+15$$
$$3x-\frac{1}{2}x=15$$
$$\frac{6}{2}x-\frac{1}{2}x=15$$
$$\frac{5}{2}x=15$$
$$\frac{2}{5}\cdot\frac{5}{2}x=\overset{3}{\cancel{15}}\cdot\frac{2}{\underset{1}{\cancel{5}}}$$
$$1x=6,\ x=6$$

Check: $3\cdot6=\frac{1}{2}\cdot6+15$

97) Let x = the number

$$x-6=2x+5$$
$$x-6+6=2x+5+6$$
$$x=2x+11$$
$$x-2x=2x-2x+11$$
$$-x=11$$
$$x=-11$$

Check: $-11-6=2(-11)+5$

Section 2.2

1) $c+5$

3) $p-31$

5) $3w$

7) $14-x$

9) Let x = Gatorade
Let $x+6.5$ = Pepsi

$$(x+6.5)+x=13.1$$
$$x+6.5+x=13.1$$
$$2x+6.5=13.1$$
$$2x+6.5-6.5=13.1-6.5$$
$$2x=6.6$$
$$\frac{2x}{2}=\frac{6.6}{2}$$
$$x=3.3$$

Gatorade has $x=3.3$ tsp and Pepsi has $x+6.5=3.3+6.5=9.8$ tsp.

11) Let x = Greece
Let $\frac{1}{2}x$ = Thailand

$$x+\frac{1}{2}x=24$$
$$\frac{2}{2}x+\frac{1}{2}x=24$$
$$\frac{3}{2}x=24$$
$$\frac{2}{3}\cdot\frac{3}{2}x=24\cdot\frac{2}{3}$$
$$1x=16,\ x=16$$

Greece won $x=16$ medals and Thailand won $\frac{1}{2}x=\frac{1}{2}\cdot16=8$ medals.

Section 2.2: Applications of Linear Equations

13) Let x = Ohio River

Let $x - 70$ = Columbia River

$x + (x - 70) = 2550$

$x + x - 70 = 2550$

$2x - 70 = 2550$

$2x - 70 + 70 = 2550 + 70$

$2x = 2620$

$\dfrac{2x}{2} = \dfrac{2620}{2}$

$x = 1310$

Ohio River is x = 1310 mi and Columbia River is $x - 70 = 1310 - 70 = 1240$ mi.

15) Let $x = 1^{st}$ piece

Let $x + 14 = 2^{nd}$ piece

$x + (x + 14) = 36$

$x + x + 14 = 36$

$2x + 14 = 36$

$2x + 14 - 14 = 36 - 14$

$2x = 22$

$\dfrac{2x}{2} = \dfrac{22}{2}$

$x = 11$

1st piece is x = 11 in. and 2nd piece is $x + 14 = 11 + 14 = 25$ in.

17) Let x = short piece

Let $2x$ = long piece

$x + 2x = 18$

$3x = 18$

$\dfrac{3x}{3} = \dfrac{18}{3}$

$x = 6$

Short piece is x = 6 ft. and long piece is $2x = 2 \cdot 6 = 12$ ft.

19) Let $x = 1^{st}$ integer

Let $x + 1 = 2^{nd}$ integer

Let $x + 2 = 3^{rd}$ integer

$x + (x + 1) + (x + 2) = 195$

$3x + 3 = 195$

$3x + 3 - 3 = 195 - 3$

$3x = 192$

$\dfrac{3x}{3} = \dfrac{192}{3}$

$x = 64$

The 1st integer is x = 64, the 2nd is $x + 1 = 64 + 1 = 65$, and the 3rd is $x + 2 = 64 + 2 = 66$.

21) Let $x = 1^{st}$ even integer

Let $x + 2 = 2^{nd}$ even integer

$2x = (x + 2) + 10$

$2x = x + 12$

$2x - x = x - x + 12$

$x = 12$

The 1st even integer is x = 12 and the 2nd is $x + 2 = 12 + 2 = 14$.

23) Let $x = 1^{st}$ odd integer

Let $x + 2 = 2^{nd}$ odd integer

Let $x + 4 = 3^{rd}$ odd integer

$x + (x + 2) + (x + 4) = 4(x + 4) + 5$

$3x + 6 = 4x + 16 + 5$

$3x + 6 = 4x + 21$

$3x + 6 - 6 = 4x + 21 - 6$

$3x = 4x + 15$

$3x - 4x = 4x - 4x + 15$

$-x = 15$

$x = -15$

The 1st odd integer is x = -15, the 2nd is $x + 2 = -15 + 2 = -13$, and the 3rd is $x + 4 = -15 + 4 = -11$.

25) Let $x = 1^{st}$ page number

Let $x + 1 = 2^{nd}$ page number

$x + (x + 1) = 345$

$2x + 1 = 345$

$2x + 1 - 1 = 345 - 1$

$2x = 344$

$$\frac{2x}{2} = \frac{344}{2}$$

$x = 172$

The 1st page number is $x = 172$ and the 2nd is $x + 1 = 172 + 1 = 173$.

27) Sale price = Original price − Discount

$= \$75.00 - (0.15)(\$75.00)$

$= \$75.00 - \$11.25$

$= \$63.75$

29) Sale price = Original price − Discount

$= \$399.00 - (0.25)(\$399.00)$

$= \$399.00 - \$99.75$

$= \$299.25$

31) Let $x =$ original price of camera

Sale price = Original price − Discount

$\$119 = x - 0.15x$

$\$119 = 0.85x$

$$\frac{\$119}{0.85} = \frac{0.85x}{0.85}$$

$x = \$140.00$

33) Let $x =$ original price of calendar

Sale price = Original price − Discount

$\$4.38 = x - 0.60x$

$\$4.38 = 0.40x$

$$\frac{\$4.38}{0.40} = \frac{0.40x}{0.40}$$

$x = \$10.95$

35) Let $x =$ original price of coffeemaker

Sale price = Original price − Discount

$\$22.75 = x - 0.30x$

$\$22.75 = 0.70x$

$$\frac{\$22.75}{0.70} = \frac{0.70x}{0.70}$$

$x = \$32.50$

37) Let $x =$ countries participating in 1980

$140 = x + 0.75x$

$140 = 1.75x$

$$\frac{140}{1.75} = \frac{1.75x}{1.75}$$

$x = 80$ countries participating in 1980

39) Let $x =$ Starbucks stores in 1994

$7569 = x + 16.81x$

$7569 = 17.81x$

$$\frac{7569}{17.81} = \frac{17.81x}{17.81}$$

$x = 425$ Starbucks stores in 1994

Section 2.2: Applications of Linear Equations

41) $P = \$800, R = 0.04, T = 1$

$I = PRT$

$I = (\$800)(0.04)(1)$

$I = \$32$

43)

Total interest = 6.5% interest + 8% interest

$= (4000)(0.065)(1) + (1500)(0.08)(1)$

$= 260 + 120$

$= \$380$ Total Interest

45)

$x =$ amount invested in 6% account

$15,000 - x =$ amount invested in 7% account

$960 = 0.06x + 0.07(15,000 - x)$

$100(960) = 100[0.06x + 0.07(15,000 - x)]$

$96,000 = 6x + 7(15,000 - x)$

$96,000 = 6x + 105,000 - 7x$

$96,000 = -x + 105,000$

$-9000 = -x$

$x = 9000$

Amount invested at 6% is $x = \$9000$ and at 7% is
$15,000 - x = 15,000 - 9000 = \$6000$.

47) $x =$ amount invested in 6% account

$x + 200 =$ amount invested in 5% account

$164 = 0.06x + 0.05(x + 200)$

$100(164) = 100[0.06x + 0.05(x + 200)]$

$16,400 = 6x + 5(x + 200)$

$16,400 = 6x + 5x + 1000$

$16,400 = 11x + 1000$

$15,400 = 11x$

$x = 1400$

Amount invested at 6% is $x = \$1400$ and at 5% is $x + 200 = 1400 + 200 = \$1600$.

49) $x =$ amount invested in 9.5% account

$7000 - x =$ amount invested in 7% account

$560 = 0.095x + 0.07(7000 - x)$

$1000(560) = 1000[0.095x + 0.07(7000 - x)]$

$560,000 = 95x + 70(7000 - x)$

$560,000 = 95x + 490,000 - 70x$

$560,000 = 25x + 490,000$

$70,000 = 25x$

$x = 2800$

Amount invested at 9.5% is $x = \$2800$ and at 7% is $7000 - x = 7000 - 2800 = \$4200$.

51) Let $x =$ distance Irma rode bike

Let $\dfrac{1}{2}x+1 =$ distance Irma walked

$$x+\left(\dfrac{1}{2}x+1\right)=7$$

$$x+\dfrac{1}{2}x+1=7$$

$$\dfrac{2}{2}x+\dfrac{1}{2}x+1=7$$

$$\dfrac{3}{2}x+1=7$$

$$\dfrac{3}{2}x+1-1=7-1$$

$$\dfrac{3}{2}x=6$$

$$\dfrac{2}{3}\cdot\dfrac{3}{2}x=\overset{2}{\cancel{6}}\cdot\dfrac{2}{\cancel{3}}$$
$$\underset{1}{}$$

$$x=4$$

Bike distance is $x = 4$ mi. and walk

distance is $\dfrac{1}{2}x+1=\dfrac{1}{\cancel{2}}\cdot\overset{2}{\cancel{4}}+1=3$ mi.

53) Let $x =$ Freshmen enrolled Aug. 2012

$$2600=x+0.04x$$

$$2600=1.04x$$

$$\dfrac{2600}{1.04}=\dfrac{1.04x}{1.04}$$

$$x=2500$$

Freshmen enrolled Aug. 2012

55) Let $x = 1^{st}$ piece

Let $x+5 = 2^{nd}$ piece

Let $2x = 3^{rd}$ piece

$$x+(x+5)+2x=53$$

$$4x+5=53$$

$$4x+5-5=53-5$$

$$4x=48$$

$$\dfrac{4x}{4}=\dfrac{48}{4}$$

$$x=12$$

1st piece is $x = 12$ in., 2nd piece is
$x+5=12+5=17$ in., and 3rd piece is
$2x=2\cdot12=24$ in.

57) Let $x = 1^{st}$ even integer

Let $x+2 = 2^{nd}$ even integer

Let $x+4 = 3^{rd}$ even integer

$$\dfrac{1}{6}x=\dfrac{1}{10}[(x+2)+(x+4)]-3$$

$$30\left(\dfrac{1}{6}x\right)=30\left[\dfrac{1}{10}[(x+2)+(x+4)]-3\right]$$

$$30\cdot\dfrac{1}{6}x=30\cdot\dfrac{1}{10}[(x+2)+(x+4)]-30\cdot3$$

$$5x=3[(x+2)+(x+4)]-90$$

$$5x=3(2x+6)-90$$

$$5x=6x+18-90$$

$$5x=6x-72$$

$$5x-6x=6x-6x-72$$

$$-x=-72$$

$$x=72$$

The 1st even integer is $x = 72$, the 2nd is
$x+2=72+2=74$, and the 3rd is
$x+4=72+4=76$.

Section 2.3: Geometry Applications and Solving Formulas

59)

Let $x = $ *Twilight Saga : Breaking Dawn Part 1*

$x + 100 = $ *Harry Potter and the Deathly Hallows Part 2*

$$(x+100) + x = 662$$
$$x + 100 + x = 662$$
$$2x + 100 = 662$$
$$2x + 100 - 100 = 662 - 100$$
$$2x = 562$$
$$\frac{2x}{2} = \frac{562}{2}$$
$$x = 281$$

Twilight Saga : Breaking Dawn Part 1 earned

$x = \$281$ million and

Harry Potter and the Deathly Hallows Part 2 earned

$x + 100 = 281 + 100 = \$381$ million.

61) $x = $ amount invested in 3% account

$2x = $ amount invested in 4% account

$x + 1000 = $ amount invested in 5% account

$$290 = 0.03x + 0.04(2x) + 0.05(x+1000)$$
$$100(290) = 100\left[0.03x + 0.04(2x) + 0.05(x+1000)\right]$$
$$29,000 = 3x + 4(2x) + 5(x+1000)$$
$$29,000 = 3x + 8x + 5x + 5000$$
$$29,000 = 16x + 5000$$
$$24,000 = 16x$$
$$x = 1500$$

Amount invested at 3% is $x = \$1500$, at 4% is $2x = 2 \cdot 1500 = \$3000$, and at 5% is $x + 1000 = 1500 + 1000 = \$2500$.

63) Let $x = $ Zoe's salary last year

$$\$40,144 = x + 0.04x$$
$$\$40,144 = 1.04x$$
$$\frac{\$40,144}{1.04} = \frac{1.04x}{1.04}$$
$$x = \$38,600$$

Section 2.3

1) Let $l = $ length of pool

$$V = lwh$$
$$1700 = l \cdot 17 \cdot 4$$
$$1700 = 68l$$
$$\frac{1700}{68} = \frac{68l}{68}$$
$$l = 25 \text{ ft}$$

3) Let $w = $ width of printed area

$$A = lw$$
$$48 = 8 \cdot w$$
$$48 = 8w$$
$$\frac{48}{8} = \frac{8w}{8}$$
$$w = 6 \text{ in.}$$

5) Let $A = $ area of circular clock face

$$A = \pi r^2$$
$$A \approx 3.14(11.5)^2$$
$$A \approx 3.14 \cdot 132.25$$
$$A \approx 415.265$$
$$A \approx 415 \text{ ft}^2$$

7) Let $w =$ width of lane

$$A = lw$$
$$228 = 19 \cdot w$$
$$228 = 19w$$
$$\frac{228}{19} = \frac{19w}{19}$$
$$w = 12 \text{ ft}$$

9) Let $h =$ height of can

$$V = \pi r^2 h$$
$$24\pi = \pi(2)^2 h$$
$$24\pi = 4\pi h$$
$$\frac{24\pi}{4\pi} = \frac{4\pi h}{4\pi}$$
$$h = 6 \text{ in.}$$

11)
$$x + 102 + (2x) = 180$$
$$3x + 102 = 180$$
$$3x + 102 - 102 = 180 - 102$$
$$3x = 78$$
$$x = 26$$
$$m\angle A = 26°, m\angle B = 2 \cdot 26 = 52°$$

13)
$$(2x + 15) + (6x - 11) + (8x) = 180$$
$$16x + 4 = 180$$
$$16x + 4 - 4 = 180 - 4$$
$$16x = 176$$
$$x = 11$$
$$m\angle A = 2 \cdot 11 + 15 = 37°, m\angle B = 6 \cdot 11 - 11 = 55°$$
$$m\angle C = 8 \cdot 11 = 88°$$

15)
$$9x + 5 = 10x - 2$$
$$9x + 5 - 5 = 10x - 2 - 5$$
$$9x = 10x - 7$$
$$9x - 10x = 10x - 10x - 7$$
$$-x = -7$$
$$x = 7$$

$$9x + 5 = 10x - 2$$
$$9 \cdot 7 + 5 = 10 \cdot 7 - 2$$
$$63 + 5 = 70 - 2$$
$$68° = 68°$$

17)
$$9x - 75 = 6x$$
$$9x - 9x - 75 = 6x - 9x$$
$$-75 = -3x$$
$$x = 25$$

$$9x - 75 = 6x$$
$$9 \cdot 25 - 75 = 6 \cdot 25$$
$$225 - 75 = 150$$
$$150° = 150°$$

19)
$$(10x + 13) + (4x - 1) = 180$$
$$14x + 12 = 180$$
$$14x + 12 - 12 = 180 - 12$$
$$14x = 168$$
$$x = 12$$
$$10x + 13 = 10 \cdot 12 + 13 = 120 + 13 = 133°$$
$$4x - 1 = 4 \cdot 12 - 1 = 48 - 1 = 47°$$

21)
$$(3x + 19) + (5x + 1) = 180$$
$$8x + 20 = 180$$
$$8x + 20 - 20 = 180 - 20$$
$$8x = 160$$
$$x = 20$$
$$3x + 19 = 3 \cdot 20 + 19 = 60 + 19 = 79°$$
$$5x + 1 = 5 \cdot 20 + 1 = 100 + 1 = 101°$$

Section 2.3: Geometry Applications and Solving Formulas

23) If x = measure of an angle then its supplement is $180 - x$.

25)

Let x = the measure of the angle

Let $180 - x$ = the measure of the supplement

$$10x = (180 - x) + 7$$
$$10x = 180 - x + 7$$
$$10x = 187 - x$$
$$10x + x = 187 - x + x$$
$$11x = 187$$
$$x = 17°$$

27)

Let x = the measure of the angle

Let $90 - x$ = the measure of the complement

Let $180 - x$ = the measure of the supplement

$$4(90 - x) = 2(180 - x) - 40$$
$$360 - 4x = 360 - 2x - 40$$
$$-4x = -2x - 40$$
$$-4x + 2x = -2x + 2x - 40$$
$$-2x = -40, x = 20°$$
$$90 - x = 90 - 20 = 70°$$
$$180 - x = 180 - 20 = 160°$$

29) Let x = measure of the angle

Let $90 - x$ = measure of the complement

Let $180 - x$ = measure of the supplement

$$x + 3(90 - x) = (180 - x) + 55$$
$$x + 270 - 3x = 180 - x + 55$$
$$-2x + 270 = 235 - x$$
$$-2x + 270 - 235 = 235 - 235 - x$$
$$-2x + 35 = -x$$
$$-2x + 2x + 35 = -x + 2x$$
$$x = 35°$$

31)

Let x = measure of the angle

Let $180 - x$ = measure of the supplement

$$3x + 2(180 - x) = 400$$
$$3x + 360 - 2x = 400$$
$$x + 360 = 400$$
$$x + 360 - 360 = 400 - 360$$
$$x = 40°$$

33) $\quad I = P \cdot r \cdot t, I = 240, P = 3000, r = 0.04$

$$240 = 3000 \cdot 0.04t$$
$$240 = 120t$$
$$t = 2$$

35) $\quad V = lwh, V = 96, l = 8, h = 3$

$$96 = 8 \cdot w \cdot 3$$
$$96 = 24w$$
$$w = 4$$

37) $\quad P = 2l + 2w, P = 50, w = 7$

$$50 = 2l + 2 \cdot 7$$
$$50 = 2l + 14$$
$$50 - 14 = 2l + 14 - 14$$
$$36 = 2l$$
$$l = 18$$

39) $V = \dfrac{1}{3}\pi r^2 h, V = 54\pi, r = 9$

$54\pi = \dfrac{1}{3}\pi \cdot (9)^2 \cdot h$

$54\pi = \dfrac{1}{3}\pi \cdot 81 \cdot h$

$54\pi = \overset{27}{\cancel{81}} \cdot \dfrac{1}{\underset{1}{\cancel{3}}}\pi h$

$54\pi = 27\pi h$

$\dfrac{54\pi}{27\pi} = \dfrac{27\pi h}{27\pi}$

$h = 2$

41) $S = 2\pi r^2 + 2\pi rh, S = 120\pi, r = 5$

$120\pi = 2\pi \cdot (5)^2 + 2\pi \cdot 5 \cdot h$

$120\pi = 2\pi \cdot 25 + 2\pi \cdot 5 \cdot h$

$120\pi = 50\pi + 10\pi h$

$120\pi - 50\pi = 50\pi - 50\pi + 10\pi h$

$70\pi = 10\pi h$

$\dfrac{70\pi}{10\pi} = \dfrac{10\pi h}{10\pi}$

$h = 7$

43) $A = \dfrac{1}{2}h(b_1 + b_2), A = 790, b_1 = 29, b_2 = 50$

$790 = \dfrac{1}{2}h(29 + 50)$

$790 = \dfrac{1}{2}h \cdot 79$

$790 = \dfrac{79}{2}h$

$\dfrac{2}{79} \cdot 790 = \dfrac{79}{2}h \cdot \dfrac{2}{79}$

$\dfrac{2}{\underset{1}{\cancel{79}}} \cdot \overset{10}{\cancel{790}} = \dfrac{79}{2}h \cdot \dfrac{2}{79}$

$h = 20$

45)   a) $x + 12 = 35$

$\boxed{x} + 12 = 35$

$\boxed{x} + 12 - 12 = 35 - 12$

$x = 23$

b) $x + n = p$

$\boxed{x} + n = p$

$\boxed{x} + n - n = p - n$

$x = p - n$

c) $x + q = v$

$\boxed{x} + q = v$

$\boxed{x} + q - q = v - q$

$x = v - q$

47) a) $5n = 30$

$5\boxed{n} = 30$

$\dfrac{5\boxed{n}}{5} = \dfrac{30}{5}$

$n = 6$

b) $yn = c$

$y\boxed{n} = c$

$\dfrac{y\boxed{n}}{y} = \dfrac{c}{y}$

$n = \dfrac{c}{y}$

c) $wn = d$

$w\boxed{n} = d$

$\dfrac{w\boxed{n}}{w} = \dfrac{d}{w}$

$n = \dfrac{d}{w}$

49)

a) $\dfrac{c}{3}=7$

$\dfrac{\boxed{c}}{3}=7$

$3\cdot\dfrac{\boxed{c}}{3}=7$

$c=2\text{'}$

b) $\dfrac{c}{u}=r$

$\dfrac{\boxed{c}}{u}=r$

$u\cdot\dfrac{\boxed{c}}{u}=r\cdot u$

$c=ru$

c) $\dfrac{c}{x}=t$

$\dfrac{\boxed{c}}{x}=t$

$x\cdot\dfrac{\boxed{c}}{x}=t\cdot x$

$c=tx$

51) a) $8d-7=17$

$8\boxed{d}-7=17$

$8\boxed{d}-7+7=17+7$

$8\boxed{d}=24$

$\dfrac{8\boxed{d}}{8}=\dfrac{24}{8}$

$d=3$

b) $kd-a=z$

$k\boxed{d}-a=z$

$k\boxed{d}-a+a=z+a$

$k\boxed{d}=z+a$

$\dfrac{k\boxed{d}}{k}=\dfrac{z+a}{k}$

$d=\dfrac{z+a}{k}$

53) a) $6z+19=4$

$6\boxed{z}+19=4$

$6\boxed{z}+19-19=4-$

$6\boxed{z}=-15$

$\dfrac{6\boxed{z}}{6}=\dfrac{-15}{6}$

$z=-\dfrac{5}{2}$

b) $yz+t=w$

$y\boxed{z}+t=w$

$y\boxed{z}+t-t=w-t$

$y\boxed{z}=w-t$

$\dfrac{y\boxed{z}}{y}=\dfrac{w-t}{y}$

$z=\dfrac{w-t}{y}$

55) $F=ma$ for m

$F=\boxed{m}a$

$\dfrac{F}{a}=\dfrac{\boxed{m}a}{a}$

$m=\dfrac{F}{a}$

57) $n=\dfrac{c}{v}$ for c

$n=\dfrac{\boxed{c}}{v}$

$v\cdot n=\dfrac{\boxed{c}}{v}\cdot v$

$c=vn$

59) $E=\sigma T^4$ for σ

$E=\boxed{\sigma}T^4$

$\dfrac{E}{T^4}=\dfrac{\boxed{\sigma}T^4}{T^4}$

$\sigma=\dfrac{E}{T^4}$

61) $V=\dfrac{1}{3}\pi r^2 h$ for h

$V=\dfrac{1}{3}\pi r^2\boxed{h}$

$3\cdot V=\dfrac{1}{3}\pi r^2\boxed{h}\cdot 3$

$3V=\pi r^2\boxed{h}$

$\dfrac{3V}{\pi r^2}=\dfrac{\pi r^2\boxed{h}}{\pi r^2}$

$h=\dfrac{3V}{\pi r^2}$

63) $R = \dfrac{E}{I}$ for E

$$R = \dfrac{\boxed{E}}{I}$$

$$I \cdot R = \dfrac{\boxed{E}}{I} \cdot I$$

$$E = IR$$

65) $I = PRT$ for R

$$I = P\boxed{R}T$$

$$\dfrac{I}{PT} = \dfrac{P\boxed{R}T}{PT}$$

$$R = \dfrac{I}{PT}$$

67) $P = 2l + 2w$ for l

$$P = 2\boxed{l} + 2w$$

$$P - 2w = 2\boxed{l} + 2w - 2w$$

$$P - 2w = 2\boxed{l}$$

$$\dfrac{P - 2w}{2} = \dfrac{2\boxed{l}}{2}$$

$$l = \dfrac{P - 2w}{2}$$

69) $H = \dfrac{D^2 N}{2.5}$ for N

$$H = \dfrac{D^2 \boxed{N}}{2.5}$$

$$2.5 \cdot H = \dfrac{D^2 \boxed{N}}{2.5} \cdot 2.5$$

$$\dfrac{2.5H}{D^2} = \dfrac{D^2 \boxed{N}}{D^2}$$

$$N = \dfrac{2.5H}{D^2}$$

71) $A = \dfrac{1}{2}h(b_1 + b_2)$ for b_2

$$A = \dfrac{1}{2}h(b_1 + \boxed{b_2})$$

$$2 \cdot A = \dfrac{1}{2}h(b_1 + \boxed{b_2}) \cdot 2$$

$$2A = h(b_1 + \boxed{b_2})$$

$$\dfrac{2A}{h} = \dfrac{h(b_1 + \boxed{b_2})}{h}$$

$$\dfrac{2A}{h} = b_1 + \boxed{b_2}$$

$$\dfrac{2A}{h} - b_1 = b_1 - b_1 + \boxed{b_2}$$

$$b_2 = \dfrac{2A}{h} - b_1$$

73) a) $P = 2l + 2w$ for w

$$P = 2l + 2\boxed{w}$$

$$P - 2l = 2l - 2l + 2\boxed{w}$$

$$P - 2l = 2\boxed{w}$$

$$\dfrac{P - 2l}{2} = \dfrac{2\boxed{w}}{2}$$

$$w = \dfrac{P - 2l}{2}$$

b) $w = \dfrac{P - 2l}{2}, P = 28, l = 11$

$$w = \dfrac{28 - 2(11)}{2}$$

$$w = \dfrac{28 - 22}{2}$$

$$w = \dfrac{6}{2}, w = 3$$

75) a) $C = \dfrac{5}{9}(F - 32)$ for F

$$C = \frac{5}{9}(\boxed{F} - 32)$$

$$9 \cdot C = \frac{5}{9}(\boxed{F} - 32) \cdot 9$$

$$9C = 5(\boxed{F} - 32)$$

$$\frac{9C}{5} = \frac{5(\boxed{F} - 32)}{5}$$

$$\frac{9C}{5} = \boxed{F} - 32$$

$$\frac{9C}{5} + 32 = F$$

$$F = \frac{9C}{5} + 32$$

b) $F = \dfrac{9C}{5} + 32, C = 25°$

$$F = \frac{9(25)}{5} + 32$$

$$F = \frac{9(\cancel{25}^{5})}{\cancel{5}_{1}} + 32$$

$$F = 45 + 32, F = 77°$$

Section 2.4

1) a) Value in dollars b) Value in cents

$0.10 \cdot 8 = \$0.80 \qquad 10 \cdot 8 = 80¢$

3) a) Value in dollars b) Value in cents

$0.01 \cdot 217 = \$2.17 \quad 1 \cdot 217 = 217¢$

5) a) Value in dollars

$(0.25)(9) + (0.10)(7)$

$2.25 + 0.70 = \$2.95$

a) Value in cents

$(25)(9) + (10)(7)$

$225 + 70 = 295$

7) a) Value in dollars b) Value in cents

$0.25 \cdot q = 0.25q \qquad 25 \cdot q = 25q$

9) a) Value in dollars b) Value in cents

$0.10 \cdot d = 0.10d \qquad 10 \cdot d = 10d$

11) a) Value in dollars

$(0.01)(p) + (0.25)(q)$

$0.01p + 0.25q$

a) Value in cents

$(1)(p) + (25)(q)$

$p + 25q$

13) Let n = nickels, $n + 8$ = quarters

$$0.05n + 0.25(n + 8) = 4.70$$

$$100[0.05n + 0.25(n + 8)] = 100(4.70)$$

$$5n + 25(n + 8) = 470$$

$$5n + 25n + 200 = 470$$

$$30n + 200 = 470$$

$$30n + 200 - 200 = 470 - 200$$

$$30n = 270$$

$$\frac{30n}{30} = \frac{270}{30}$$

$$n = 9 \text{ nickels}$$

$$n + 8 = 9 + 8 = 17 \text{ quarters}$$

15) Let $b = \$5$ bills, $25 - b = \$1$ bills

$$5b + 1(25 - b) = 69$$
$$5b + 25 - 1b = 69$$
$$4b + 25 = 69$$
$$4b + 25 - 25 = 69 - 25$$
$$4b = 44$$
$$b = 11 \ \$5 \text{ bills}$$
$$25 - b = 25 - 11 = 14 \ \$1 \text{ bills}$$

17) Let a = adult ticket, $\dfrac{1}{2}a$ = child ticket

$$9a + 7 \cdot \frac{1}{2}a = 475$$
$$9a + \frac{7}{2}a = 475$$
$$\frac{18}{2}a + \frac{7}{2}a = 475$$
$$\frac{25}{2}a = 475$$
$$\frac{2}{25} \cdot \frac{25}{2}a = \overset{19}{\cancel{475}} \cdot \frac{2}{\underset{1}{\cancel{25}}}$$
$$a = 19 \cdot 2 = 38 \text{ adult tickets}$$
$$\frac{1}{2}a = \frac{1}{\underset{1}{\cancel{2}}} \cdot \overset{19}{\cancel{38}} = 19 \text{ child tickets}$$

19) 5% of 40 oz

$$0.05 \cdot 40 = 2 \text{ oz}$$

21) 10% of 60 mL + 4% of 40 mL

$$(0.10 \cdot 60) + (0.04 \cdot 40) = 6 + 1.6 = 7.6 \text{ mL}$$

23) Let $x = 4\%$ soln, $24 - x = 10\%$ soln

$$0.04x + 0.10(24 - x) = 0.06(24)$$
$$100\left[0.04x + 0.10(24 - x)\right] = 100\left[0.06(24)\right]$$
$$4x + 10(24 - x) = 6(24)$$
$$4x + 240 - 10x = 144$$
$$-6x + 240 = 144$$
$$-6x = -96$$
$$x = 16 \text{ oz of } 4\% \text{ soln}$$
$$24 - x = 24 - 16 = 8 \text{ oz of } 10\% \text{ soln}$$

25) Let $x = 40\%$ antifreeze

$$x + 5 = 60\% \text{ antifreeze}$$
$$0.40x + 0.70(5) = 0.60(x + 5)$$
$$100\left[0.40x + 0.70(5)\right] = 100\left[0.60(x + 5)\right]$$
$$40x + 70(5) = 60(x + 5)$$
$$40x + 350 = 60x + 300$$
$$-20x + 350 = 300$$
$$-20x = -50$$
$$\frac{-20x}{-20} = \frac{-50}{-20}$$
$$x = 2\frac{1}{2}\text{L of } 40\% \text{ antifreeze}$$

27) Let $x =$ lb of \$6 Aztec

   $5 - x =$ lb of \$8 Cinnamon

$6.00x + 8.00(5 - x) = 7.20(5)$

$100[6.00x + 8.00(5 - x)] = 100[7.20(5)]$

$600x + 800(5 - x) = 720(5)$

$600x + 4000 - 800x = 3600$

$-200x + 4000 = 3600$

$-200x = -400$

$x = 2$ lb of \$6 Aztec

$5 - x = 5 - 2 = 3$ lb of \$8 Cinnamon

29) Let $r =$ mph westbound car

   Let $r - 8 =$ mph eastbound car

t	r	d
3	r	$3r$
3	$r - 8$	$3(r - 8)$

$3r + 3(r - 8) = 414$

$3r + 3r - 24 = 414$

$6r - 24 = 414$

$6r = 438$

$r = 73$ mph westbound car

$r - 8 = 73 - 8 = 65$ mph eastbound car

31) Let $t =$ time Yvette drives

   Let $t - \dfrac{1}{6} =$ time Maureen drives

t	r	d
t	60	$60t$
$t - \dfrac{1}{6}$	72	$72(x - \dfrac{1}{6})$

$60t = 72(t - \dfrac{1}{6})$

$60t = 72t - 12$

$-12t = -12$

$t = 1$

$t - \dfrac{1}{6} = 1 - \dfrac{1}{6} = \dfrac{6}{6} - \dfrac{1}{6} = \dfrac{5}{6}$ hr

33) Let $r =$ speed of freight train

   Let $r + 20 =$ speed of passenger train

t	r	d
5	r	$5r$
5	$r + 20$	$5(r + 20)$

$5r + 5(r + 20) = 400$

$5r + 5r + 100 = 400$

$10r + 100 = 400$

$10r = 300$

$r = 30$ mph of freight train

$r + 20 = 30 + 20 = 50$ mph of passenger train

35) Let $t =$ time truck and car will be
6 miles apart

t	r	d
t	35	$35t$
t	45	$45t$

$35t + 6 = 45t$

$6 = 10t$

$\dfrac{10t}{10} = \dfrac{6}{10}$

$t = \dfrac{3}{5}$

$\dfrac{3}{\cancel{5}} \cdot \overset{12}{\cancel{60}} = 36$ minutes

37) Let $t =$ time Ajay and Rojan will be
105 mi apart

t	r	d
t	30	$30t$
t	40	$40t$

$30t + 40t = 105$

$70t = 105$

$\dfrac{70t}{70} = \dfrac{105}{70}$

$t = \dfrac{3}{2}$ or $1\dfrac{1}{2}$ hr

$\dfrac{3}{\cancel{2}} \cdot \overset{30}{\cancel{60}} = 90$ minutes $= 1$ hour 30 minutes

$3:00 + 1:30 = 4:30$ pm

39) Let $r =$ speed car is traveling

t	r	d
1	30	30
$\dfrac{1}{2}$	r	$\dfrac{1}{2}r$

$30 + \dfrac{1}{2}r = 54$

$2\left(30 + \dfrac{1}{2}r\right) = 2(54)$

$60 + r = 108$

$r = 48$ mph car is traveling

41) Let $q =$ quarters, $q + 7 =$ dimes

$0.25q + 0.10(q + 7) = 6.30$

$100[0.25q + 0.10(q + 7)] = 100(6.30)$

$25q + 10(q + 7) = 630$

$25q + 10q + 70 = 630$

$35q + 70 = 630$

$35q + 70 - 70 = 630 - 70$

$35q = 560$

$\dfrac{35n}{35} = \dfrac{560}{35}$

$q = 16$ quarters

$q + 7 = 16 + 7 = 23$ dimes

43) Let $r =$ speed of small plane

Let $2r =$ speed of jet

t	r	d
30	r	$30r$
30	$2r$	$60r$

$30r + 100 = 60r$

$100 = 30r$

$\dfrac{30r}{30} = \dfrac{100}{30}$

$r = \dfrac{10}{3}$ mi per min

$\dfrac{10}{3} \cdot 60 = \dfrac{10}{\cancel{3}} \cdot \cancel{60}^{20} = 200$ mph

$2r = 2 \cdot 200 = 400$ mph

45) Let $x = 0.08\%$ soln, $20 - x = 0.03\%$ soln

$0.0008x + 0.0003(20 - x) = 0.0005(20)$

$10,000\left[0.0008x + 0.0003(20 - x)\right]$

$= 10,000\left[0.0005(20)\right]$

$8x + 3(20 - x) = 5(20)$

$8x + 60 - 3x = 100$

$5x + 60 = 100$

$5x = 40$

$x = 8$ oz of 0.08% soln

$20 - x = 20 - 8 = 12$ oz of 0.03% soln

Chapter 2 Review

1) $\qquad 2n + 13 = 10; n = -\dfrac{3}{2}$

$\cancel{2}^{1}\left(-\dfrac{3}{\cancel{2}_{1}}\right) + 13 = 10$

$-3 + 13 = 10$

$10 = 10$ yes

3) $\qquad \dfrac{3}{2}k - 5 = 1; k = -4$

$\dfrac{3}{\cancel{2}}\left(-\cancel{4}^{2}\right) - 5 = 1$

$-6 - 5 = 1$

$-11 \neq 1$ no

5) The variables are eliminated and you get a false statement like $5 = 13$.

7) $-9z = 30$

$\dfrac{-9z}{-9} = \dfrac{30}{-9}$

$z = -\dfrac{10}{3}$ or $-3\dfrac{1}{3}$

9) $\qquad 21 = k + 2$

$21 - 2 = k + 2 - 2$

$19 = k, k = 19$

11) $\qquad -\dfrac{4}{9}w = -\dfrac{10}{7}$

$\left(-\dfrac{9}{4}\right)\left(-\dfrac{4}{9}\right)w = \left(-\dfrac{\cancel{10}^{5}}{7}\right)\left(-\dfrac{9}{\cancel{4}_{2}}\right)$

$w = \dfrac{45}{14}$

13) $21 = 0.6q$

$0.6q = 21$

$\dfrac{0.6q}{0.6} = \dfrac{21}{0.6}$

$q = 35$

15) $6 = 15 + \dfrac{9}{2}v$

$6 - 15 = 15 - 15 + \dfrac{9}{2}v$

$-9 = \dfrac{9}{2}v$

$\dfrac{2}{\cancel{9}^{1}}\left(-\cancel{9}^{1}\right) = \dfrac{2}{9} \cdot \dfrac{9}{2}v$

$v = -2$

17) $\dfrac{2}{7} - \dfrac{3}{4}k = -\dfrac{17}{14}$

$28\left(\dfrac{2}{7} - \dfrac{3}{4}k\right) = 28\left(-\dfrac{17}{14}\right)$

$28 \cdot \dfrac{2}{7} - 28 \cdot \left(-\dfrac{3}{4}\right) = 28\left(-\dfrac{17}{14}\right)$

$8 - 21k = -34$

$-21k = -42$

$k = 2$

19) $11x + 13 = 2x - 5$

$11x - 2x + 13 = 2x - 2x - 5$

$9x + 13 = -5$

$9x + 13 - 13 = -5 - 13$

$9x = -18$

$\dfrac{9x}{9} = \dfrac{-18}{9}$

$x = -2$

21) $6 - 5(4d - 3) = 7(3 - 4d) + 8d$

$6 - 20d + 15 = 21 - 28d + 8d$

$21 - 20d = 21 - 20d$

$21 - 20d + 20d = 21 - 20d + 20d$

$21 = 21$

All real numbers

23) $0.05m + 0.11(6 - m) = 0.08(6)$

$100\left[0.05m + 0.11(6 - m)\right] = 100\left[0.08(6)\right]$

$5m + 11(6 - m) = 8(6)$

$5m + 66 - 11m = 48$

$66 - 6m = 48$

$66 - 66 - 6m = 48 - 66$

$-6m = -18$

$\dfrac{-6m}{-6} = \dfrac{-18}{-6}$

$m = 3$

25) $-0.78 = -0.6t$

$\dfrac{\cancel{-0.78}^{1.3}}{\cancel{-0.6}} = \dfrac{\cancel{-0.6}}{\cancel{-0.6}}t$

$1.3 = t$

27) $0.18a + 0.1(20 - a) = 0.14(20)$

$0.18a + (2 - 0.1a) = 2.8$

$0.18a - 0.1a + 2 = 2.8$

$0.08a = 2.8 - 2$

$0.08a = 0.8$

$\dfrac{0.08}{0.08}a = \dfrac{0.8}{0.08}$

$a = 10$

29) $16 = -\dfrac{12}{5}d$

$-\dfrac{5}{\cancel{12}_{3}}\left(\cancel{16}^{4}\right) = -\dfrac{12}{5} \cdot \left(-\dfrac{5}{12}\right)d$

$\left\{-\dfrac{20}{3}\right\} = d$

31) Let x = a number

$x - 12 = 5$

$x - 12 + 12 = 5 + 12$

$x = 17$

The number is 17.

33) $26 - c$

35) Let x = the number of Clarkson CDs sold. Then the number of Aiken CDs sold = $x + 316{,}000$.

$\begin{pmatrix} \text{Clarkson} \\ \text{CDs} \end{pmatrix} + \begin{pmatrix} \text{Aiken} \\ \text{CDs} \end{pmatrix} = 910{,}000$

$x \quad + x + 316{,}000 = 910{,}000$

$2x = 594{,}000$

$x = 297{,}000$

Aiken CDs = $x + 316{,}000$

$= 297{,}000 + 316{,}000$

$= 613{,}000$

Clarkson sold 297,000 copies and Aiken sold 613,000 copies.

37)

Let x = the length of the gravel portion.
Then the length of the paved portion = $3x$.
gravel portion + paved portion = 500

$x \quad + \quad 3x \quad = 500$

$4x = 500$

$x = 125$

The gravel portion of the road is 125 ft long.

39) Let x = the vet care spending in 2007.

2008 Spending = 2007 Spending + Increase

$11 = x + x(0.085)$

$11 = 1.085x$

$10.1 \approx x$

The pet care spending was $10.1 billion in 2007.

41) h = height of triangle

$b = 9$, $A = 54$

$A = \dfrac{1}{2}bh$

$54 = \dfrac{1}{2}(9)h$

$108 = 9h$

$12 = h$

The height of the triangle is 12 cm.

43) $x + x + (x + 15) = 180$

$3x + 15 = 180$

$3x = 165$

$x = 55$

$x + 15 = 70$

$m\angle A = 55°,\ m\angle B = 55°,\ m\angle C = 70°$

45) $(3x+1)° = (4x-14)°$

$3x + 1 = 4x - 14$

$15 = x$

$(3x+1)° = (3(15)+1)° = 46°$

$(4x-14)° = (4(15)-14)° = 46°$

47)
$$p - n = x$$
$$p - n + n = x + n$$
$$p = x + n$$

49) $pV = nRT$
$$\frac{pV}{nT} = \frac{nRT}{nT}$$
$$\frac{pV}{nT} = R$$

51) $x = \#$ of dimes; $91 - x = \#$ of quarters
Dime Value + Quarter Value = Ttl Value
$$0.10x + 0.25(91 - x) = 14.05$$
$$100\big(0.10x + 0.25(91 - x)\big) = 100(14.05)$$
$$10x + 25(91 - x) = 1405$$
$$10x + 2275 - 25x = 1405$$
$$-15x = -870$$
$$x = 58$$

quarters $= 91 - x$
$$= 91 - 58$$
$$= 33$$
There are 58 dimes and 33 quarters.

53) $t =$ the amount of time traveling for Peter
$t - \dfrac{1}{4} =$ the amount of time
traveling for Mitchell

	d	= r	·	t
Peter	30t	30		t
Mitchell	$40\left(t - \dfrac{1}{4}\right)$	40		$t - \dfrac{1}{4}$

Peter's distance = Mitchell's distance

$$30t = 40\left(t - \frac{1}{4}\right)$$
$$30t = 40t - 10$$
$$-10t = -10$$
$$t = 1$$
Mitchell's Time $= t - \dfrac{1}{4}$
$$= 1 - \frac{1}{4}$$
$$= \frac{3}{4}; \frac{3}{4}\,\text{hr}$$
$$= 45\,\text{min}$$
It will take Mitchell 45 minutes
to catch Peter.

Chapter 2 Test

1) $7p + 16 = 30$
$$7p = 14$$
$$p = 2$$

3) $\dfrac{5}{8}(3k + 1) - \dfrac{1}{4}(7k + 2) = 1$
$$8\left[\frac{5}{8}(3k + 1) - \frac{1}{4}(7k + 2)\right] = 8(1)$$
$$5(3k + 1) - 2(7k + 2) = 8$$
$$15k + 5 - 14k - 4 = 8$$
$$k + 1 = 8$$
$$k = 7$$

5) $8(3 + n) = 5(2n - 3)$
$$24 + 8n = 10n - 15$$
$$24 + 15 = 10n - 8n$$
$$39 = 2n$$
$$\frac{39}{2} = \frac{2}{2}n$$
$$\left\{\frac{39}{2}\right\} = n$$

33

Chapter 2: Test

7) Let x = the number

$$2x - 9 = 33$$
$$2x - 9 + 9 = 33 + 9$$
$$2x = 42$$
$$\frac{2x}{2} = \frac{42}{2}$$
$$x = 21$$

Check: $2(21) - 9 = 33$

9) Let P = Perimeter of the airplane tray

Let w = width and l = length.

$$P = 2l + 2w$$
Then, $l = w + 5$
$$2(w + 5) + 2w = 50$$
$$2w + 10 + 2w = 50$$
$$4w = 50 - 10$$
$$\frac{4}{4}w = \frac{40}{4}$$
$$w = 10 \text{ in.}$$

Then length, $l = w + 5 = 10 + 5 = 15$ in.

Dimensions are 10 in. × 15 in.

11) Let x = number of drive-in theaters in Wisconsin in 1967

$$9 = x - 0.82x$$
$$9 = 0.18x$$
$$\frac{0.18x}{0.18} = \frac{9}{0.18}$$
$$x = 50 \text{ drive-in theaters}$$

13) Let x be speed of the eastbound car.

Let $(x + 6)$ be the speed of the westbound car.

$$\text{Speed} = \frac{\text{Distance}}{\text{Time}}$$
$$= \frac{345}{2.5}$$
$$= 138$$

$$x + (x + 6) = 138$$
$$2x + 6 = 138$$
$$2x = 132$$
$$x = 66$$

Eastbound car: 66 mph; westbound car: 72 mph

15) $S = 2\pi r^2 + 2\pi rh$ for h

$$S = 2\pi r^2 + 2\pi r\boxed{h}$$
$$S - 2\pi r^2 = 2\pi r^2 - 2\pi r^2 + 2\pi r\boxed{h}$$
$$S - 2\pi r^2 = 2\pi r\boxed{h}$$
$$\frac{S - 2\pi r^2}{2\pi r} = \frac{2\pi r\boxed{h}}{2\pi r}$$
$$h = \frac{S - 2\pi r^2}{2\pi r}$$

17) $m\angle A = x°$ $m\angle B = (4x + 11)°$

$m\angle C = (x + 13)°$

$$m\angle A + m\angle B + m\angle C = 180$$
$$x + (4x + 11) + (x + 13) = 180$$
$$6x + 24 = 180$$
$$6x = 156$$
$$x = 26$$

$$m\angle A = x° = 26°$$
$$m\angle B = (4x + 11)° = (104 + 11)° = 115°$$
$$m\angle C = (x + 13)° = (26 + 13)° = 39°$$

Cumulative Review: Chapters 1–2

1) $\dfrac{5}{12}-\dfrac{7}{9}=\dfrac{15}{36}-\dfrac{28}{36}=-\dfrac{13}{36}$

3) $52-12\div4+3\cdot5$

$\qquad52-3+15$

$\qquad\quad49+15$

$\qquad\qquad64$

5) $-3^{4}=-1(3)(3)(3)(3)=-81$

7) Given the set of numbers

$\left\{-13.7,\dfrac{19}{7},0,8,\sqrt{17},0.\overline{61},\sqrt{81},-2\right\}$

The rational numbers are

$\left\{-13.7,\dfrac{19}{7},0,8,0.\overline{61},\sqrt{81},-2\right\}$

9) Given the set of numbers

$\left\{-13.7,\dfrac{19}{7},0,8,\sqrt{17},0.\overline{61},\sqrt{81},-2\right\}$

The whole numbers are $\{0,8\}$

11) $9+(4+1)=(9+4)+1$

Associative Property

13) $5\times\dfrac{1}{5}=1$

Inverse Property

15) $-31=\dfrac{4}{7}z+9$

$-31-9=\dfrac{4}{7}z+9-9$

$\qquad-40=\dfrac{4}{7}z$

$\dfrac{7}{4}\cdot\dfrac{4}{7}z=-40\cdot\dfrac{7}{4}$

$z=-\overset{10}{\cancel{40}}\cdot\dfrac{7}{\underset{1}{\cancel{4}}}$

$z=-70$

17) $\dfrac{1}{2}(t+1)-\dfrac{1}{3}=\dfrac{17}{12}+\dfrac{1}{4}(2t-5)$

$12\left[\dfrac{1}{2}(t+1)-\dfrac{1}{3}\right]=12\left[\dfrac{17}{12}+\dfrac{1}{4}(2t-5)\right]$

$\qquad6(t+1)-4=17+3(2t-5)$

$\qquad\quad6t+6-4=17+6t-15$

$\qquad\qquad6t+2=2+6t$

$\qquad\qquad\quad2=2$

All real numbers

19) Let x = supplement

$\qquad x+59=180$

$\quad x+59-59=180-59$

$\qquad\qquad x=121°$

21) $A=l\cdot w$ and $P=2l+2w$

$A=15\text{ cm}\cdot7\text{ cm}=105\text{ cm}^{2}$

$P=2(15\text{ cm})+2(7\text{ cm})$

$\quad=30\text{ cm}+14\text{ cm}=44\text{ cm}$

23) Let $x =$ name brand drugs

Let $72 - x =$ generic drugs

$$2x + x = 72$$
$$3x = 72$$
$$x = 24 \text{ name brand drugs}$$
$$72 - x = 72 - 24 = 48 \text{ generic drugs}$$

25) Let x be the original price of the food.

Lorenzo purchased a bag of dog food = $28
30% off the original price

Sale price = Original price – Amount of discount
$$28 = x - 0.30x$$
$$28 = 0.70x$$
$$\frac{28}{0.70} = \frac{0.70}{0.70}x$$
$$x = 40$$

The original price of the dog food is $40.

Chapter 3: Linear Inequalities and Absolute Value

Section 3.1

1) You use parentheses when there is a $<$ or $>$ symbol or when you use ∞ or $-\infty$.

3)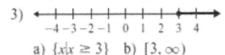

a) $\{x | x \geq 3\}$ b) $[3, \infty)$

5)

a) $\{c | c < -1\}$ b) $(-\infty, -1)$

7)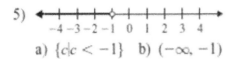

a) $\left\{ w \,\middle|\, w > -\dfrac{11}{3} \right\}$ b) $\left(-\dfrac{11}{3}, \infty \right)$

9)

a) $\{n | 1 \leq n \leq 4\}$ b) $[1, 4]$

11)

a) $\{a | -2 < a < 1\}$ b) $(-2, 1)$

13)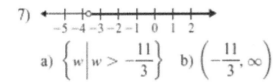

a) $\left\{ z \,\middle|\, \dfrac{1}{2} < z \leq 3 \right\}$ b) $\left(\dfrac{1}{2}, 3 \right]$

15) $n - 8 \leq -3$
$$n - 8 + 8 \leq -3 + 8$$
$$n \leq 5$$

a) $\{n | n \leq 5\}$ b) $(-\infty, 5]$

17) $y + 5 \geq 1$
$$y + 5 - 5 \geq 1 - 5$$
$$y \geq -4$$

a) $\{y | y \geq -4\}$ b) $[-4, \infty)$

19) $3c > 12$
$$\frac{3c}{3} > \frac{12}{3}$$
$$c > 4$$

a) $\{c | c > 4\}$ b) $(4, \infty)$

21) $15k < -55$
$$\frac{15k}{15} < \frac{-55}{15}$$
$$k < -\frac{11}{3}$$

a) $\left\{ k \,\middle|\, k < -\dfrac{11}{3} \right\}$ b) $\left(-\infty, -\dfrac{11}{3} \right)$

23) $-4b \leq 32$
$$\frac{-4b}{-4} \geq \frac{32}{-4}$$
$$b \geq -8$$

a) $\{b | b \geq -8\}$ b) $[-8, \infty)$

25) $-14w > -42$
$$\frac{-14w}{-14} < \frac{-42}{-14}$$
$$w < 3$$

a) $\{w | w < 3\}$ b) $(-\infty, 3)$

27) $\dfrac{1}{3}x < -2$

$3 \cdot \dfrac{1}{3}x < -2 \cdot 3$

$x < -6$

a) $\{x | x < -6\}$ b) $(-\infty, -6)$

29) $-\dfrac{2}{5}p \geq 4$

$\left(-\dfrac{5}{2}\right) \cdot \left(-\dfrac{2}{5}\right)p \leq \overset{2}{\cancel{4}} \cdot \left(-\dfrac{5}{\underset{1}{\cancel{2}}}\right)$

$p \leq -10$

a) $\{p | p \leq -10\}$ b) $(-\infty, -10]$

31) $8z + 19 > 11$

$8z + 19 - 19 > 11 - 19$

$8z > -8$

$\dfrac{8z}{8} > \dfrac{-8}{8}$

$z > -1$

$(-1, \infty)$

$(-\infty, 4]$

33) $12 - 7t \geq 15$

$12 - 12 - 7t \geq 15 - 12$

$-7t \geq 3$

$\dfrac{-7t}{-7} \leq \dfrac{3}{-7}$

$t \leq -\dfrac{3}{7}$

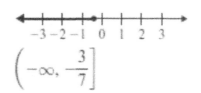

$\left(-\infty, -\dfrac{3}{7}\right]$

35) $-23 - w < -20$

$-23 + 23 - w < -20 + 23$

$-w < 3$

$w > -3$

$(-3, \infty)$

37) $7a + 4(5 - a) \leq 4 - 5a$

$7a + 20 - 4a \leq 4 - 5a$

$3a + 20 \leq 4 - 5a$

$3a + 20 - 20 \leq 4 - 5a - 20$

$3a \leq -16 - 5a$

$3a + 5a \leq -16 - 5a + 5a$

$8a \leq -16$

$\dfrac{8a}{8} \leq \dfrac{-16}{8}$

$a \leq -2$

$(-\infty, -2]$

39) $9c + 17 > 14c - 3$

$9c + 17 - 17 > 14c - 3 - 17$

$9c > 14c - 20$

$9c - 14c > 14c - 14c - 20$

$-5c > -20$

$\dfrac{-5c}{-5} < \dfrac{-20}{-5}$

$c < 4$

$(-\infty, 4)$

38

41) $\frac{8}{3}(2k+1) > \frac{1}{6}k + \frac{8}{3}$

$6\left[\frac{8}{3}(2k+1)\right] > 6\left[\frac{1}{6}k + \frac{8}{3}\right]$

$16(2k+1) > k+16$

$32k+16 > k+16$

$32k+16-16 > k+16-16$

$32k > k+0$

$32k-k > k-k+0$

$31k > 0$

$\frac{31k}{31} > \frac{0}{31}$

$k > 0$

$(0, \infty)$

43) $0.04x + 0.12(10-x) \geq 0.08(10)$

$100[0.04x + 0.12(10-x)] \geq 100[0.08(10)]$

$4x + 12(10-x) \geq 8(10)$

$4x + 120 - 12x \geq 80$

$-8x + 120 \geq 80$

$-8x + 120 - 120 \geq 80 - 120$

$-8x \geq -40$

$\frac{-8x}{-8} \leq \frac{-40}{-8}$

$x \leq 5$

$(-\infty, 5]$

45) $-8 \leq a - 5 \leq -4$

$-8+5 \leq a-5+5 \leq -4+5$

$-3 \leq a \leq 1$

$[-3, 1]$

47) $9 < 6n < 18$

$\frac{9}{6} < \frac{6n}{6} < \frac{18}{6}$

$\frac{3}{2} < n < 3$

$\left(\frac{3}{2}, 3\right)$

49) $-19 \leq 7p + 9 \leq 2$

$-19-9 \leq 7p+9-9 \leq 2-9$

$-28 \leq 7p \leq -7$

$\frac{-28}{7} \leq \frac{7p}{7} \leq \frac{-7}{7}$

$-4 \leq p \leq -1$

$[-4, -1]$

51) $-6 \leq 4c - 13 < -1$

$-6+13 \leq 4c-13+13 < -1+13$

$7 \leq 4c < 12$

$\frac{7}{4} \leq \frac{4c}{4} < \frac{12}{4}$

$\frac{7}{4} \leq c < 3$

$\left[\frac{7}{4}, 3\right)$

53) $2 < \dfrac{3}{4}u + 8 < 11$

$$4 \cdot 2 < 4\left(\dfrac{3}{4}u + 8\right) < 11 \cdot 4$$

$$8 < 3u + 32 < 44$$

$$8 - 32 < 3u + 32 - 32 < 44 - 32$$

$$-24 < 3u < 12$$

$$\dfrac{-24}{3} < \dfrac{3u}{3} < \dfrac{12}{3}$$

$$-8 < u < 4$$

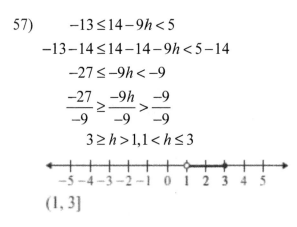

$(-8, 4)$

55) $-\dfrac{1}{2} \le \dfrac{5d+2}{6} \le 0$

$$\overset{3}{\cancel{6}}\left(-\dfrac{1}{\cancel{2}}\right) \le 6\left(\dfrac{5d+2}{6}\right) \le 0 \cdot 6$$

$$-3 \le 5d + 2 \le 0$$

$$-3 - 2 \le 5d + 2 - 2 \le 0 - 2$$

$$-5 \le 5d \le -2$$

$$\dfrac{-5}{5} \le \dfrac{5d}{5} \le \dfrac{-2}{5}$$

$$-1 \le d \le -\dfrac{2}{5}$$

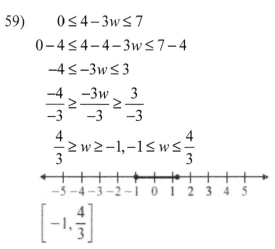

$\left[-1, -\dfrac{2}{5}\right]$

57) $-13 \le 14 - 9h < 5$

$$-13 - 14 \le 14 - 14 - 9h < 5 - 14$$

$$-27 \le -9h < -9$$

$$\dfrac{-27}{-9} \ge \dfrac{-9h}{-9} > \dfrac{-9}{-9}$$

$$3 \ge h > 1, 1 < h \le 3$$

$(1, 3]$

59) $0 \le 4 - 3w \le 7$

$$0 - 4 \le 4 - 4 - 3w \le 7 - 4$$

$$-4 \le -3w \le 3$$

$$\dfrac{-4}{-3} \ge \dfrac{-3w}{-3} \ge \dfrac{3}{-3}$$

$$\dfrac{4}{3} \ge w \ge -1, -1 \le w \le \dfrac{4}{3}$$

$\left[-1, \dfrac{4}{3}\right]$

61) $k + 11 > 4$

$$k + 11 - 11 > 4 - 11$$

$$k > -7$$

$$(-7, \infty)$$

63) $-12p \ge -16$

$$\dfrac{-12p}{-12} \le \dfrac{-16}{-12}$$

$$p \le \dfrac{4}{3}$$

$$\left(-\infty, \dfrac{4}{3}\right]$$

65) $5(2b-3)-7b > 5b+9$

$\quad 10b-15-7b > 5b+9$

$\quad\quad 3b-15 > 5b+9$

$\quad 3b-3b-15 > 5b+9-3b$

$\quad\quad -15 > 2b+9$

$\quad -15-9 > 2b+9-9$

$\quad\quad -24 > 2b$

$\quad\quad \dfrac{2b}{2} < \dfrac{-24}{2}$

$\quad\quad b < -12$

$\quad\quad (-\infty, -12)$

67) $-12 < \dfrac{8}{5}t+12 \le 6$

$\quad 5(-12) < 5\left(\dfrac{8}{5}t+12\right) \le 6 \cdot 5$

$\quad -60 < 8t+60 \le 30$

$\quad -60-60 < 8t+60-60 \le 30-60$

$\quad -120 < 8t \le -30$

$\quad \dfrac{-120}{8} < \dfrac{8t}{8} \le \dfrac{-30}{8}$

$\quad -15 < t \le -\dfrac{15}{4}$

$\quad\quad \left(-15, -\dfrac{15}{4}\right]$

69) $\dfrac{5}{4}(k+4)+\dfrac{1}{4} \ge \dfrac{5}{6}(k+3)-1$

$\quad 12\left[\dfrac{5}{4}(k+4)+\dfrac{1}{4}\right] \ge 12\left[\dfrac{5}{6}(k+3)-1\right]$

$\quad 15(k+4)+3 \ge 10(k+3)-12$

$\quad 15k+60+3 \ge 10k+30-12$

$\quad 15k+63 \ge 10k+18$

$\quad 15k+63-63 \ge 10k+18-63$

$\quad 15k \ge 10k-45$

$\quad 15k-10k \ge 10k-10k-45$

$\quad 5k \ge -45$

$\quad \dfrac{5k}{5} \ge \dfrac{-45}{5}$

$\quad k \ge -9$

$\quad\quad [-9, \infty)$

71) $\quad 4 < 4-7y \le 18$

$\quad 4-4 < 4-4-7y \le 18-4$

$\quad 0 < -7y \le 14$

$\quad \dfrac{0}{-7} > \dfrac{-7y}{-7} \ge \dfrac{14}{-7}$

$\quad 0 > y \ge -2, -2 \le y < 0$

$\quad\quad [-2, 0)$

73) Let $x = 15$ min intervals over 75 minutes

$\quad 19+5x \le 50$

$\quad 19-19+5x \le 50-19$

$\quad 5x \le 31$

$\quad \dfrac{5x}{5} \le \dfrac{31}{5}$

$\quad x \le 6.2$

$\quad 75+15(6) = 75+90 = 165$

$\quad 165 \div 60 = 2.75 = 2$ hr 45 min

75) Let x = fifths of a mile

$$2 + 0.25x \le 12$$

$$2 - 2 + 0.25x \le 12 - 2$$

$$0.25x \le 10$$

$$\frac{0.25x}{0.25} \le \frac{10}{0.25}$$

$$x \le 40$$

$$\frac{1}{5} \cdot 40 = \frac{1}{\cancel{5}} \cdot \overset{8}{\cancel{40}} = 8 \text{ mi}$$

77) Let x = grade on 3rd test

$$\frac{87 + 94 + x}{3} \ge 90$$

$$3\left(\frac{87 + 94 + x}{3}\right) \ge 3(90)$$

$$87 + 94 + x \ge 270$$

$$181 + x \ge 270$$

$$181 - 181 + x \ge 270 - 181$$

$$x \ge 89$$

Section 3.2

1) $A \cap B$ means "A intersect B." $A \cap B$ is the set of all numbers which are in set A and set B.

3) $\{8\}$

5) $\{2, 4, 5, 6, 7, 8, 9, 10\}$

7) $\varnothing$

9) $\{1, 2, 3, 4, 5, 6, 8, 10\}$

11)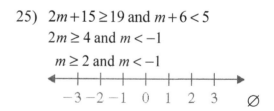
$[-3, 2]$

13)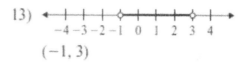
$(-1, 3)$

15)
$[3, \infty)$

17)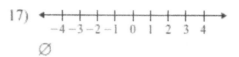
$\varnothing$

19)
$[2, 5]$

21) $b - 7 > -9$ and $8b < 24$

$b > -2$ and $b < 3$

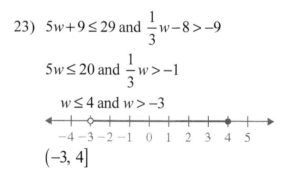

$(-2, 3)$

23) $5w + 9 \le 29$ and $\dfrac{1}{3}w - 8 > -9$

$5w \le 20$ and $\dfrac{1}{3}w > -1$

$w \le 4$ and $w > -3$

$(-3, 4]$

25) $2m + 15 \ge 19$ and $m + 6 < 5$

$2m \ge 4$ and $m < -1$

$m \ge 2$ and $m < -1$

$\varnothing$

27) $r - 10 > -10$ and $3r - 1 > 8$

$r > 0$ and $3r > 9$

$r > 0$ and $r > 3$

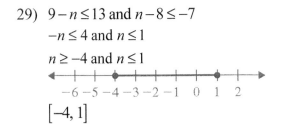

$(3, \infty)$

29) $9 - n \leq 13$ and $n - 8 \leq -7$

$-n \leq 4$ and $n \leq 1$

$n \geq -4$ and $n \leq 1$

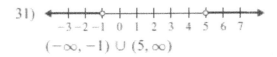

$[-4, 1]$

31)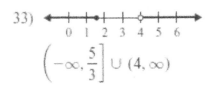

$(-\infty, -1) \cup (5, \infty)$

33)

$\left(-\infty, \dfrac{5}{3} \right] \cup (4, \infty)$

35)

$(1, \infty)$

37)

$(-\infty, \infty)$

39)

$(-\infty, -1) \cup (3, \infty)$

41) $6m \leq 21$ or $m - 5 > 1$

$m \leq \dfrac{7}{2}$ or $m > 6$

$\left(-\infty, \dfrac{7}{2} \right] \cup (6, \infty)$

43) $3t + 4 > -11$ or $t + 19 > 17$

$3t > -15$ or $t > -2$

$t > -5$ or $t > -2$

$(-5, \infty)$

45) $-2v - 5 \leq 1$ or $\dfrac{7}{3} v < -14$

$-2v \leq 6$ or $v < -6$

$v \geq -3$ or $v < -6$

$(-\infty, -6) \cup [-3, \infty)$

47) $c + 3 \geq 6$ or $\dfrac{4}{5} c \leq 10$

$c \geq 3$ or $c \leq \dfrac{25}{2}$

$(-\infty, \infty)$

49) $7 - 6n \geq 19$ or $n + 14 < 11$

$-6n \geq 12$ or $n < -3$

$n \leq -2$ or $n < -3$

$(-\infty, -2]$

51) $4n + 7 \leq 9$ and $n + 6 \geq 1$

$4n \leq 2$

$n \leq \dfrac{1}{2}$ and $n \geq -5$

$\left[-5, \dfrac{1}{2} \right]$

Section 3.3: Absolute Value Equations and Inequalities

53) $\dfrac{4}{3}x+5<2 \quad$ or $x+3\ge 8$

$\qquad \dfrac{4}{3}x<-3$

$\qquad x<-\dfrac{9}{4} \quad$ or $\quad x\ge 5$

$\qquad \left(-\infty,-\dfrac{9}{4}\right)\cup[5,\infty)$

55) $\dfrac{8}{3}w<-16$ or $w+9>-4$

$\qquad w<-6 \quad$ or $\quad w>-13$

$\qquad (-\infty,\infty)$

56) $5y-2>8 \quad$ and $\dfrac{3}{4}y+2<11$

$\qquad 5y>10 \quad$ and $\quad \dfrac{3}{4}y<9$

$\qquad y>2 \quad$ and $\quad y<12$

$\qquad (2,12)$

57) $7-r>7$ and $0.3r<6$

$\qquad -r>0$

$\qquad r<0$ and $\quad r<20$

$\qquad (-\infty,0)$

59) $3-2k>11 \quad$ and $\dfrac{1}{2}k+5\ge 1$

$\qquad -2k>8 \quad$ and $\quad \dfrac{1}{2}k\ge -4$

$\qquad k<-4 \quad$ and $\quad k\ge -8$

$\qquad [-8,-4)$

61) $\{$Liliane Bettancourt, Alice Walton$\}$

63) $\left\{\begin{array}{l}\text{Liliane Bettancourt, J.K.Rowling,}\\ \text{Oprah Winfrey}\end{array}\right\}$

Section 3.3

1) Answers may vary.

3) $|q|=6$

$\qquad \{-6,6\}$

5) $|q-5|=3$

$\qquad q-5=3 \quad$ or $q-5=-3$

$\qquad q=8 \quad$ or $\quad q=2$

$\qquad \{2,8\}$

7) $|4t-5|=7$

$\qquad 4t-5=7 \quad$ or $4t-5=-7$

$\qquad 4t=12$ or $\quad 4t=-2$

$\qquad t=3 \quad$ or $\quad t=-\dfrac{1}{2}$

$\qquad \left\{-\dfrac{1}{2},3\right\}$

9) $|12c+5|=1$

$\qquad 12c+5=1 \quad$ or $12c+5=-1$

$\qquad 12c=-4$ or $\quad 12c=-6$

$\qquad c=-\dfrac{1}{3}$ or $\quad c=-\dfrac{1}{2}$

$\qquad \left\{-\dfrac{1}{2},-\dfrac{1}{3}\right\}$

11) $\left|\dfrac{2}{3}b+3\right|=13$

$\qquad \dfrac{2}{3}b+3=13 \quad$ or $\dfrac{2}{3}b+3=-13$

$\qquad \dfrac{2}{3}b=10$ or $\quad \dfrac{2}{3}b=-16$

$\qquad b=15$ or $\quad b=-24$

$\qquad \{-24,15\}$

13) $\left|4 - \dfrac{3}{5}d\right| = 6$

$4 - \dfrac{3}{5}d = 6$ or $4 - \dfrac{3}{5}d = -6$

$-\dfrac{3}{5}d = 2$ or $-\dfrac{3}{5}d = -10$

$d = -\dfrac{10}{3}$ or $d = \dfrac{50}{3}$

$\left\{-\dfrac{10}{3}, \dfrac{50}{3}\right\}$

15) $|m - 5| = -3$, $\varnothing$

Absolute value is distance, which is always positive, not negative.

17) $|z - 6| + 4 = 20$

$|z - 6| = 16$

$z - 6 = 16$ or $z - 6 = -16$

$z = 22$ or $\quad z = -10$

$\{-10, 22\}$

19) $|2a + 5| + 8 = 13$

$|2a + 5| = 5$

$2a + 5 = 5$ or $2a + 5 = -5$

$2a = 0$ or $\quad 2a = -10$

$a = 0$ or $\quad a = -5$

$\{-5, 0\}$

21) $|w + 14| = 0$

$w + 14 = 0$

$w = -14$

$\{-14\}$

23) $|8n + 11| = -1$, $\varnothing$

Absolute value is distance, which is always positive, not negative.

25) $|5b + 3| + 6 = 19$

$|5b + 3| = 13$

$5b + 3 = 13$ or $5b + 3 = -13$

$5b = 10$ or $\quad 5b = -16$

$b = 2$ or $\quad b = -\dfrac{16}{5}$

$\left\{-\dfrac{16}{5}, 2\right\}$

27) $|3m - 1| + 5 = 2$

$|3m - 1| = -3$

$\varnothing$, Absolute value is distance, which always positive, not negative.

29) $|s + 9| = |2s + 5|$

$s + 9 = 2s + 5$ or $s + 9 = -(2s + 5)$

$-s = -4$ or $s + 9 = -2s - 5$

$s = 4$ or $3s = -14$

$s = 4$ or $s = -\dfrac{14}{3}$

$\left\{-\dfrac{14}{3}, 4\right\}$

31) $|3z + 2| = |6 - 5z|$

$3z + 2 = 6 - 5z$ or $3z + 2 = -(6 - 5z)$

$8z = 4$ or $3z + 2 = -6 + 5z$

$z = \dfrac{1}{2}$ or $-2z = -8$

$z = \dfrac{1}{2}$ or $z = 4$

$\left\{\dfrac{1}{2}, 4\right\}$

Section 3.3: Absolute Value Equations and Inequalities

33) $\left|\frac{3}{2}x-1\right|=|x|$

$\frac{3}{2}x-1=x$ or $\frac{3}{2}x-1=-(x)$

$\frac{1}{2}x=1$ or $\frac{3}{2}x-1=-x$

$x=2$ or $\frac{5}{2}x=1$

$x=2$ or $x=\frac{2}{5}$

$\left\{\frac{2}{5},2\right\}$

35) $\left|\frac{1}{4}t-\frac{5}{2}\right|=\left|5-\frac{1}{2}t\right|$

$\frac{1}{4}t-\frac{5}{2}=5-\frac{1}{2}t$ or $\frac{1}{4}t-\frac{5}{2}=-\left(5-\frac{1}{2}t\right)$

$4\left(\frac{1}{4}t-\frac{5}{2}\right)=4\left(5-\frac{1}{2}t\right)$ or $4\left(\frac{1}{4}t-\frac{5}{2}\right)=4\left[-\left(5-\frac{1}{2}t\right)\right]$

$t-10=20-2t$ or $t-10=-4\left(5-\frac{1}{2}t\right)$

$3t=30$ or $t-10=-20+2t$

$t=10$ or $-t=-10$

$t=10$ or $t=10$

$\{10\}$

37) $|x|=9$, may vary

39) $|x|=\frac{1}{2}$, may vary

41) $[-1,5]$

42) $(-\infty,2)\cup(9,\infty)$

45) $\left(-\infty,-\frac{9}{2}\right]\cup\left[\frac{3}{5},\infty\right)$

47) $|m|\le 7$

$-7\le m\le 7$

$[-7,7]$

49) $|3k|<12$

$-12<3k<12$

$-4<k<4$

$(-4,4)$

51) $|w-2|<4$

$-4<w-2<4$

$-2<w<6$

$(-2,6)$

53) $|3r+10|\le 4$

$-4\le 3r+10\le 4$

$-14\le 3r\le -6$

$-\frac{14}{3}\le r\le -2$

$\left[-\frac{14}{3},-2\right]$

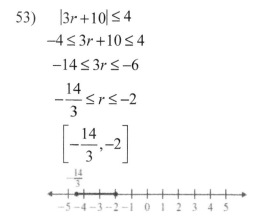

55) $|7 - 6p| \leq 3$

$\quad -3 \leq 7 - 6p \leq 3$

$\quad -10 \leq -6p \leq -4$

$\quad \dfrac{5}{3} \geq p \geq \dfrac{2}{3}$

$\quad \dfrac{2}{3} \leq p \leq \dfrac{5}{3}$

$\quad \left[\dfrac{2}{3}, \dfrac{5}{3}\right]$

57) $|5q + 11| < 0$

$\quad \varnothing$, Absolute value is distance, which is positive, not less than 0, which is negative.

59) $|2x + 7| \leq -12$

$\quad \varnothing$, Absolute value is distance, which is positive, not negative.

61) $|8c - 3| + 15 < 20$

$\quad |8c - 3| < 5$

$\quad -5 < 8c - 3 < 5$

$\quad -2 < 8c < 8$

$\quad -\dfrac{1}{4} < c < 1$

$\quad \left(-\dfrac{1}{4}, 1\right)$

63) $\left|\dfrac{3}{2}h + 6\right| - 2 \leq 10$

$\quad \left|\dfrac{3}{2}h + 6\right| \leq 12$

$\quad -12 \leq \dfrac{3}{2}h + 6 \leq 12$

$\quad -18 \leq \dfrac{3}{2}h \leq 6$

$\quad -12 \leq h \leq 4$

$\quad [-12, 4]$

65) $\quad |t| \geq 7$

$\quad t \geq 7$ or $t \leq -7$

$\quad (-\infty, -7] \cup [7, \infty)$

67) $\quad |d + 10| \geq 4$

$\quad d + 10 \geq 4$ or $d + 10 \leq -4$

$\quad d \geq -6$ or $d \leq -14$

$\quad (-\infty, -14] \cup [-6, \infty)$

69) $\quad |4v - 3| \geq 9$

$\quad 4v - 3 \geq 9$ or $4v - 3 \leq -9$

$\quad 4v \geq 12$ or $4v \leq -6$

$\quad v \geq 3$ or $v \leq -\dfrac{3}{2}$

$\quad \left(-\infty, -\dfrac{3}{2}\right] \cup [3, \infty)$

71) $|17 - 6x| > 5$

$17 - 6x > 5$ or $17 - 6x < -5$

$-6x > -12$ or $-6x < -22$

$x < 2$ or $x > \dfrac{11}{3}$

$(-\infty, 2) \cup \left(\dfrac{11}{3}, \infty\right)$

79) $-3 + \left|\dfrac{5}{6}n + \dfrac{1}{2}\right| \geq 1$

$\left|\dfrac{5}{6}n + \dfrac{1}{2}\right| \geq 4$

$\dfrac{5}{6}n + \dfrac{1}{2} \geq 4$ or $\dfrac{5}{6}n + \dfrac{1}{2} \leq -4$

$5n + 3 \geq 24$ or $5n + 3 \leq -24$

$5n \geq 21$ or $5n \leq -27$

$n \geq \dfrac{21}{5}$ or $n \leq -\dfrac{27}{5}$

$\left(-\infty, -\dfrac{27}{5}\right] \cup \left[\dfrac{21}{5}, \infty\right)$

73) $|8k + 5| \geq 0$

$8k + 5 \geq 0$ or $8k + 5 \leq 0$

$8k \geq -5$ or $8k \leq -5$

$k \geq -\dfrac{5}{8}$ or $k \leq -\dfrac{5}{8}$

$(-\infty, \infty)$

81) The absolute value of a quantity is always 0 or positive; it cannot be less than 0.

83) The absolute value of a quantity is always 0 or positive, so for any real number, x, the quantity $|2x + 1|$ will be greater than -3.

75) $|z - 3| \geq -5$

$z - 3 \geq -5$ or $z - 3 \leq 5$

$z \geq -2$ or $z \leq 8$

$(-\infty, \infty)$

85) $|2v + 9| > 3$

$2v + 9 > 3$ or $2v + 9 < -3$

$2v > -6$ or $2v < -12$

$v > -3$ or $v < -6$

$(-\infty, -6) \cup (-3, \infty)$

77) $|2m - 1| + 4 > 5$

$|2m - 1| > 1$

$2m - 1 > 1$ or $2m - 1 < -1$

$2m > 2$ or $2m < 0$

$m > 1$ or $m < 0$

$(-\infty, 0) \cup (1, \infty)$

87) $3 = |4t + 5|$

$4t + 5 = 3$ or $4t + 5 = -3$

$4t = -2$ or $4t = -8$

$t = -\dfrac{1}{2}$ or $t = -2$

$\left\{-2, -\dfrac{1}{2}\right\}$

89) $9 \le |7 - 8q|$

$7 - 8q \ge 9$ or $7 - 8q \le -9$

$-8q \ge 2$ or $-8q \le -16$

$q \le -\dfrac{1}{4}$ or $q \ge 2$

$\left(-\infty, -\dfrac{1}{4}\right] \cup [2, \infty)$

91) $2(x - 8) + 10 < 4x$

$2x - 16 + 10 < 4x$

$2x - 6 < 4x$

$-6 < 2x$

$2x > -6$

$x > -3$

$(-3, \infty)$

93) $|6y + 5| \le -9$

$\varnothing$, Absolute value is distance, which is always positive, not negative.

95) $\left|\dfrac{4}{3}x + 1\right| = \left|\dfrac{5}{3}x + 8\right|$

$\dfrac{4}{3}x + 1 = \dfrac{5}{3}x + 8$ or $\dfrac{4}{3}x + 1 = -\left(\dfrac{5}{3}x + 8\right)$

$4x + 3 = 5x + 24$ or $4x + 3 = -5x - 24$

$-x = 21$ or $9x = -27$

$x = -21$ or $x = -3$

$\{-21, -3\}$

97) $|4 - 9t| + 2 = 1$

$|4 - 9t| = -1$

$\varnothing$, Absolute value is distance, which is always positive, not negative.

99) $-\dfrac{3}{5} \ge \dfrac{5}{2}a - \dfrac{1}{2}$

$-6 \ge 25a - 5$

$-1 \ge 25a$

$a \le -\dfrac{1}{25}$

$\left(-\infty, -\dfrac{1}{25}\right]$

101) $|6k + 17| > -4$

$6k + 17 > -4$ or $6k + 17 < 4$

$6k > -21$ or $6k < -13$

$k > -\dfrac{7}{2}$ or $k < -\dfrac{13}{6}$

$(-\infty, \infty)$

103) $5 \ge |c + 8| - 2$

$7 \ge |c + 8|$

$-7 \le c + 8 \le 7$

$-15 \le c \le -1$

$[-15, -1]$

105) $|5h - 8| > 7$

$5h - 8 > 7$ or $5h - 8 < -7$

$5h > 15$ or $5h < 1$

$h > 3$ or $h < \dfrac{1}{5}$

$\left(-\infty, \dfrac{1}{5}\right) \cup (3, \infty)$

107) $|a - 128| \le 0.75$

$-0.75 \le a - 128 \le 0.75$

$127.25 \le a \le 128.75$

There is between 127.25 oz and 128.75 oz of milk in the container.

109) $|b-38| \le 5$

$-5 \le b-38 \le 5$

$33 \le b \le 43$

Emmanuel will spend between $33 and $43 on his daughter's birthday gift.

Chapter 3 Review

1) $z+6 \ge 14$

$z+6-6 \ge 14-6$

$z \ge 8$

$[8, \infty)$

3) $w+8 > 5$

$w+8-8 > 5-8$

$w > -3$

$(-3, \infty)$

5) $5x-2 \le 18$

$5x-2+2 \le 18+2$

$5x \le 20$

$\dfrac{1}{5} \cdot 5x \le \dfrac{1}{5} \cdot 20$

$x \le 4$

$(-\infty, 4]$

7) $-15 < 4p-7 \le 5$

$-15+7 < 4p-7+7 \le 5+7$

$-8 < 4p \le 12$

$\dfrac{-8}{4} < \dfrac{4p}{4} \le \dfrac{12}{4}$

$-2 < p \le 3$

$(-2, 3]$

9) $3(3c+8)-7 > 2(7c+1)-5$

$3(3c+8)-7 > 2(7c+1)-5$

$9c+24-7 > 14c+2-5$

$9c+17 > 14c-3$

$9c+17-14c > 14c-3-14c$

$-5c+17 > -3$

$-5c+17-17 > -3-17$

$-5c > -20$

$-\dfrac{1}{5}(-5c) < -\dfrac{1}{5}(-20)$

$c < 4$

$(-\infty, 4)$

11)

$-3 < \dfrac{3}{4}a-6 \le 0$

$-3+6 < \dfrac{3}{4}a-6+6 \le 0+6$

$3 < \dfrac{3}{4}a \le 6$

$\dfrac{4}{3}(3) < \dfrac{4}{3}\left(\dfrac{3}{4}a\right) \le \dfrac{4}{3}(6)$

$4 < a \le 8$

$(4, 8]$

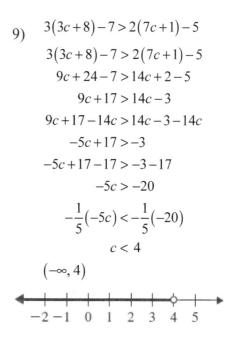

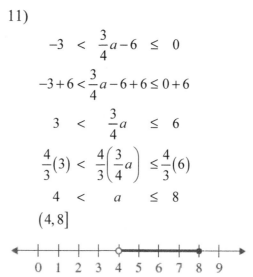

13) Let x be Gia's score on the fourth test. Then

$$\frac{94+88+91+x}{4} \geq 90$$

$$4\left(\frac{273+x}{4}\right) \geq 4 \cdot 90$$

$$273+x \geq 360$$

$$273+x-273 \geq 360-273$$

$$x \geq 87$$

Gia must make at least an 87 on her fourth test.

15) $a+6 \leq 9$ and $7a-2 \geq 5$

$a \leq 3$ and $7a \geq 7$

$a \leq 3$ and $a \geq 1$

$[1, 3]$

17) $8-y < 9$ or $\frac{1}{10}y > \frac{3}{5}$

$-y < 1$ or $y > 6$

$y > -1$ or $y > 6$

$(-1, \infty)$

19) $\{\text{Toyota}\}$

21) $|m| = 9$

$m = 9$ or $m = -9$ $\{-9, 9\}$

23) $|7t+3| = 4$

$7t+3 = 4$ or $7t+3 = -4$

$7t = 1$ $7t = -7$

$t = \frac{1}{7}$ or $t = -1$ $\left\{-1, \frac{1}{7}\right\}$

25) $|8p+11| - 7 = -3$

$|8p+11| = 4$

$8p+11 = 4$ or $8p+11 = -4$

$8p = -7$ $8p = -15$

$p = -\frac{7}{8}$ or $p = -\frac{15}{8}$

$\left\{-\frac{15}{8}, -\frac{7}{8}\right\}$

27) $\left|4-\frac{5}{3}x\right| = \frac{1}{3}$

$4-\frac{5}{3}x = \frac{1}{3}$ or $4-\frac{5}{3}x = -\frac{1}{3}$

$12-5x = 1$ $12-5x = -1$

$-5x = -11$ $-5x = -13$

$x = \frac{11}{5}$ or $x = \frac{13}{5}$

$\left\{\frac{11}{5}, \frac{13}{5}\right\}$

29) $|7r-6| = |8r+2|$

$7r-6 = 8r+2$ or $7r-6 = -8r-2$

$15r = 4$

$-8 = r$ or $r = \frac{4}{15}$

$\left\{-8, \frac{4}{15}\right\}$

31) $\varnothing$, the absolute value of a quantity cannot be negative.

33) $|9d+4| = 0$

$9d+4 = 0$

$9d = -4$

$d = -\frac{4}{9}$ $\left\{-\frac{4}{9}\right\}$

35) $|a| = 4$

37) $\quad |c| \le 3$

$-3 \le c \le 3 \; [-3,3]$

$|w+1| < 11$

$-11 < w+1 < 11$

$-12 < w < 10 \quad (-12,10)$

39) $|4t| > 8 \qquad\qquad (-\infty,-2) \cup (2,\infty)$

$4t > 8 \;$ or $\; 4t < -8$

$t > 2 \qquad\quad t < -2$

41) $|12r+5| \ge 7$

$12r+5 \ge 7 \;$ or $\; 12r+5 \le -7$

$12r \ge 2 \qquad\quad 12r \le -12$

$r \ge \dfrac{1}{6} \qquad\quad r \le -1$

$(-\infty,-1] \cup \left[\dfrac{1}{6},\infty\right)$

43) $\quad |4-a| < 9$

$-9 < 4-a < 9$

$-13 < -a < 5$

$13 > a > -5 \quad (-5,13)$

45) $|4c+9| - 8 \le -2$

$|4c+9| \le 6$

$-6 \le 4c+9 \le 6$

$-15 \le 4c \le -3$

$-\dfrac{15}{4} \le c \le -\dfrac{3}{4} \quad \left[-\dfrac{15}{4},-\dfrac{3}{4}\right]$

47)

$|5y+12| - 15 \ge -8$

$|5y+12| \ge 7$

$5y+12 \ge 7 \;$ or $\; 5y+12 \le -7$

$5y \ge -5 \qquad\qquad 5y \le -19$

$y \ge -1 \qquad\qquad y \le -\dfrac{19}{5}$

$\left(-\infty,-\dfrac{19}{5}\right] \cup [-1,\infty)$

49) $(-\infty,\infty)$

51) $|12s+1| \le 0$

$12s+1 = 0$

$12s = -1$

$s = -\dfrac{1}{12} \quad \left\{-\dfrac{1}{12}\right\}$

Chapter 3 Test

1) $r + 7 \le 2$

$r + 7 - 7 \le 2 - 7$

$r \le -5$

$(-\infty, -5]$

3) $9 - 3(2x - 1) < 4x + 5(x + 2) - 8$

$9 - 6x + 3 < 4x + 5x + 10 - 8$

$12 - 6x < 9x + 2$

$12 - 12 - 6x < 9x + 2 - 12$

$-6x < 9x - 10$

$-6x - 9x < 9x - 9x - 10$

$-15x < -10$

$\dfrac{-15x}{-15} > \dfrac{-10}{-15}$

$x > \dfrac{2}{3}$

$\left(\dfrac{2}{3}, \infty\right)$

5) $-1 < \dfrac{w - 5}{4} \le \dfrac{1}{2}$

$-4 < w - 5 \le 2$

$1 < w \le 7$

$(1, 7]$

7) Let x = number of hours able to keep forklift

$46 + 9x \le 100$

$9x \le 54$

$x \le 6$ hr

9) $3n + 5 > 12$ or $\dfrac{1}{4}n < -2$

$3n > 7$

$n > \dfrac{7}{3}$ or $n < -8$

$(-\infty, -8) \cup \left(\dfrac{7}{3}, \infty\right)$

11) $6 - p < 10$ or $p - 7 < 2$

$-p < 4$ or $p < 9$

$p > -4$ or $p < 9$

$(-\infty, \infty)$

13) $|d + 6| - 3 = 7$

$|d + 6| = 10$

$d + 6 = 10$ or $d + 6 = -10$

$d = 4$ or $d = -16$

$\{-16, 4\}$

15) $\left|\dfrac{1}{2}n - 1\right| = -8$

$\varnothing$, Absolute value is distance, which is positive, not negative.

17) $|2z - 7| \le 9$

$-9 \le 2z - 7 \le 9$

$-2 \le 2z \le 16$

$-1 \le z \le 8$

$[-1, 8]$

19) $|10 - 3w| < -2$

$\varnothing$, Absolute value is distance, which is positive, not negative.

Cumulative Review: Chapters 1–3

1)

$$\frac{3}{8} - \frac{5}{6} \quad \text{LCD is 24.}$$

$$\frac{3}{8} - \frac{5}{6} = \frac{3}{8} \cdot \frac{3}{3} - \frac{5}{6} \cdot \frac{4}{4}$$

$$= \frac{9}{24} - \frac{20}{24}$$

$$= \frac{9 - 20}{24}$$

$$= -\frac{11}{24}$$

3) $\quad 26 - 14 \div 2 + 5 \cdot 7 = 26 - 7 + 35 = 54$

5)

$$-39 - |7 - 15| = -39 - |7 - 15|$$
$$= -39 - |-8|$$
$$= -39 - (8)$$
$$= -47$$

7) the integers: $\{-5, 0, 9\}$

9) the whole numbers: $\{0, 9\}$

11) No. For example, $10 - 3 \neq 3 - 10$.

13) $\quad 8t - 17 = 10t + 6$

$$8t - 17 - 6 = 10t + 6 - 6$$
$$8t - 23 = 10t$$
$$8t - 8t - 23 = 10t - 8t$$
$$-23 = 2t$$
$$t = -\frac{23}{2} \quad \left\{-\frac{23}{2}\right\}$$

15) $\quad 3(7w - 5) - w = -7 + 4(5w - 2)$

$$21w - 15 - w = -7 + 20w - 8$$
$$20w - 15 = 20w - 15$$
$$-15 = -15$$

All real numbers

17) $\quad -\frac{1}{2}c + \frac{1}{5}(2c - 3) = \frac{3}{10}(2c + 1) - \frac{3}{4}c$

$$20\left[-\frac{1}{2}c + \frac{1}{5}(2c - 3)\right] = 20\left[\frac{3}{10}(2c + 1) - \frac{3}{4}c\right]$$

$$-10c + 4(2c - 3) = 6(2c + 1) - 15c$$
$$-10c + 8c - 12 = 12c + 6 - 15c$$
$$2c - 12 = -3c + 6$$
$$2c - 12 - 6 = -3c + 6 - 6$$
$$2c - 18 = -3c$$
$$c = 18 \quad \{18\}$$

19)

$$7 - k > 1$$
$$-7 + 7 - k > -7 + 1$$
$$-k > -6$$
$$k < 6$$

$$(-\infty, 6)$$

21) $\quad -14 < 6y + 10 < 3$

$$-24 < 6y < -7$$
$$-4 < y < -\frac{7}{6}$$
$$\left(-4, -\frac{7}{6}\right)$$

23) $\left|\dfrac{1}{2}m+7\right|\leq 11$

$-11\leq \dfrac{1}{2}m+7\leq 11$

$-22\leq m+14\leq 22$

$-36\leq m\leq 8$

$[-36,8]$

25) $\dfrac{87+76+x}{3}\geq 80$

$\dfrac{163+x}{3}\geq 80$

$163+x\geq 240$

$x\geq 240-163$

$x\geq 77$ At least 77

Section 4.1

1)

    a)   28.8 million people watched the Season 3 finale.

    b)   36.4 million

    c)   Season 6

    d) $(1, 22.8)$

3)

    A. $(5,1)$; quadrant I

    B. $(2,-3)$; quadrant IV

    C. $(-2,4)$; quadrant II

    D. $(-3,-4)$; quadrant III

    E. $(3,0)$; no quadrant

    F. $(0,-2)$; no quadrant

5, 7)

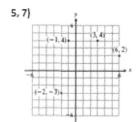

9)

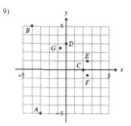

11) positive

13) negative

15) zero

17) yes

19) yes

21) no

23) yes

25) no

27) A linear equation with one variable is either a horizontal (zero slope) or vertical line (undefined slope). A linear equation with two variables is a line that has a slope. $x = -1$ and $x + y = 3$

29) A line and every point on the line is a solution to the equation.

31) It is the point where the graph intersects the y-axis. To find the y-intercept, let $x = 0$ and solve for y.

33)

x	y
0	−1
1	2
2	5
−1	−4

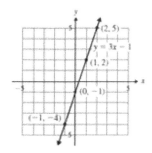

35)

x	y
0	4
−3	6
3	2
6	0

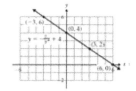

37)

x	y
0	$\frac{1}{2}$
−3	0
5	4
−1	1

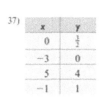
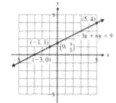

39)

x	y
0	−4
−3	−4
−1	−4
2	−4

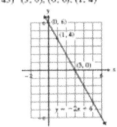

47) $(−1, 0), (0, 4), (1, 8)$

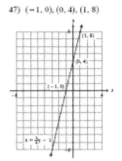

41)

   a) $(3,−5),(6,−3),(−3,−9)$

   b) $\left(1,-\dfrac{19}{3}\right),\left(5,-\dfrac{11}{3}\right),\left(-2,-\dfrac{25}{3}\right)$

   c)   The x-values in part a) are multiples of the denominator of $\dfrac{2}{3}$. So, when you multiply $\dfrac{2}{3}$ by a multiple of 3 the fraction is eliminated.

49) $(−3, 0), (0, −2), \left(1, -\dfrac{8}{3}\right)$

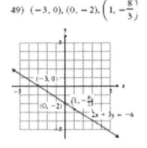

43) $(3, 0), (0, 6), (1, 4)$

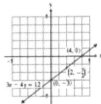

51) $(0, 0), (1, −1), (−1, 1)$

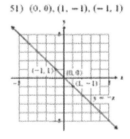

45) $(4, 0), (0, −3), \left(2, -\dfrac{3}{2}\right)$

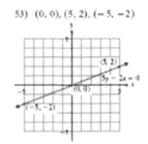

53) $(0, 0), (5, 2), (−5, −2)$

55) $(5, 0), (5, 2), (5, -1)$
Answers may vary.

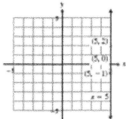

57) $(0, 0), (1, 0), (-2, 0)$
Answers may vary.

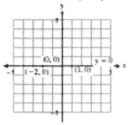

59) $(0, -3), (1, -3), (-3, -3)$
Answers may vary.

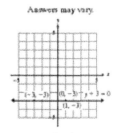

61) $(8, 0), \left(0, \frac{8}{3}\right), (2, 2)$

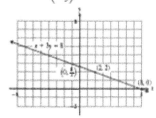

63)
   a) $y = 0$
   b) $x = 0$

65) $\quad (x_1, y_1); (x_2, y_2)$
$(1, 3)$ and $(7, 9)$

$$\text{Midpoint} = \left(\frac{x_1 + x_2}{2}, \frac{y_1 + y_2}{2}\right)$$

$$= \left(\frac{1+7}{2}, \frac{3+9}{2}\right) = (4, 6)$$

67) $\quad (x_1, y_1); (x_2, y_2)$
$(-5, 2)$ and $(-1, -8)$

$$\text{Midpoint} = \left(\frac{x_1 + x_2}{2}, \frac{y_1 + y_2}{2}\right)$$

$$= \left(\frac{-5+(-1)}{2}, \frac{2+(-8)}{2}\right)$$

$$= (-3, -3)$$

69)

$(x_1, y_1); (x_2, y_2)$
$(-3, -7)$ and $(1, -2)$

$$\text{Midpoint} = \left(\frac{x_1 + x_2}{2}, \frac{y_1 + y_2}{2}\right)$$

$$= \left(\frac{-3+1}{2}, \frac{-7+(-2)}{2}\right) = \left(-1, -\frac{9}{2}\right)$$

71)

$(x_1, y_1); (x_2, y_2)$
$(4, 0)$ and $(-3, -5)$

$$\text{Midpoint} = \left(\frac{x_1 + x_2}{2}, \frac{y_1 + y_2}{2}\right)$$

$$= \left(\frac{4+(-3)}{2}, \frac{0+(-5)}{2}\right) = \left(\frac{1}{2}, -\frac{5}{2}\right)$$

73) $(x_1, y_1); (x_2, y_2)$

$\left(\dfrac{3}{2}, -1\right)$ and $\left(\dfrac{5}{2}, \dfrac{7}{2}\right)$

$\text{Midpoint} = \left(\dfrac{x_1 + x_2}{2}, \dfrac{y_1 + y_2}{2}\right)$

$= \left(\dfrac{\dfrac{3}{2} + \dfrac{5}{2}}{2}, \dfrac{-1 + \dfrac{7}{2}}{2}\right) = \left(\dfrac{\dfrac{8}{2}}{2}, \dfrac{\dfrac{5}{2}}{2}\right)$

$= \left(2, \dfrac{5}{4}\right)$

75) $(x_1, y_1); (x_2, y_2)$

$(-6.2, 1.5)$ and $(4.8, 5.7)$

$\text{Midpoint} = \left(\dfrac{x_1 + x_2}{2}, \dfrac{y_1 + y_2}{2}\right)$

$= \left(\dfrac{-6.2 + 4.8}{2}, \dfrac{1.5 + 5.7}{2}\right)$

$= \left(-\dfrac{1.4}{2}, \dfrac{7.2}{2}\right) = (-0.7, 3.6)$

77)
a) $(16, 800), (17, 1300), (18, 1800),$
    $(19, 1900)$

b)
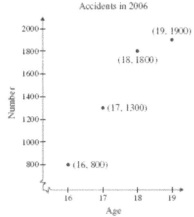

c) There were 1800 18-year-old drivers in fatal motor accidents in 2006.

79)
a)

x	y
1	120
3	160
4	180
6	220

(1, 120), (3, 160), (4, 180), (6, 220)

b)

Cost of Moon Jump

c) The cost of renting the moon jump for 4 hours is $180

d) 9 hours

81)
a)

x	y
0	0
10	15
20	30
60	90

(0, 0), (10, 15), (20, 30), (60, 90)

b) $(0, 0)$: Before engineers began working $(x = 0$ days$)$, the tower did not move toward vertical $(y = 0)$.

$(10, 15)$: After 10 days of working, the tower has moved 15 mm toward vertical.

$(20, 30)$: After 20 days of working, the tower was moved 30 mm toward vertical.

Section 4.2: Slope of a Line and Slope-Intercept Form

$(60, 90)$: After 60 days of working, the tower was moved 90 mm toward vertical.

c)

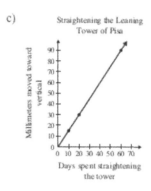

Straightening the Leaning Tower of Pisa

d) 300 days

83)

a) Answers will vary.

b)

29.86in. 28.86in. 26.36in. 24.86in.; yes

c)

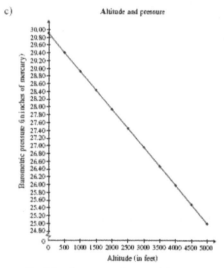

Altitude and pressure

d) No, because the problem states that the equation applies to altitudes 0 ft – 5000 ft.

Section 4.2

1) The slope of a line is the ratio of vertical change to horizontal change. It is $\dfrac{\text{change in } y}{\text{change in } x}$ or $\dfrac{\text{rise}}{\text{run}}$ or

$\dfrac{y_2 - y_1}{x_2 - x_1}$ where (x_1, y_1), and (x_2, y^2) are points on the line.

3) It slants downward from left to right.

5) undefined

7) a) $m = \dfrac{3}{4}$

b) $m = \dfrac{-1-2}{1-5} = \dfrac{-3}{-4} = \dfrac{3}{4}$

9) a) $m = \dfrac{2}{-6} = -\dfrac{1}{3}$

b) $m = \dfrac{-2-(-4)}{-3-3} = \dfrac{2}{-6} = -\dfrac{1}{3}$

11) a) $m = -5$

b) $m = \dfrac{-1-4}{0-(-1)} = \dfrac{-5}{1} = -5$

13) a) $m = 0$

b) $m = \dfrac{-5-(-5)}{3-(-1)} = \dfrac{-5+5}{3+1} = \dfrac{0}{4} = 0$

15)

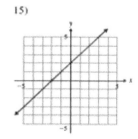

17) Let $(x_1, y_1) = (3, 2)$ and $(x_2, y_2) = (9, 5)$

$m = \dfrac{5-2}{9-3} = \dfrac{3}{6} = \dfrac{1}{2}$

19) Let $(x_1, y_1) = (-2, 8)$ and $(x_2, y_2) = (2, 4)$

$$m = \frac{4-8}{2-(-2)} = \frac{-4}{4} = -1$$

21) Let $(x_1, y_1) = (9, 2)$ and $(x_2, y_2) = (0, 4)$

$$m = \frac{4-2}{0-9} = \frac{2}{-9} = -\frac{2}{9}$$

23) Let $(x_1, y_1) = (3, 5)$ and $(x_2, y_2) = (-1, 5)$

$$m = \frac{5-5}{-1-3} = \frac{0}{-4} = 0$$

25) Let $(x_1, y_1) = (3, 2)$ and $(x_2, y_2) = (3, -1)$

$$m = \frac{-1-2}{3-3} = \frac{-3}{0}; \text{undefined}$$

27) Let $(x_1, y_1) = \left(\frac{3}{8}, -\frac{1}{3}\right)$ and $(x_2, y_2) = \left(\frac{1}{2}, \frac{1}{4}\right)$

$$m = \frac{\frac{1}{4} - \left(-\frac{1}{3}\right)}{\frac{1}{2} - \frac{3}{8}} = \frac{\frac{1}{4} + \frac{1}{3}}{\frac{1}{2} - \frac{3}{8}} = \frac{\frac{3}{12} + \frac{4}{12}}{\frac{4}{8} - \frac{3}{8}} = \frac{\frac{7}{12}}{\frac{1}{8}}$$

$$= \frac{7}{12} \div \frac{1}{8} = \frac{7}{\cancel{12}_3} \cdot \frac{\cancel{8}^2}{1} = \frac{14}{3}$$

29) Let $(x_1, y_1) = (-1.7, -1.2)$ and $(x_2, y_2) = (2.8, -10.2)$

$$m = \frac{-10.2 - (-1.2)}{2.8 - (-1.7)} = \frac{-10.2 + 1.2}{2.8 + 1.7} = \frac{9}{4.5} = 2$$

31) a) $m = \dfrac{10}{12} = \dfrac{5}{6}$

b) $m = \dfrac{8}{12} = \dfrac{2}{3}$

2 ft. = 24 in.

$m = \dfrac{8}{24} = \dfrac{1}{3}; 4-12$ pitch

33) $m = \dfrac{4}{40} = 0.10 = 10\%$

Yes the slope of the driveway will be 10%. The slope must be below 12% to meet city requirements.

35) a) $\$22,000$

b)   negative

c)   The value of the car is decreasing over time.

d) $m = -2000$; the value of the car is decreasing by $2000 per year.

37)

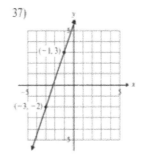

39)

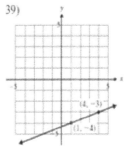

41)

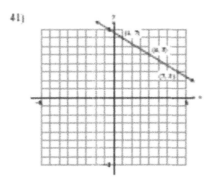

43)

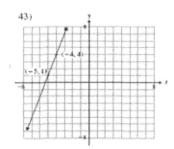

45)

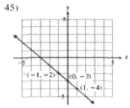

47)

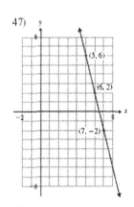

49)

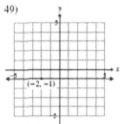

51)

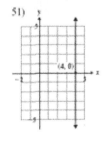

53)

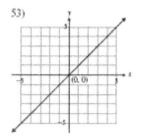

55) The slope is m and the intercept is $(0, b)$.

57) $m = \dfrac{2}{5}; (0, -6)$

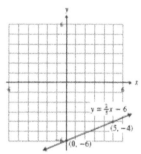

59) $m = -\dfrac{5}{3}; (0, 4)$

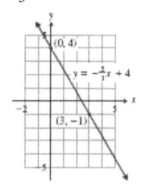

61) $m = \dfrac{3}{4}$, y-int: $(0, 1)$

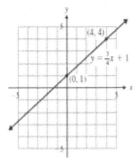

69) $m = 0; (0, -2)$

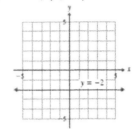

63) $m = 4$, y-int: $(0, -2)$

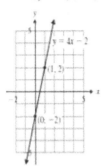

71) $x + 3y = -6$

$$3y = -x - 6$$

$$y = -\frac{1}{3}x - 2$$

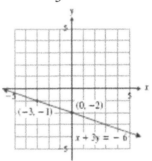

73) $12x - 8y = 32$

$$-8y = -12x + 32$$

$$\frac{-8}{-8}y = \frac{-12}{-8}x + \frac{32}{-8}$$

$$y = \frac{3}{2}x - 4$$

65) $m = -1$, y-int: $(0, 5)$

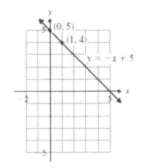

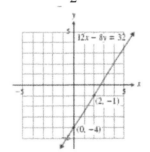

67) $m = \dfrac{3}{2}$, y-int: $\left(0, \dfrac{1}{2} \right)$

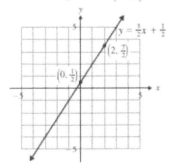

75) $x + 9 = 2$

$x = 2 - 9$

$x = -7$

This cannot be written in slope-intercept form.

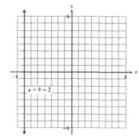

77)

$y = \frac{5}{2}x + 3$

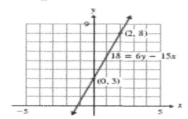

79) $y = 0$

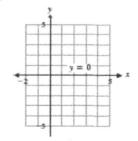

81) a) $(0, 34,000)$; if Dave has zero in sales, his income is $34,000

b) $m = 0.05$; Dave earns $0.05 for every $1 dollar in sales.

c) $I = 0.05s + 34,000$

$I = 0.05 \cdot 80,000 + 34,000$

$I = 4000 + 34,000$

$I = \$38,000$

83) a) $(0, 40.53)$; 40.53 gallons of whole milk were consumed in 1945.

b) $m = -0.59$; whole milk consumption decreased by 0.59 gallons for every year after 1945 for each person.

c) $y = -0.59 \cdot 55 + 40.53$

$y = 8.08$ gallons

85) a) $(0, 68,613)$; in 2000, the average annual salary of pharmacist was $68,613.

b) $m = 3986$; the average annual salary of a pharmacists is increasing by $3986 per year.

c) $y = 3986 \cdot 4 + 68,613$

$y = \$84,557$

$y = 3986 \cdot 14 + 68,613$

$y = \$124.17$

d) answers may vary

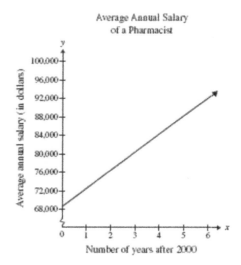

87) $y = -4x + 7$

89) $y = \frac{8}{5}x - 6$

91) $y = \frac{1}{3}x + 5$

93) $y = -x$

95) $y = -2$

Section 4.3

1) Substitute the slope and y-intercept into $y = mx + b$.

3) $y = -7x + 2$

5) If $m = 1$ and $b = -3$, then
$$y = 1 \cdot x + (-3)$$
$$y = x - 3$$
$$x - y = 3$$

7) If $m = -\dfrac{1}{3}$ and $b = -4$, then
$$y = -\frac{1}{3} \cdot x + (-4)$$
$$y = -\frac{1}{3}x - 4$$
$$3y = -x - 12$$
$$x + 3y = -12$$

9) $y = x$

11) a) $y_2 - y_1 = m(x_2 - x_1)$
   b) Substitute the point and the slope into the point-slope formula.

13) $y - 6 = 5(x - 1)$
$$y - 6 = 5x - 5$$
$$y = 5x - 5 + 6$$
$$y = 5x + 1$$

15) $y - 4 = -[x - (-9)]$
$$y - 4 = -x - 9$$
$$y = -x - 9 + 4$$
$$y = -x - 5$$

17) $y - (-1) = 4[x - (-2)]$
$$y + 1 = 4(x + 2)$$
$$y + 1 = 4x + 8$$
$$-4x + y = 8 - 1$$
$$-4x + y = 7$$
$$4x - y = -7$$

19) $y - (-5) = \dfrac{1}{6}[x - (-4)]$
$$6(y + 5) = \cancel{6}\left[\left(\frac{1}{\cancel{6}}\right)(x + 4)\right]$$
$$6y + 30 = x + 4$$
$$6y = x + 4 - 30$$
$$6y = x - 26$$
$$y = \frac{1}{6}x - \frac{13}{3}$$

21) $y - 0 = -\dfrac{5}{9}(x - 6)$
$$y = -\frac{5}{9}x + \frac{30}{9}$$
$$9y = -5x + 30$$
$$5x + 9y = 30$$

23) Find the slope and use it with one of the points in the point-slope equation.

25) $m = \dfrac{8 - 4}{7 - 3} = \dfrac{4}{4} = 1$
$$y - 4 = 1(x - 3)$$
$$y - 4 = x - 3$$
$$y = x - 3 + 4$$
$$y = x + 1$$

Section 4.3: Writing an Equation of a Line

27) $m = \dfrac{3-4}{1-(-2)} = \dfrac{-1}{1+2} = -\dfrac{1}{3}$

$y - 3 = -\dfrac{1}{3}(x-1)$

$y - 3 = -\dfrac{1}{3}x + \dfrac{1}{3}$

$y = -\dfrac{1}{3}x + \dfrac{1}{3} + \dfrac{3}{1}$

$y = -\dfrac{1}{3}x + \dfrac{1}{3} + \dfrac{9}{3}$

$y = -\dfrac{1}{3}x + \dfrac{10}{3}$

29) $m = \dfrac{-2-(-5)}{3-(-1)} = \dfrac{-2+5}{3+1} = \dfrac{3}{4}$

$y - (-2) = \dfrac{3}{4}(x-3)$

$4(y+2) = \cancel{4}\left[\dfrac{3}{\cancel{4}}(x-3)\right]$

$4y + 8 = 3(x-3)$

$4y + 8 = 3x - 9$

$-3x + 4y = -9 - 8$

$-3x + 4y = -17$

$3x - 4y = 17$

31) $m = \dfrac{-3-1}{6-4} = \dfrac{-4}{2} = -2$

$y - 1 = -2(x-4)$

$y - 1 = -2x + 8$

$2x + y = 8 + 1$

$2x + y = 9$

33) $m = \dfrac{7.2-4.2}{3.1-2.4} = \dfrac{3}{.6} = 5.0$

$y - 4.2 = 5.0(x-2.5)$

$y - 4.2 = 5.0x - 12.5$

$y = 5.0x - 12.5 + 4.2$

$y = 5.0x - 8.3$

35) If $m = \dfrac{-1-(-4)}{2-0} = \dfrac{3}{2}$ and $b = -4$,

then $y = \dfrac{3}{2}x - 4$.

37) $m = \dfrac{-1-3}{-1-(-5)} = \dfrac{-4}{4} = -1$

$y - 3 = -1[x-(-5)]$

$y - 3 = -(x+5)$

$y - 3 = -x - 5$

$y = -x - 2$

39) $y = 5$

41) $y - (-1) = 4[x-(-4)]$

$y + 1 = 4(x+4)$

$y + 1 = 4x + 16$

$y = 4x + 16 - 1$

$y = 4x + 15$

43) $y = \dfrac{8}{3}x - 9$

45) $m = \dfrac{1-(-2)}{-5-(-1)} = \dfrac{1+2}{-5+1} = -\dfrac{3}{4}$

$y - (-2) = -\dfrac{3}{4}[x-(-1)]$

$y + 2 = -\dfrac{3}{4}(x+1)$

$4(y+2) = 4\left(-\dfrac{3}{4}x - \dfrac{3}{4}\right)$

$4y + 8 = -3x - 3$

$4y = -3x - 11$

$y = -\dfrac{3}{4}x - \dfrac{11}{4}$

47) $x = 3$

49) $y = -8$

51) $m = \dfrac{0-2}{6-0} = \dfrac{-2}{6} = -\dfrac{1}{3}$

$y - 0 = -\dfrac{1}{3}(x - 6)$

$y = -\dfrac{1}{3}x + \dfrac{1}{3} \cdot 6$

$y = -\dfrac{1}{3}x + 2$

53) $y = x$

55) The slopes are negative reciprocals.

57) perpendicular

59) $m_1 = \dfrac{2}{9}$

$4x - 18y = 9$

$-18y = -4x + 9$

$y = \dfrac{-4}{-18}x + \dfrac{9}{-18}$

$y = \dfrac{2}{9}x - \dfrac{1}{2}$

$m_2 = \dfrac{2}{9}$

parallel

61) $-3x + 2y = -10$

$2y = 3x - 10$

$y = \dfrac{3}{2}x - \dfrac{10}{2}$

$m_1 = \dfrac{3}{2}$

$3x - 4y = -2$

$-4y = -3x - 2$

$y = \dfrac{-3}{-4}x + \dfrac{-2}{-4}$

$y = \dfrac{3}{4}x + \dfrac{1}{2}$

$m_2 = \dfrac{3}{4}$

neither

63) $m_1 = 1$

$x + y = 7$

$y = -x + 7$

$m_2 = -1$

perpendicular

65) $4x - 3y = 18$

$-3y = -4x + 18$

$y = \dfrac{-4}{-3}x + \dfrac{18}{-3}$

$y = \dfrac{4}{3}x - 6$

$m_1 = \dfrac{4}{3}$

$-8x + 6y = 5$

$6y = 8x + 5$

$y = \dfrac{8}{6}x + \dfrac{5}{6}$

$m_1 = \dfrac{8}{6} = \dfrac{4}{3}$

Parallel

67) $m_1 = $ undefined;

$m_2 = $ undefined;

parallel

69) $L_1 : m_1 = \dfrac{-13-2}{6-1} = \dfrac{-15}{5} = -3$

$L_2 : m_2 = \dfrac{-10-5}{3-(-2)} = \dfrac{-15}{5} = -3$

parallel

71) $L_1 : m_1 = \dfrac{8-(-7)}{2-(-1)} = \dfrac{15}{3} = 5$

$L_2 : m_2 = \dfrac{4-2}{0-10} = \dfrac{2}{-10} = -\dfrac{1}{5}$

perpendicular

73) $L_1 : m_1 = \dfrac{3-(-1)}{7-5} = \dfrac{4}{2} = 2$

$L_2 : m_2 = \dfrac{5-0}{4-(-6)} = \dfrac{5}{10} = \dfrac{1}{2}$

neither

75) $y = 4x+9; (0,2)$; slope-intercept form

$y-2 = 4(x-0)$

$y-2 = 4x$

$y = 4x+2$

77) $y = \dfrac{1}{2}x-5; (4,5)$; standard form

$y-5 = \dfrac{1}{2}(x-4)$

$y-5 = \dfrac{1}{2}x - \dfrac{1}{\cancel{2}} \cdot \cancel{4}^{2}$

$2(y-5) = 2(\dfrac{1}{2}x-2)$

$2y-10 = x-4$

$-x+2y = -4+10$

$-x+2y = 6 \text{ or } x-2y = -6$

79) $4x+3y = -6; (-9,4)$; standard form

$4x+3y = -6$

$3y = -4x-6$

$y = -\dfrac{4}{3}x - \dfrac{6}{3}$

$m = -\dfrac{4}{3}$

$y-4 = -\dfrac{4}{3}[x-(-9)]$

$3(y-4) = \cancel{3}(\dfrac{-4}{\cancel{3}})(x+9)$

$3y-12 = -4(x+9)$

$3y-12 = -4x-36$

$4x+3y = -36+12$

$4x+3y = -24$

81) $x+5y = 10; (15,7)$; slope-intercept form

$x+5y = 10$

$5y = -x+10$

$y = -\dfrac{1}{5}x+2; \; m = -\dfrac{1}{5}$

$y-7 = -\dfrac{1}{5}(x-15)$

$y-7 = -\dfrac{1}{5}x+3$

$y = -\dfrac{1}{5}x+10$

83) $m_{\text{perpindicular}} = -\dfrac{3}{2}$

$y-(-3) = -\dfrac{3}{2}(x-6)$

$y+3 = -\dfrac{3}{2}x + \dfrac{18}{2}$

$y+3 = -\dfrac{3}{2}x+9$

$y = -\dfrac{3}{2}x+9-3$

$y = -\dfrac{3}{2}x+6$

85) $m_{\text{perpindicular}} = \dfrac{1}{5}$

$$y - 0 = \frac{1}{5}(x - 10)$$

$$5(y) = 5\left(\frac{1}{\cancel{5}}x - \frac{10}{\cancel{5}}\right)$$

$$5y = x - 10$$

$$-x + 5y = 10$$

$$x - 5y + = 10$$

87) $\quad x + y = 9$

$$y = -x + 9$$

$$m = -1$$

$$m_{\text{perpindicular}} = 1$$

$$y - (-5) = 1[x - (-5)]$$

$$y + 5 = x + 5$$

$$y = x + \cancel{5} - \cancel{5}$$

$$y = x$$

89) $-15y = -24x + 10$

$$y = \frac{-\cancel{24}^{8}}{-\cancel{15}_{5}}x + \frac{\cancel{10}^{2}}{-\cancel{15}_{3}}$$

$$y = \frac{8}{5}x - \frac{2}{3}$$

$$m = \frac{8}{5}$$

$$m_{\text{perpindicular}} = -\frac{5}{8}$$

$$y - (-7) = -\frac{5}{8}(x - 16)$$

$$y + 7 = -\frac{5}{8}x + \frac{5 \cdot \cancel{16}^{2}}{\cancel{8}}$$

$$y + 7 = -\frac{5}{8}x + 10$$

$$8(y + 7) = 8\left(-\frac{5}{8}x + 10\right)$$

$$8y + 56 = -5x + 80$$

$$5x + 8y = 80 - 56$$

$$5x + 8y = 24$$

91) $\quad -6y = -2x - 3$

$$y = \frac{-2}{-6}x + \frac{-3}{-6}$$

$$y = \frac{1}{3}x + \frac{1}{2}$$

$$m = \frac{1}{3}$$

$$m_{\text{perpindicular}} = -3$$

$$y - 2 = -3(x - 2)$$

$$y - 2 = -3x + 6$$

$$y = -3x + 6 + 2$$

$$y = -3x + 8$$

93) $\quad y = 2x + 1$

$$m = 2$$

$$m_{\text{parallel}} = 2$$

$$y - (-3) = 2(x - 1)$$

$$y + 3 = 2x - 2$$

$$y = 2x - 2 - 3$$

$$y = 2x - 5$$

95) $x = -1$

97) $x = 2$

Section 4.3: Writing an Equation of a Line

99) $-6y = -21x + 2$

$$y = \frac{-21}{-6}x + \frac{2}{-6}$$

$$y = \frac{7}{2}x - \frac{1}{3}$$

$$m = \frac{7}{2}$$

$$m_{\text{perpindicular}} = -\frac{2}{7}$$

$$y - (-1) = -\frac{2}{7}(x-4)$$

$$y + 1 = -\frac{2}{7}(x-4)$$

$$7(y+1) = 7(-\frac{2}{7})(x-4)$$

$$7y + 7 = -2x + 8$$

$$7y = -2x + 1$$

$$y = -\frac{2}{7}x + \frac{1}{7}$$

101) $y = -\frac{5}{2}$

103) a) Use (8,10) and (9.5, 10.5) to find m.

$$m = \frac{10.5 - 10}{9.5 - 8} = \frac{0.5}{1.5} = \frac{1}{3}$$

Now use $m = \frac{1}{3}$ and (8, 10) in the point-slope formula.

$$L - 10 = \frac{1}{3}(S-8)$$

$$L - 10 = \frac{1}{3}S - \frac{8}{3}$$

$$L = \frac{1}{3}S - \frac{8}{3} + \frac{30}{3}$$

$$L = \frac{1}{3}S + \frac{22}{3} \text{ or } 3L - S = 22$$

b) $3L - S = 22$

$$3(11.5) - S = 22$$

$$34.5 - 22 = S$$

$$12.5 = S$$

105)

a)

For 2001, $x = 2001 - 1998 = 3$.
So, use $m = 8700$ and (3, 1,284,000) in the point-slope formula.

$$y - 1,284,000 = 8700(x-3)$$

$$y - 1,284,000 = 8700x - 26,100$$

$$y = 8700x - 26,100 + 1,284,000$$

$$y = 8700x + 1,257,900$$

b) The population of Main is increasing by 8700 people per year.

c) For 1998, $x = 0$ and $y = 1,257,900$.
   For 2002, $x = 2002 - 1998 = 4$.

$$y = 8700(4) + 1,257,900 = 1,292,700$$

d) Let $y = 1,431,900$ and solve the equation for x.

$$1,431,900 = 8700x + 1,257,900$$

$$1,431,900 - 1,257,900 = 8700x$$

$$174,000 = 8700x$$

$$x = \frac{174,000}{8700} = 20$$

$x = 20$ corresponds to the year $1998 + 20 = 2018$.

107) a) Use (0, 124) and (5, 92) to

   find *m*.

   $$m = \frac{92-124}{5-0} = \frac{-32}{5} = -6.4$$

   Now use $m = -6.4$ and (0, 124)

   in the slope-intercept form.

   $$y = -6.4x + 124$$

   b) The number of farms with milk
   cows is decreasing by 6.4 thousand
   (6400) per year.
   For 2004, $x = 2004 - 1997 = 7$.

   $$y = -6.4(7) + 124 = 79.2$$

   79.2 thousand, or 79,200

   c) For 2004, $x = 2004 - 1997 = 7$.

   $$y = -6.4(7) + 124 = 79.2$$

   79.2 thousand, or 79,200

109) a) Use (0, 6479) and (3, 43,435)

   to find *m*.

   $$m = \frac{43,435 - 6479}{3-0} = \frac{36,956}{3} \approx$$

   12,318.7

   Now use $m = 12,318.7$ and

   (0, 6479) in the slope-

   intercept form.

   $$y = 12,318.7x + 6479$$

   b) The number of registered hybrid
   vehicles is increasing by 12,318.7
   per year.

   c) For 2002, $x = 2002 - 2000 = 2$

   $$y = 12,318.7(2) + 6479 =$$

   31,116.4

   This is slightly lower than the actual
   value.

d) For 2010, $x = 2010 - 2000 = 10$.

   $$y = 12,318.7(10) + 6479 =$$

   129,666

Section 4.4

1)
In solution set: $(0,-2), (3,-1), (-2,-2)$
Not in solution set: $(0,0), (1,1), (-3,0)$

3)
In solution set: $(0,0), (1,-5), (3,2)$
Not in solution set: $(-4,0), (-2,1), (-3,0)$

5)
In solution set: $(0,-2), (-3,0), (-1,-1)$
Not in solution set: $(0,1), (2,0), (3,2)$

7) dotted

9)
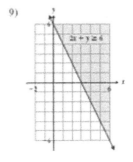

11)
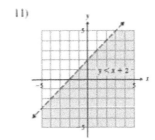

Section 4.4: Linear and Compound Linear Inequalities in Two Variables

13)

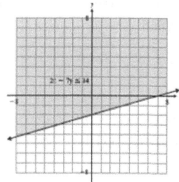

15)

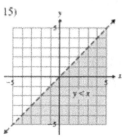

17)

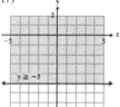

19) below

21)

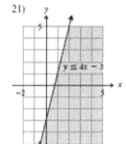

23)

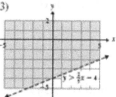

25)

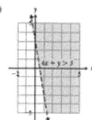

27)

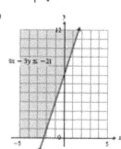

29)

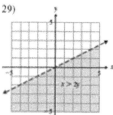

31) slope-intercept form because the equation
is already solved for y.

33)

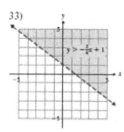

35)

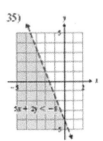

37)

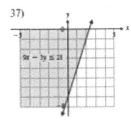

39)

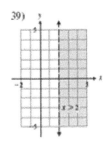

41)

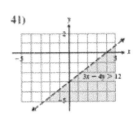

43) No; (3,5) satisfies $x - y \geq -6$ but
not $2x + y \leq 7$.
Since the inequality contains <u>and</u>, it
must satisfy <u>both</u> inequalities.

$$x - y \geq -6$$
$$3 - 5 \geq -6$$
$$-2 \geq -6$$
Is a solution

$$2x + y < 7$$
$$2 \cdot 3 + 5 < 7$$
$$6 + 5 < 7$$
$$11 < 7$$
Not a solution

45)

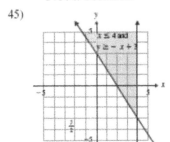

47)

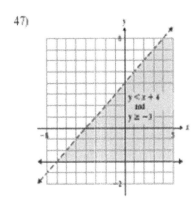

49)

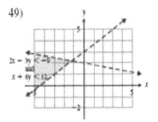

51)

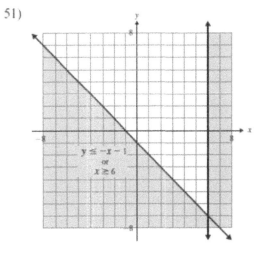

73

53)

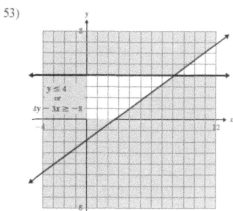

$y \leq 4$
or
$4y - 3x \geq -8$

55)

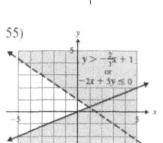

$y > -\frac{2}{3}x + 1$
or
$-2x + 5y \leq 0$

57)

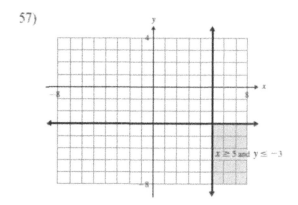

$x \geq 5$ and $y \leq -3$

59)

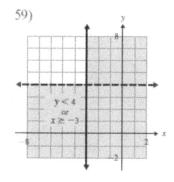

$y < 4$
or
$x \geq -3$

61)

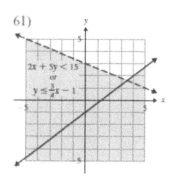

$2x + 5y < 15$
or
$y \leq \frac{3}{4}x - 1$

63)

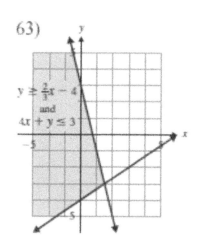

$y \geq \frac{2}{3}x - 4$
and
$4x + y \leq 3$

Section 4.5

1) a) Any set of ordered pairs.
   b) For every x value there is exactly one y value.
   c) Answers may vary.

3) Domain: $\{5, -2, 1, -8\}$

   Range: $\{13, 6, 4, -3\}$

   Function

   Not a function
5) Domain: $\{9, 25, 1\}$

   Range: $\{-1, -3, 1, 5, 7\}$
   Not a function

7) Domain: $\{-1, 2, 5, 8\}$

   Range: $\{-7, -3, 12, 19\}$

   Not a Function

9) Domain: $(-\infty, \infty)$

   Range: $(-\infty, \infty)$

   Function

11) Domain : $(-\infty, 4]$

    Range : $(-\infty, \infty)$

    Not a function

13) Domain: $(-\infty, \infty)$

    Range: $(-\infty, 6]$

    Function

15) yes

17) yes

19) no

21) no

23) $(-\infty, \infty)$; function

25) $(-\infty, \infty)$; function

27) $[0, \infty)$; Not a function

29) $(-\infty, 0) \cup (0, \infty)$; function

31) $(-\infty, -4) \cup (-4, \infty)$; function

33) $(-\infty, 5) \cup (5, \infty)$; function

35) $\left(-\infty, \dfrac{3}{5}\right) \cup \left(\dfrac{3}{5}, \infty\right)$; function

37) $\left(-\infty, -\dfrac{4}{3}\right) \cup \left(-\dfrac{4}{3}, \infty\right)$; function

39) $(-\infty, 3) \cup (3, \infty)$; function

41) $(-\infty, \infty)$; function

43) y is a function; and y is a function of x

45) a) $y = 5 \cdot 3 - 8$

   $y = 15 - 8$

   $y = 7$

   b) $f(3) = 5 \cdot 3 - 8$

   $f(3) = 15 - 8$

   $f(3) = 7$

47) $f(5) = -4 \cdot 5 + 7$

   $f(5) = -20 + 7$

   $f(5) = -13$

49) $f(0) = -4 \cdot 0 + 7$

   $f(0) = 0 + 7$

   $f(0) = 7$

51) $g(4) = 4^2 + 9 \cdot 4 - 2$

$g(4) = 16 + 36 - 2$

$g(4) = 50$

53) $g(-1) = (-1)^2 + 9(-1) - 2$

$g(-1) = 1 - 9 - 2$

$g(-1) = -10$

55) $g\left(-\dfrac{1}{2}\right) = \left(-\dfrac{1}{2}\right)^2 + 9\left(-\dfrac{1}{2}\right)(-2)$

$g\left(-\dfrac{1}{2}\right) = \dfrac{1}{4} - \dfrac{9}{2} - \dfrac{2}{1}$

$g\left(-\dfrac{1}{2}\right) = \dfrac{1}{4} - \dfrac{18}{4} - \dfrac{8}{4}$

$g\left(-\dfrac{1}{2}\right) = \dfrac{25}{4}$

57) $f(6) - g(6) = -4 \cdot 6 + 7 - [6^2 + 9 \cdot 6 - 2]$

$f(6) - g(6) = -24 + 7 - [36 + 54 - 2]$

$f(6) - g(6) = -24 + 7 - [88]$

$f(6) - g(6) = -24 + 7 - 88$

$f(6) - g(6) = -105$

59) $f(-1) = 10; f(4) = -5$

61) $f(-1) = 6; f(4) = 2$

63) $f(-1) = 7; f(4) = 3$

65) $f(x) = -3x - 2$

$10 = -3x - 2$

$12 = -3x$

$-4 = x$

67) $g(x) = \dfrac{2}{3}x + 1$

$3 \cdot 5 = 3\left(\dfrac{2}{3}x + 1\right)$

$15 = 2x + 3$

$12 = 2x$

$6 = x$

69) Substitute $k + 6$ for x.

$= 4k + 24 - 5 = 4k + 19$

71)

a) $f(c) = -7c + 2$

b) $f(t) = -7t + 2$

c) $f(a + 4) = -7(a + 4) + 2$

$f(a + 4) = -7a - 28 + 2$

$f(a + 4) = -7a - 26$

d) $f(z - 9) = -7(z - 9) + 2$

$f(z - 9) = -7z + 63 + 2$

$f(z - 9) = -7z + 65$

e) $g(k) = k^2 - 5k + 12$

f) $g(m) = m^2 - 5m + 12$

g) $f(x + h) = -7(x + h) + 2$

$f(x + h) = -7x - 7h + 2$

h) $f(x + h) - f(x) = -7(x + h)$

$+ 2 - (-7x + 2)$

$f(x + h) - f(x) = -7x - 7h + 2 + 7x - 2$

$f(x + h) - f(x) = -7h$

73)

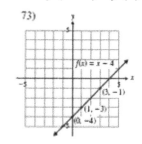

76

75)

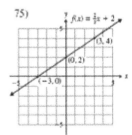

77)

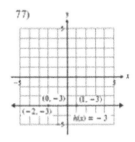

79) x-int: $(-1, 0)$; y-int: $(0, 3)$

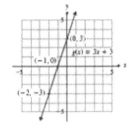

81) x-int: $(4, 0)$; y-int: $(0, 2)$

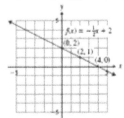

83) intercept: $(0, 0)$

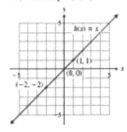

85) $m = -4$; y-int: $(0, -1)$

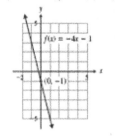

87) $m = \dfrac{3}{5}$; y-int: $(0, -2)$

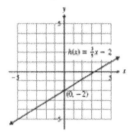

89) $m = 2$; y-int: $\left(0, \dfrac{1}{2}\right)$

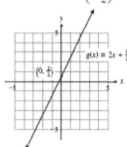

91)

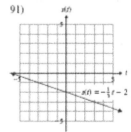

93)

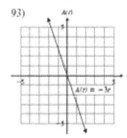

95)

a) $D(2) = 54 \cdot 2 = 108$ miles

b) $D(4) = 54 \cdot 4 = 216$ miles

c) $\qquad 135 = 54t$

$2.5 \text{ hours} = t$

d)

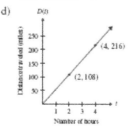

97)

a) $E(10) = 7.5 \cdot 10 = 75$; when Jenelle works 10 hours she earns $75.

b) $E(15) = 7.5 \cdot 15 = 112.5$; when Jenelle works 15 hours she earns $112.5.

c) $210 = 7.5t$

$28 = t$

Jenelle must work 28 hours to earn $210.

99) a) $D(12) = 21.13 \cdot 12 = 253.56 MB$

b) $1 \text{ min} = 60 \text{ sec}$

$D(60) = 21.13 \cdot 60 = 1267.8 MB$

c) $623.70 = 17.82t$

$35 = t$

d)

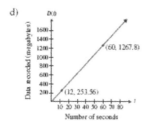

101) a) $F(2) = 30 \cdot 2 = 60$

b) $\qquad 30_s = 105$

$3.5 \text{ sec} = s$

103) a) $S(50) = 44.1 \cdot 50 = 2205$; the number of samples read after 50 sec is 2,205,000.

b) $S(180) = 44.1 \cdot 180 = 7938$; the number of samples read after 180 sec is 7,938,000.

c) $2646 = 44.2t$

$60 \text{ sec} = t$

It will take 60 sec to read 2,646,000 samples.

105) a) 2 hr; 400 mg

b) after about 30 minutes and after 6 hours

c) 200 mg

d) $A(8) = 50$; after about 8 hours there will be about 50mg of ibuprofen in Sasha's bloodstream.

Chapter 4 Review

1) $4 \cdot 1 - (-5) = 9$

$4 + 5 = 9$

$9 = 9$

yes

3) $3 = \dfrac{5}{4} \cdot 2 + \dfrac{1}{2}$

$3 = \dfrac{\cancel{10}^{5}}{\cancel{4}} + \dfrac{1}{2}$

$3 = \dfrac{5}{2} + \dfrac{1}{2}$

$3 = \dfrac{6}{2}$

$3 = 3$

yes

5) $y = -6 \cdot (-3) + 10$

$y = 18 + 10$

$y = 28$

7) -8

9)

x	y
0	-11
3	-8
-1	-12
-5	-16

11)

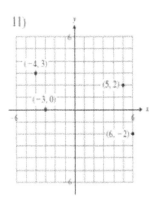

13)

a)

x	y
1	0.10
2	0.20
7	0.70
10	1.00

(1, 0.10), (2, 0.20), (7, 0.70), (10, 1.00)

b)

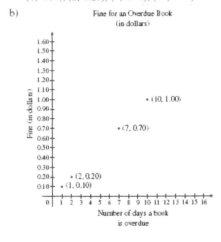

c) If the book is 14 days overdue, then the fine is $1.40.

15)

x	y
0	3
1	1
2	-1
-2	7

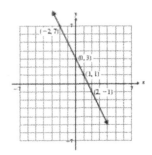

17) (6, 0), (0, -3); (2, -2) may vary.

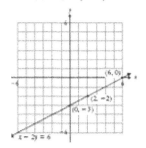

19) (24, 0), (0, 4); (12, 2) may vary.

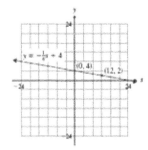

21) (5, 0), (5, 1) may vary, (5, 2) may vary.

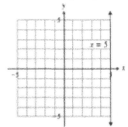

23) $(3,8) \text{ and } (5,2)$

$$\text{Midpoint} = \left(\frac{x_1 + x_2}{2}, \frac{y_1 + y_2}{2} \right)$$

$$= \left(\frac{3+5}{2}, \frac{8+2}{2} \right)$$

$$= \left(\frac{8}{2}, \frac{10}{2} \right)$$

$$= (4,5)$$

25) $(7,-3) \text{ and } (6,-4)$

$$\text{Midpoint} = \left(\frac{x_1 + x_2}{2}, \frac{y_1 + y_2}{2} \right)$$

$$= \left(\frac{7+6}{2}, \frac{-3-4}{2} \right)$$

$$= \left(\frac{13}{2}, -\frac{7}{2} \right)$$

27) $m = -\dfrac{2}{5}$

29) $m = \dfrac{2-7}{-4-1} = \dfrac{-5}{-5} = 1$

31) $m = \dfrac{-8-5}{3-(-2)} = \dfrac{-13}{3+2} = -\dfrac{13}{5}$

33) $m = \dfrac{7-(-1)}{-\dfrac{5}{2}-\dfrac{3}{2}} = \dfrac{7+1}{-\dfrac{8}{2}} = \dfrac{8}{-4} = -2$

35) $m = \dfrac{4-0}{9-9} = \dfrac{4}{0}$; undefined

37) a)   In 2008, one share of the stock was worth $32.
     b)   The slope is positive, so the value of the stock is increasing over time.
     c) $m = 3$; the value of one share of stock is increasing by $3.00 per year.

39)

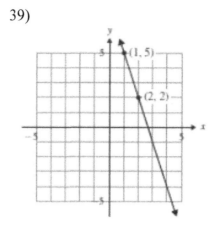

41)

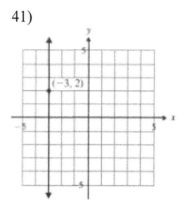

43) $m = 1; \; y-\text{int} : (0,-3)$

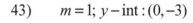

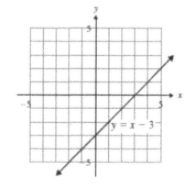

80

45) $m=-\dfrac{3}{4}; y-\text{int}:(0,1)$

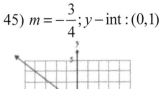

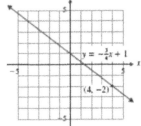

47) $-3y=-x-6$

$$y=\dfrac{-1}{-3}x-\dfrac{6}{-3}$$

$$y=\dfrac{1}{3}x+2$$

$$m=\dfrac{1}{3}; y-\text{int}:(0,2)$$

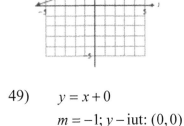

49) $\quad y=x+0$

$$m=-1; y-\text{iut}:(0,0)$$

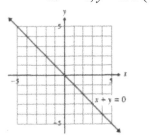

51)
a) $(0,5920.1)$; In 1998 the amount of money spent for personal consumption was $5920.1 billion.
b) It has been increasing by $371.5 billion per year.

c) Estimate from graph: $7400 billion
Estimate from equation:
$$y=371.5(4)+5920.1=\$7406.1 \text{ billion}$$

53) $y-5=7(x-2)$

$$y-5=7x-14$$

$$y=7x-14+5$$

$$y=7x-9$$

55) $y=-\dfrac{4}{9}x+2$

57) $m=\dfrac{-2-(-6)}{-9-3}=\dfrac{4}{-12}=-\dfrac{1}{3}$

$$y-(-6)=-\dfrac{1}{3}(x-3)$$

$$y+6=-\dfrac{1}{3}x+\dfrac{3}{3}$$

$$y=-\dfrac{1}{3}x+1-6$$

$$y=-\dfrac{1}{3}x-5$$

59) $y=9$

61) $m=\dfrac{7-2}{8-(-2)}=\dfrac{5}{10}=\dfrac{1}{2}$

$$y-2=\dfrac{1}{2}[x-(-2)]$$

$$2(y-2)=2\left(\dfrac{1}{2}x+1\right)$$

$$2y-4=x+2$$

$$-x+2y=2+4$$

$$x-2y=-6$$

63) $y - 1 = -3\left(x - \dfrac{4}{3}\right)$

$\quad y - 1 = -3x + 4$

$\quad\quad y = -3x + 4 + 1$

$\quad\quad y = -3x + 5$

$\quad 3x + y = 5$

65) $\quad\quad y = 6x + 0$

$\quad -6x + y = 0$

$\quad\quad 6x - y = 0$

67) $m = \dfrac{-5 - 1}{-7 - 1} = \dfrac{-6}{-8} = \dfrac{3}{4}$

$\quad 4(y - 1) = 4\left[\dfrac{3}{4}(x - 1)\right]$

$\quad\quad 4y - 4 = 3(x - 1)$

$\quad\quad 4y - 4 = 3x - 3$

$\quad -3x + 4y = -3 + 4$

$\quad -3x + 4y = 1$

$\quad\quad 3x - 4y = -1$

69)

a. 2001 corresponds to $x = 0$ and
2004 corresponds to $x = 2004 - 2001 = 3$
use $(0, 944.2)$ and $(3, 1502.9)$ to find m.

$\quad m = \dfrac{1502.9 - 944.2}{3 - 0} = \dfrac{588.7}{3} \approx 186.2$

now use $m = 186.2$ and $(0, 944.2)$ in the
slope intercept formula.

$\quad y = 186.2x + 944.2$

b. The number of worldwide wireless
subscribers is increasing by 186.
2 million per year.

c. 1316.6 million; this is slightly less than
the number given on the chart.

71) $-4y = -9x - 1$

$\quad y = \dfrac{-9}{-4}x - \dfrac{1}{-4}$

$\quad y = \dfrac{9}{4}x + \dfrac{1}{4}$

$\quad 12y = 27x + 2$

$\quad y = \dfrac{27}{12}x + \dfrac{2}{12}$

$\quad y = \dfrac{9}{4}x + \dfrac{1}{6}$

parallel

73) perpendicular

75) $y = -4x + 9$

neither

77) $\quad\quad m = 5$

$\quad y - (-4) = 5[x - (-2)]$

$\quad\quad y + 4 = 5x + 10$

$\quad\quad y = 5x + 6$

79) $-4y = -x + 9$

$\quad y = \dfrac{1}{4}x - \dfrac{9}{4}; \quad$ so $m = \dfrac{1}{4}$

$\quad y - 3 = \dfrac{1}{4}(x - 5)$

$\quad 4y - 12 = x - 5$

$\quad -x + 4y = -5 + 12$

$\quad -x + 4y = 7$

$\quad x - 4y = -7$

81) $\quad\quad \perp m = 2$

$\quad y - 5 = 2(x - 6)$

$\quad\quad y = 2x - 12 + 5$

$\quad\quad y = 2x - 7$

83) $-11y = -2x + 15$

$$y = \frac{2}{11}x - \frac{15}{11}$$

$$m_{\text{perpindicular}} = -\frac{11}{2}$$

$$y - (-7) = -\frac{11}{2}(x - 2)$$

$$y + 7 = -\frac{11}{2}x + 11$$

$$y = -\frac{11}{2}x + 11 - 7$$

$$y = -\frac{11}{2}x + 4$$

85) $x = 2$

87) $x = -1$

89)

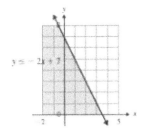

91)

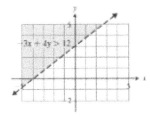

93)

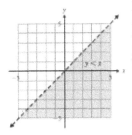

95)

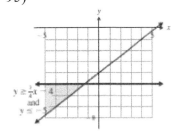

97)

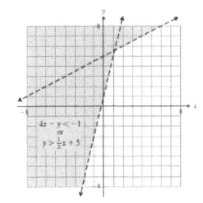

99)

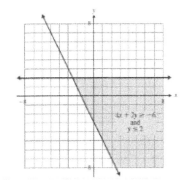

101) Domain: $\{-3, 5, 12\}$
Range: $\{1, 3, -3, 4\}$
Not a function

103) Domain: $\{\text{Beagle, Siamese, Parrot}\}$
Range: $\{\text{Dog, Cat, Bird}\}$
Function

105) Domain: $[0, 4]$
Range: $[0, 2]$
Not a function

107) $(-\infty, -3) \cup (-3, \infty)$; function

109) $[0, \infty)$; Not a function

111) $\left(-\infty, \dfrac{2}{7}\right] \cup \left[\dfrac{2}{7}, \infty\right)$; function

113) $f(3) = 27; f(-2) = -8$

115)
a) $f(4) = 5 \cdot 4 - 12 = 20 - 12 = 8$
b) $f(4) = 5 \cdot (-3) - 12 = -15 - 12 = -27$
c) $g(3) = 3^2 + 6 \cdot 3 + 5 = 9 + 18 + 5 = 32$
d) $g(0) = 0^2 + 6 \cdot 0 + 5 = 0 + 0 + 5 = 5$
e) $f(a) = 5a - 12$
f) $g(t) = t^2 + 6t + 5$
g)

$$f(k+8) = 5(k+8) - 12 = 5k + 40 - 12 = 5k + 28$$

h)

$$f(c-2) = 5(c-2) - 12 = 5c - 10 - 12 = 5k - 22$$

i) $f(x+h) = 5(x+h) - 12 = 5x + 5h - 12$

j)

$$f(x+h) + f(x) = 5(x+h) - 12 - (5x - 12)$$
$$f(x+h) + f(x) = 5x + 5h - 12 - 5x + 12$$
$$f(x+h) + f(x) = 5h$$

117) $\quad 2\left(\dfrac{11}{2}\right) = 2\left(\dfrac{3}{2}x + 5\right)$
$$11 = 3x + 10$$
$$11 - 10 = 3x$$
$$1 = 3x$$
$$\dfrac{1}{3} = x$$

119)

a)

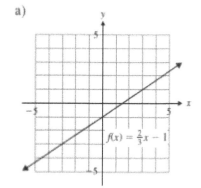

b)

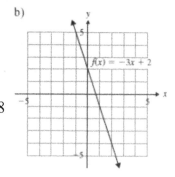

121)

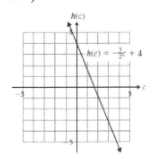

123)
 a) $f(2) = 480 \cdot 2 = 960 \ MB$

   $f(6) = 480 \cdot 6 = 2880 \ MB$

 b) $1200 = 480t$

   $2.5 \sec = t$

Chapter 4 Test

1) $5 \cdot 9 + 3 \cdot (-13) = 6$

   $45 - 39 = 6$

    $6 = 6$

     yes

4) a) $2x - 3 \cdot 0 = 12$

    $2x = 12$

    $x = 6$

   intercept: $(6,0)$

 b) $2 \cdot 0 - 3y = 12$

    $y = -4$

   intercept: $(0,-4)$

 c) $2x - 3 \cdot 2 = 12$

   $2x - 6 = 12$

    $2x = 18$

    $x = 9$

     $(2,9)$

d)
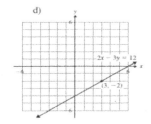

5)
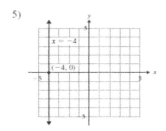

7) a) $m = \dfrac{-14 - (-5)}{4 - (-8)} = \dfrac{-14 + 5}{4 + 8} = \dfrac{-9}{12} = -\dfrac{3}{4}$

 b) $m = \dfrac{2 - 2}{3 - 9} = \dfrac{0}{-6} = 0$

9)

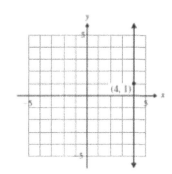

11) $2(y - 8) = \cancel{2}\left(\dfrac{1}{\cancel{2}}\right)(x - 3)$

   $2y - 16 = x - 3$

   $-x + 2y = -3 + 16$

   $-x + 2y = 13$

    $x - 2y = -13$

13) $\perp m = \dfrac{1}{3}$

$$y - (-5) = \dfrac{1}{3}(x - 12)$$

$$y + 5 = \dfrac{1}{3}x - 4$$

$$y = \dfrac{1}{3}x - 4 - 5$$

$$y = \dfrac{1}{3}x - 9$$

15)

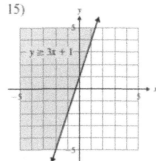

17)

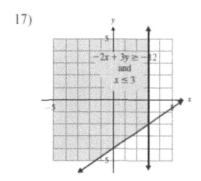

19) Domain: $\{-2, 1, 3, 8\}$

　　Range: $\{-5, -1, 1, 4\}$

　　function

21) a. $(-\infty, \infty)$

　　b. yes

23) $f(2) = -3$

25) $f(6) = -4 \cdot 6 + 2$

　　$f(6) = -24 + 2$

　　$f(6) = -22$

27) $g(t) = t^2 - 3t + 7$

29)

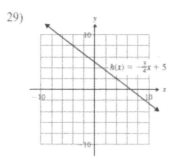

Cumulative Review Chapters 1–4

1) $\dfrac{252}{840} = \dfrac{3}{10}$

3) $-2^6 = -(2)(2)(2)(2)(2)(2) = 64$

5) $3 - \dfrac{2}{5} = \dfrac{15}{5} - \dfrac{2}{5} = \dfrac{13}{5}$

7) $12 - 5(2n + 9) = 3n + 2(n + 6)$

　　$12 - 10n - 45 = 3n + 2n + 12$

　　$-10n - 33 = 5n + 12$

　　$-10n - 5n = 12 + 33$

　　$-15n = 45$

　　$n = -3 \qquad \{-3\}$

9) $t + zw = r$

　　$zw = r - t$

　　$w = \dfrac{r - t}{z}$

11) $19 - 4x > 25$

　　$-4x > 6$

　　$x < -\dfrac{6}{4}$

　　$x < -\dfrac{3}{2} \qquad \left(-\infty, -\dfrac{3}{2}\right)$

13) cc = chocolate chip

cm = chunky monkey

$cc = cm - cm(0.10)$

$270 = cm - 0.10cm$

$270 = 0.9cm$

$300 = cm$

15) Lynette's age $= 3x - 7$

daughter's age $= x$

$x + 3x - 7 = 57$

$4x = 64$

$x = 16$

daugher's age $= 16$

Lynette's age $= 41$

17)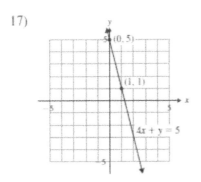

19)

$(x_1, y_1) = (4, -12); \quad m = \dfrac{1}{3}; \quad m_{\text{perpendicular}} = -3$

$y - y_1 = m(x - x_1)$

$y + 12 = -3(x - 4)$

$y + 12 = -3x + 12$

$y = -3x + 12 - 12$

$y = -3x$

20) a) yes   b) $(-\infty, 7) \cup (7, \infty)$

21) $f(x) = 8x + 3$

$f(-5) = 8(-5) + 3 = -37$

23)

$f(x) = 8x + 3$

$f(t + 2) = 8(t + 2) + 3 = 8t + 16 + 3 = 8t + 19$

25)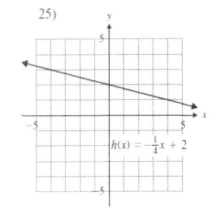

Section 5.1

1) Is $(-2,5)$ a solution to the system of equations?

$$3x + 2y = 4$$
$$3(-2) + 2 \cdot 5 = 4$$
$$-6 + 10 = 4$$
$$4 = 4$$
$$\text{True}$$
$$4x - y = -3$$
$$4(-2) - 5 = -3$$
$$-8 - 5 = -3$$
$$-13 = -3$$
$$\text{False}$$

No, $(-2,5)$ is not a solution.

3) Is $(-1,-2)$ a solution to the system of equations?

$$y = 5x - 7$$
$$-2 = 5(-1) - 7$$
$$-2 = -5 - 7$$
$$-2 = -12$$
$$\text{False}$$
$$3x + 9 = y$$
$$3(-1) + 9 = -2$$
$$-3 + 9 = -2$$
$$6 = -2$$
$$\text{False}$$

No, $(-1,-2)$ is not a solution to the system of equations.

5) The lines are parallel.

7) dependent

9) $(3,1)$

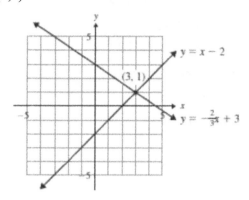

11) $(-1,-3)$

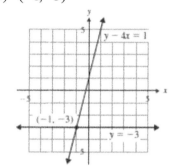

13) $\varnothing$; inconsistent system

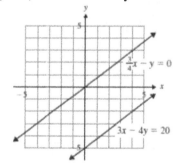

15) $(1,-1)$

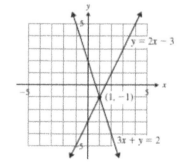

17) $\{(x,y)\mid y=-3x+1\}$

dependent equations

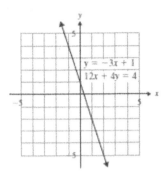

19) Answers may vary.

21) Answers may vary.

23) B: $(-2,1)$ is in quadrant II.

25) The slopes are different.

27) $y=\dfrac{3}{2}x+\dfrac{7}{2}$

$-9x+6y=21$

$6y=9x+21$

$y=\dfrac{9}{6}x+\dfrac{21}{6}$

$y=\dfrac{3}{2}x+\dfrac{7}{2}$

Infinite number of solutions

29) $5x-2y=-11 \qquad x+6y=18$

$-2y=-5x-11 \qquad 6y=-x+18$

$y=\dfrac{-5}{-2}x-\dfrac{11}{-2} \qquad y=-\dfrac{1}{6}x+\dfrac{18}{6}$

$y=\dfrac{5}{2}x+\dfrac{11}{2} \qquad y=-\dfrac{1}{6}x+3$

One solution

31) $x+y=10 \qquad -9x-9y=2$

$\qquad y=-x+10 \qquad -9y=9x+2$

$\qquad\qquad\qquad y=\dfrac{9}{-9}x+\dfrac{2}{-9}$

$\qquad\qquad\qquad y=-x-\dfrac{2}{9}$

No solution

33) a) after 2001 \qquad b) 2001; 5.3 million

c) $1999-2001$ \qquad d) $1999-2001$

35) $(x,y)=(3,-2)$

37) $(x,y)=(-5,-1)$

39) $(x,y)=(-0.5,-1.25)$

41) It is the only variable with a coefficient of 1.

43) $\begin{cases} 2x-y=5 \\ x=y+6 \end{cases}$

$2x-y=5$

$2(y+6)-y=5$

$2y+12-y=5$

$y+12=5$

$y=5-12$

$y=-7$

$x=y+6$

$x=-7+6$

$x=-1$

$(x,y)=(-1,-7)$

45) $\begin{cases} x - 6y = 4 & \Rightarrow & x = 6y + 4 \\ 2x + 5y = 8 & & 2x + 5y = 8 \end{cases}$

$2x + 5y = 8$

$2(6y + 4) + 5y = 8$

$12y + 8 + 5y = 8$

$17y = 8 - 8$

$17y = 0$

$y = 0$

$x = 6y + 4$

$x = 6 \cdot 0 + 4$

$x = 4$

$(x, y) = (4, 0)$

47) $\begin{cases} 2x - y = 3 & \Rightarrow & y = 2x - 3 \\ 9y - 18x = 5 & & 9y - 18x = 5 \end{cases}$

$9y - 18x = 5$

$9(2x - 3) - 18x = 5$

$18x - 27 - 18x = 5$

$-27 = 5$

$\varnothing$; inconsistent

49)

$\begin{cases} 5x - y = 8 & & 5x - y = 8 \\ 4y = 10 - x & \Rightarrow & 4y - 10 = -x \Rightarrow -4y + 10 = x \end{cases}$

$5x - y = 8$

$5(-4y + 10) - y = 8$

$-20y + 50 - y = 8$

$-21y + 50 = 8$

$-21y = 8 - 50$

$-21y = -42$

$y = 2$

$-4y + 10 = x$

$-4(2) + 10 = x$

$-8 + 10 = x$

$x = 2$

$(x, y) = (2, 2)$

51) $\begin{cases} 10x + y = -5 & \Rightarrow & y = -10x - 5 \\ -5x + 2y = 10 & & -5x + 2y = 10 \end{cases}$

$-5x + 2y = 10 \qquad y = -10\left(-\dfrac{4}{5}\right) - 5$

$-5x + 2(-10x - 5) = 10 \qquad y = 8 - 5$

$-5x - 20x - 10 = 10 \qquad y = 3$

$-25x - 10 = 10$

$-25x = 20$

$x = \dfrac{20}{-25} \qquad (x, y) = \left(-\dfrac{4}{5}, 3\right)$

$x = -\dfrac{4}{5}$

53) $\begin{cases} 6x + y = -6 \\ -12x - 2y = 12 \end{cases}$ $\Rightarrow y = -6x - 6$

$-12x - 2y = 12$

$$-12x - 2y = 12$$
$$-12x - 2(-6x - 6) = 12$$
$$-12x + 12x + 12 = 12$$
$$12 = 12$$

$\{(x, y) \mid 6x + y = -6\}$; dependent

55)

$\begin{cases} 2x - y = 6 \\ 3y = -18 - x \end{cases}$ $\Rightarrow -y = -2x + 6 \Rightarrow y = 2x - 6$

$3y = -18 - x$

$$3y = -18 - x$$
$$3(2x - 6) = -18 - x$$
$$6x - 18 = -18 - x$$
$$6x + x = -18 + 18$$
$$7x = 0$$
$$x = 0$$
$$y = 2x - 6$$
$$y = 2(0) - 6$$
$$y = -6$$

$(x, y) = (0, -6)$

57) Multiply the equation by the LCD of the fractions to eliminate the fractions.

59) $\begin{cases} \dfrac{1}{4}x - \dfrac{1}{2}y = 1 \\ \dfrac{2}{3}x + \dfrac{1}{6}y = \dfrac{25}{6} \end{cases}$

$\dfrac{1}{4}x - \dfrac{1}{2}y = 1$ $\qquad$ $\dfrac{2}{3}x + \dfrac{1}{6}y = \dfrac{25}{6}$

$4\left(\dfrac{1}{4}x - \dfrac{1}{2}y\right) = 4 \cdot 1$ $\quad$ $6\left(\dfrac{2}{3}x + \dfrac{1}{6}y\right) = 6\left(\dfrac{25}{6}\right)$

$x - 2y = 4$ $\qquad\qquad$ $4x + y = 25$

$x = 2y + 4$

$\begin{cases} x = 2y + 4 \\ 4x + y = 25 \end{cases}$

$4(2y + 4) + y = 25$ $\qquad$ $x = 2(1) + 4$

$8y + 16 + y = 25$ $\qquad$ $x = 2 + 4$

$9y + 16 = 25$ $\qquad\quad$ $x = 6$

$9y = 9$

$y = 1$ $\qquad$ $(x, y) = (6, 1)$

61) $\begin{cases} -\dfrac{2}{15}x - \dfrac{1}{3}y = \dfrac{2}{3} \\ \dfrac{2}{3}x + \dfrac{5}{3}y = \dfrac{1}{2} \end{cases}$

$-\dfrac{2}{15}x - \dfrac{1}{3}y = \dfrac{2}{3}$ $\qquad$ $\dfrac{2}{3}x + \dfrac{5}{3}y = \dfrac{1}{2}$

$15\left(-\dfrac{2}{15}x - \dfrac{1}{3}y\right) = 15\left(\dfrac{2}{3}\right)$ $\quad$ $6\left(\dfrac{2}{3}x + \dfrac{5}{3}y\right) = 6 \cdot \dfrac{1}{2}$

$-2x - 5y = 10$ $\qquad\qquad$ $4x + 10y = 3$

$-5y = 2x + 10$

$y = -\dfrac{2}{5}x - 2$

$\begin{cases} 4x + 10y = 3 \\ \quad y = -\dfrac{2}{5}x - 2 \end{cases}$

$4x + 10\left(-\dfrac{2}{5}x - 2\right) = 3$

$4x - 4x - 20 = 3$

$-20 \neq 3$

$\varnothing$, inconsistent

63) $\begin{cases} \dfrac{3}{4}x+\dfrac{1}{2}y=6 \\ x=3y+8 \end{cases}$

$\dfrac{3}{4}x+\dfrac{1}{2}y=6 \qquad x=3y+8$

$4\left(\dfrac{3}{4}x+\dfrac{1}{2}y\right)=4\cdot 6$

$3x+2y=24$

$\begin{cases} 3x+2y=24 \\ x=3y+8 \end{cases}$

$3(3y+8)+2y=24 \qquad x=3\cdot 0+8$

$9y+24+2y=24 \qquad x=0+8$

$11y+24=24 \qquad x=8$

$11y=0$

$y=0 \qquad (x,y)=(8,0)$

65) $\begin{cases} 0.01x+0.10y=-0.11 \\ 0.02x-0.01y=0.20 \end{cases}$

$100(0.01x+0.10y)=100(-0.11) \qquad 100(0.02x-0.01y)=100(0.20)$

$x+10y=-11 \qquad\qquad 2x-y=20$

$x=-10y-11$

$\begin{cases} 2x-y=20 \\ x=-10y-11 \end{cases}$

$2(-10y-11)-y=20 \qquad x=-10(-2)-11$

$-20y-22-y=20 \qquad x=20-11$

$-21y=42 \qquad x=9$

$y=-2 \qquad (x,y)=(9,-2)$

67) $\begin{cases} 4x-3y=-5 \\ -4x+5y=11 \end{cases}$

$4x-3y=-5$

$\underline{-4x+5y=11}$

$2y=6$

$y=3$

$4x-3(3)=-5$

$4x-9=-5$

$4x=4$

$x=1 \qquad (x,y)=(1,3)$

69) $-8x+5y=-6 \qquad 4x-7y=3$

$\qquad\qquad\qquad\qquad 2(4x-7y)=2(3)$

$\qquad\qquad\qquad\qquad 8x-14y=6$

$\begin{cases} -8x+5y=-6 \\ 8x-14y=6 \end{cases}$

$-8x+5y=-6$

$\underline{8x-14y=6}$

$-9y=0$

$y=0$

$-8x+5(0)=-6$

$-8x=-6$

$x=\dfrac{-6}{-8}=\dfrac{3}{4} \qquad (x,y)=(\dfrac{3}{4},0)$

71) $5x-6y=-2 \qquad 10x-12y=7$

$-2(5x-6y)=-2(-2)$

$-10x+12y=4$

$\begin{cases} -10x+12y=4 \\ 10x-12y=7 \end{cases}$

$-10x+12y=4$

$\underline{10x-12y=7}$

$\qquad\qquad 0=11 \text{ False}$

$\varnothing$; inconsistent

73) $x-6y=-5 \qquad\qquad -24y+4x=-20$

$-4(x-6y)=-4(-5) \quad 4x-24y=-20$

$-4x+24y=20$

$\begin{cases} -4x+24y=20 \\ 4x-24y=-20 \end{cases}$

$-4x+24y=20$

$\underline{4x-24y=-20}$

$0=0 \text{ True}$

$\{(x,y)\,|\,x-6y=-5\}$; dependent

75)

$9x-7y=-14 \qquad\qquad 4x+3y=6$

$4(9x-7y)=4(-14) \quad -9(4x+3y)=-9(6)$

$36x-28y=-56 \qquad -36x-27y=-54$

$\begin{cases} 36x-28y=-56 \\ -36x-27y=-54 \end{cases}$

$36x-28y=-56$

$\underline{-36x-27y=-54}$

$\qquad\qquad -55y=-110$

$\qquad\qquad\quad y=2$

$4x+3(2)=6$

$4x+6=6$

$4x=0$

$x=0 \qquad (x,y)=(0,2)$

77) $7x+2y=12 \qquad\qquad 24-14x=4y$

$2(7x+2y)=2(12) \quad -14x-4y=-24$

$14x+4y=24$

$\begin{cases} 14x+4y=24 \\ -14x-4y=-24 \end{cases}$

$14x+4y=24$

$\underline{-14x-4y=-24}$

$\qquad\qquad 0=0 \text{ True}$

$\{(x,y)\,|\,7x+2y=12\}$; dependent

79) $\dfrac{x}{4}+\dfrac{y}{2}=-1 \qquad\qquad \dfrac{3}{8}x+\dfrac{5}{3}y=-\dfrac{7}{12}$

$4\left(\dfrac{x}{4}+\dfrac{y}{2}\right)=4(-1) \quad 24\left(\dfrac{3}{8}x+\dfrac{5}{3}y\right)=24\left(-\dfrac{7}{12}\right)$

$x+2y=-4 \qquad\qquad 9x+40y=-14$

$-9(x+2y)=-9(-4)$

$-9x-18y=36$

$\begin{cases} -9x-18y=36 \\ 9x+40y=-14 \end{cases}$

$-9x-18y=36$

$\underline{9x+40y=-14}$

$\qquad\qquad 22y=22$

$\qquad\qquad\quad y=1$

$$-9x - 18(1) = 36$$
$$-9x - 18 = 36$$
$$-9x = 54$$
$$x = -6 \qquad (x, y) = (-6, 1)$$

81) $\dfrac{x}{2} - \dfrac{y}{5} = \dfrac{1}{10} \qquad\qquad \dfrac{x}{3} + \dfrac{y}{4} = \dfrac{5}{6}$

$10\left(\dfrac{x}{2} - \dfrac{y}{5}\right) = 10\left(\dfrac{1}{10}\right) \qquad 12\left(\dfrac{x}{3} + \dfrac{y}{4}\right) = 12\left(\dfrac{5}{6}\right)$

$5x - 2y = 1 \qquad\qquad 4x + 3y = 10$

$3(5x - 2y) = 3(1) \qquad 2(4x + 3y) = 2(10)$

$15x - 6y = 3 \qquad\qquad 8x + 6y = 20$

$$\begin{cases} 15x - 6y = 3 \\ 8x + 6y = 20 \end{cases}$$

$$\begin{aligned} 15x - 6y &= 3 \\ \underline{8x + 6y} &= \underline{20} \\ 23x &= 23 \\ x &= 1 \end{aligned}$$

$$15(1) - 6y = 3$$
$$15 - 6y = 3$$
$$-6y = -12$$
$$y = 2 \qquad (x, y) = (1, 2)$$

83) $0.1x + 2y = -0.8 \qquad\qquad 0.03x + 0.10y = 0.26$

$10(0.1x + 2y) = 10(-0.8) \quad 100(0.03x + 0.10y) = 100(0.26)$

$x + 20y = -8 \qquad\qquad 3x + 10y = 26$

$-3(x + 20y) = -3(-8)$

$-3x - 60y = 24$

$$\begin{cases} -3x - 60y = 24 \\ 3x + 10y = 26 \end{cases}$$

$$\begin{aligned} -3x - 60y &= 24 \\ \underline{3x + 10y} &= \underline{26} \\ -50y &= 50 \\ y &= -1 \end{aligned}$$

$$-3x - 60(-1) = 24$$
$$-3x + 60 = 24$$
$$-3x = -36$$
$$x = 12 \quad (x, y) = (12, -1)$$

85) $-0.4x + 0.2y = 0.1$ $\qquad 0.6x - 0.3y = 1.5$

$10(-0.4x + 0.2y) = 10(0.1)$ $10(0.6x - 0.3y) = 10(1.5)$

$-4x + 2y = 1$ $\qquad\qquad\quad 6x - 3y = 15$

$3(-4x + 2y) = 3(1)$ $\qquad\quad 2(6x - 3y) = 2(15)$ $\qquad -12x + 6y = 3$

$-12x + 6y = 3$ $\qquad\qquad 12x - 6y = 30$ $\qquad\quad \underline{12x - 6y = 30}$

$\qquad\qquad\qquad\qquad\qquad\qquad\qquad\qquad\qquad\qquad 0 = 33 \qquad \varnothing$, inconsistent

$$\begin{cases} -12x + 6y = 3 \\ 12x - 6y = 30 \end{cases}$$

$$-30x + 15y = -6$$
$$\underline{-30x + 15y = -6}$$
$$\qquad 0 = 0 \quad \text{True}$$

$\{(x, y) \mid x - 0.5y = 0.2\}$; dependent

87) a) Rent for Less: $y = 0.40(60) = \$24$

$\qquad$ Frugal: $y = 0.30(60) + 12 = \$30$

b) Rent for Less: $y = 0.40(160) = \$64$

$\qquad$ Frugal: $y = 0.30(160) + 12 = \$60$

c) $\begin{cases} y = 0.40x \\ y = 0.30x + 12 \end{cases}$

$y = 0.40x$ $\qquad\qquad\qquad y = 0.30x + 12$

$0.30x + 12 = 0.40x$ $\qquad y = 0.40(120)$

$0.10x = 12$ $\qquad\qquad\qquad y = 48$

$x = 120$ $\qquad\qquad\qquad (x, y) = (120, 48)$

If the car is driven 120 mi, the cost would be the same from each company: \$48.

d)

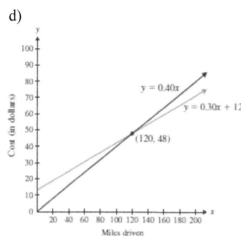

If a car is driven less than 120 mi, it is cheaper to rent from Rent-for-Less. If a car is driven more than 120 mi, it is cheaper to rent from Frugal Rentals. If a car is driven exactly 120 mi, the cost is the same from each company.

89) $(x, y) = (-3, 7)$

91) $(x, y) = \left(5, -\dfrac{3}{2}\right)$

93) $(x, y) = (8, 0)$

95) $\varnothing$; inconsistent

97) $(x, y) = (-6, 2)$

99) $(x, y) = (-3, 3)$

101) $(x, y) = \left(-\dfrac{3}{2}, 4\right)$

103)   a) The variables are eliminated, and you get a false statement.

   b) The variables are eliminated, and you get a true statement.

105)   a) 3

   b) c can be any real number except 3.

107) $-x + by = -7; (2, -1)$
$$-2 + b(-1) = -7$$
$$-2 - b = -7$$
$$-b = -5$$
$$b = 5$$

109) $\begin{cases} ax + by = 5 \\ 2ax - by = 1 \end{cases}$

$ax + by = 5$
$\underline{2ax - by = 1}$
$3ax = 6$

$x = \dfrac{6}{3a} = \dfrac{2}{a}$

$(x, y) = \left(\dfrac{2}{a}, \dfrac{3}{b}\right)$

$ax + by = 5$
$a\left(\dfrac{2}{a}\right) + by = 5$
$2 + by = 5$
$by = 3$
$y = \dfrac{3}{b}$

111) $x - 3by = 2$ $3x + by = -4$
$-3(x - 3by) = -3(2)$
$-3x + 9by = -6$

$\begin{cases} -3x + 9by = -6 \\ 3x + by = -4 \end{cases}$

$$-3x+9by=-6$$
$$\underline{3x+by=-4}$$
$$10by=-10$$

$$y=\frac{-10}{10b}=-\frac{1}{b}$$

$$x-3by=2$$
$$x-3b\left(-\frac{1}{b}\right)=2$$
$$x+3=2$$
$$x=-1 \qquad (x,y)=(-1,-\frac{1}{b})$$

113) $\begin{cases} ax+by=c \\ -ax+by=c \end{cases}$

$$ax+by=c$$
$$\underline{-ax+by=c}$$
$$2by=2c$$
$$y=\frac{2c}{2b}=\frac{c}{b}$$

$$ax+by=c$$
$$ax+b\left(\frac{c}{b}\right)=c$$
$$ax+c=c$$
$$ax=0$$
$$x=0$$

$$(x,y)=\left(0,\frac{c}{b}\right)$$

115) $\begin{cases} \dfrac{1}{x}+\dfrac{2}{y}=-\dfrac{7}{8} \\ \dfrac{1}{x}-\dfrac{1}{y}=\dfrac{5}{8} \end{cases}$

$$u+2v=-\frac{7}{8} \qquad\qquad u-v=\frac{5}{8}$$

$$-8(u+2v)=-8\left(-\frac{7}{8}\right) \qquad 8(u-v)=8\left(\frac{5}{8}\right)$$

$$-8u-16v=7 \qquad\qquad 8u-8v=5$$

$\begin{cases} -8u-16v=7 \\ 8u-8v=5 \end{cases}$

$$-8u-16v=7 \qquad\qquad -8u-16v=7$$

$$\underline{8u-8v=5} \qquad\qquad -8u-16\left(-\frac{1}{2}\right)=7$$

$$-24v=12 \qquad\qquad -8u+8=7$$

$$v=-\frac{12}{24}=-\frac{1}{2} \qquad\qquad -8u=-1$$

$$u=\frac{1}{8}$$

$$(u,v)=\left(\frac{1}{8},-\frac{1}{2}\right)$$

$\therefore$ by substitution,

because $u=\dfrac{1}{x}$ and $v=\dfrac{1}{y}$

then $\dfrac{1}{8}=\dfrac{1}{x}$ and $-\dfrac{1}{2}=\dfrac{1}{y}$.

So, $(x,y)=(8,-2)$.

117) $\begin{cases} -\dfrac{1}{x} + \dfrac{2}{y} = -\dfrac{7}{3} \\ \dfrac{2}{x} - \dfrac{3}{y} = 5 \end{cases}$

$-u + 2v = -\dfrac{7}{3}$ $2u - 3v = 5$

$-3(-u + 2v) = -3\left(-\dfrac{7}{3}\right)$ $-2(2u - 3v) = -2(5)$

$3u - 6v = 7$ $-4u + 6v = -10$

$\begin{cases} 3u - 6v = 7 \\ -4u + 6v = -10 \end{cases}$

$\begin{array}{ll} 3u - 6v = 7 & 3(3) - 6v = 7 \\ \underline{-4u + 6v = -10} & 9 - 6v = 7 \\ -u = -3 & -6v = -2 \\ u = 3 & v = \dfrac{2}{6} \\ & v = \dfrac{1}{3} \end{array}$

$(u, v) = \left(3, \dfrac{1}{3}\right)$

$\therefore$ by substitution,

because $u = \dfrac{1}{x}$ and $v = \dfrac{1}{y}$

then $3 = \dfrac{1}{x}$ and $\dfrac{1}{3} = \dfrac{1}{y}$.

So, $(x, y) = \left(\dfrac{1}{3}, 3\right)$.

119) $f(x) = x - 3, g(x) = -\dfrac{1}{4}x + 2$

$x = 4$

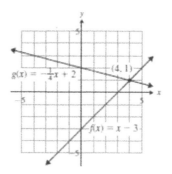

121) $f(x) = 3x + 3, g(x) = x + 1$

$x = -1$

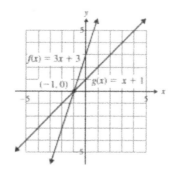

Section 5.2

1) $4x + 3y - 7z = -6$
$x - 2y + 5z = -3$
$-x + y + 2z = 7$
$$(-2, 3, 1)$$

$4x + 3y - 7z = -6$
$4(-2) + 3(3) - 7(1) = -6$
$-8 + 9 - 7 = -6$
$-6 = -6$

$$x-2y+5z=-3$$
$$-2-2(3)+5(1)=-3$$
$$-2-6+5=-3$$
$$-3=-3$$

$$-x+y+2z=7 \qquad \text{yes}$$
$$-(-2)+3+2(1)=7$$
$$2+3+2=7$$
$$7=7$$

3) $\quad -x+y-2z=2$
$$3x-y+5z=4$$
$$2x+3y-z=7$$
$$(0,6,2)$$

$$-x+y-2z=2$$
$$-(0)+6-2(2)=2$$
$$6-4=2$$
$$2=2$$

$$3x-y+5z=4$$
$$3(0)-6+5(2)=4$$
$$-6+10=4$$
$$4=4$$

$$2x+3y-z=7 \qquad \text{no}$$
$$2(0)+3(6)-2=7$$
$$18-2=7$$
$$16 \neq 7$$

5) Answers may vary.

7) $\boxed{1} \quad x+3y+z=3$
$\boxed{2}\ 4x-2y+3z=7$
$\boxed{3}\ -2x+y-z=-1$
$\boxed{1} \quad x+3y+z=3$
$\boxed{3}\ \underline{-2x+y-z=-1}$
$\boxed{A} \qquad -x+4y=2$

$-3\times\boxed{1} \quad -3x-9y-3z=-9$
$\boxed{2} \quad \underline{4x-2y+3z=7}$
$\boxed{B} \qquad x-11y=-2$

$\boxed{A} \quad -x+4y=2 \qquad \boxed{A}\,-x+4y=2$
$\boxed{B} \quad \underline{x-11y=-2} \quad -x+4(0)=2$
$\qquad\quad -7y=0 \qquad\qquad -x=2$
$\qquad\qquad y=0 \qquad\qquad\quad x=-2$

$\boxed{1} \quad x+3y+z=3$
$\qquad -2+3(0)+z=3$
$\qquad\qquad -2+z=3$
$\qquad\qquad\quad z=5$
$$(-2,0,5)$$

9) $\boxed{1} \quad 5x+3y-z=-2$
$\boxed{2}\,-2x+3y+2z=3$
$\boxed{3} \quad x+6y+z=-1$

$\boxed{1}\ 5x+3y-z=-2$
$\boxed{3}\ \underline{x+6y+z=-1}$
$\boxed{A} \qquad 6x+9y=-3$
$2\times\boxed{1}\ 10x+6y-2z=-4$
$\boxed{2}\ \underline{-2x+3y+2z=3}$
$\boxed{B} \qquad 8x+9y=-1$
$-1\times\boxed{A}\ -6x-9y=3$
$\boxed{B} \qquad \underline{8x+9y=-1}$
$\qquad\qquad 2x=2$
$\qquad\qquad\ x=1$

$\boxed{A}$ $6x+9y=-3$

$6(1)+9y=-3$

$6+9y=-3$

$9y=-9$

$y=-1$

$\boxed{1}$ $5x+3y-z=-2$

$5(1)+3(-1)-z=-2$

$5-3-z=-2$

$2-z=-2$

$-z=-4$

$z=4$

$(1,-1,4)$

11) $\boxed{1}$ $\quad 3a+5b-3c=-4$

$\boxed{2}$ $\quad a-3b+c=6$

$\boxed{3}$ $-4a+6b+2c=-6$

$2\times\boxed{1}$ $\quad 6a+10b-6c=-8$

$3\times\boxed{3}$ $\underline{-12a+18b+6c=-18}$

$\boxed{A}$ $\quad -6a+28b=-26$

$\boxed{1}$ $\quad 3a+5b-3c=-4$

$3\times\boxed{2}$ $\underline{3a-9b+3c=18}$

$\boxed{B}$ $\quad 6a-4b=14$

$\boxed{A}$ $-6a+28b=-26$

$\boxed{B}$ $\underline{\quad 6a-4b=14}$

$24b=-12, b=-\dfrac{1}{2}$

$\boxed{A}$ $-6a+28b=-26$

$-6a+28\left(-\dfrac{1}{2}\right)=-26$

$-6a-14=-26$

$-6a=-12, a=2$

$\boxed{1}$ $\quad 3a+5b-3c=-4$

$3(2)+5\left(-\dfrac{1}{2}\right)-3c=-4$

$6-\dfrac{5}{2}-3c=-4$

$\dfrac{7}{2}-3c=-4$

$-3c=-\dfrac{15}{2}, c=\dfrac{5}{2}$

$\left(2,-\dfrac{1}{2},\dfrac{5}{2}\right)$

13) $\boxed{1}$ $\quad a-5b+c=-4$

$\boxed{2}$ $\quad 3a+2b-4c=-3$

$\boxed{3}$ $\quad 6a+4b-8c=9$

$-2\times\boxed{2}$ $\quad -6a-4b+8c=6$

$\boxed{3}$ $\quad \underline{6a+4b-8c=9}$

$\boxed{A}$ $\quad\quad 0\neq15$

$\varnothing$, inconsistent

15) $\boxed{1}$ $\quad -15x-3y+9z=3$

$\boxed{2}$ $\quad\quad 5x+y-3z=-1$

$\boxed{3}$ $\quad 10x+2y-6z=-2$

$\boxed{1}$ $-15x-3y+9z=3$

$3\times\boxed{2}$ $\quad\underline{15x+3y-9z=-3}$

$\boxed{A}$ $\quad\quad\quad 0=0$

dependent, $\{(x,y,z)\,|\,5x+y-3z=-1\}$

17) $\boxed{1}$ $-3a+12b-9c=-3$

$\boxed{2}$ $5a-20b+15c=5$

$\boxed{3}$ $-a+4b-3c=-1$

$\boxed{1}$ $-3a+12b-9c=-3$

$-3\times\boxed{3}$ $\underline{3a-12b+9c=3}$

$\boxed{A}$ $0=0$

dependent, $\{(a,b,c)\,|\,-a+4b-3c=-1\}$

19) $\boxed{1}$ $5x-2y+z=-5$

$\boxed{2}$ $x-y-2z=7$

$\boxed{3}$ $4y+3z=5$

$\boxed{1}$ $5x-2y+z=-5$

$-5\times\boxed{2}$ $\underline{-5x+5y+10z=-35}$

$\boxed{A}$ $3y+11z=-40$

$-11\times\boxed{3}$ $-44y-33z=-55$

$3\times\boxed{A}$ $\underline{9y+33z=-120}$

$-35y=-175$

$y=5$

$\boxed{3}$ $4y+3z=5$

$4(5)+3z=5$

$20+3z=5$

$3z=-15$

$z=-5$

$\boxed{2}$ $x-y-2z=7$

$x-5-2(-5)=7$

$x-5+10=7$

$x+5=7$

$x=2$

$(2,5,-5)$

21) $\boxed{1}$ $a+15b=5$

$\boxed{2}$ $4a+10b+c=-6$

$\boxed{3}$ $-2a-5b-2c=-3$

$2\times\boxed{2}$ $8a+20b+2c=-12$

$\boxed{3}$ $\underline{-2a-5b-2c=-3}$

$\boxed{A}$ $6a+15b=-15$

$-6\times\boxed{1}$ $-6a-90b=-30$

$\boxed{A}$ $\underline{6a+15b=-15}$

$-75b=-45$

$b=\dfrac{3}{5}$

$\boxed{1}$ $a+15b=5$

$a+15\left(\dfrac{3}{5}\right)=5$

$a+9=5$

$a=-4$

$\boxed{3}$ $-2a-5b-2c=-3$

$-2(-4)-5\left(\dfrac{3}{5}\right)-2c=-3$

$8-3-2c=-3$

$5-2c=-3$

$-2c=-8$

$c=4$

$\left(-4,\dfrac{3}{5},4\right)$

23) $\boxed{1}\ x+2y+3z=4$

  $\boxed{2}\quad -3x+y=-7$

  $\boxed{3}\quad 4y+3z=-10$

$\boxed{1}\quad x+2y+3z=4$

  $x=-2y-3z+4$

$\boxed{2}\qquad\quad -3x+y=-7$

  $-3(-2y-3z+4)+y=-7$

  $6y+9z-12+y=-7$

  $7y+9z=5$

$-3\times\boxed{3}\ \ -12y-9z=30$

$\boxed{A}\quad\underline{\ \ 7y+9z=5\ }$

  $-5y=35$

  $y=-7$

$\boxed{2}\quad -3x+y=-7$

  $-3x+(-7)=-7$

  $-3x-7=-7$

  $-3x=0$

  $x=0$

$\boxed{3}\quad 4y+3z=-10$

  $4(-7)+3z=-10$

  $-28+3z=-10$

  $3z=18$

  $z=6$

  $(0,-7,6)$

25) $\boxed{1}\ -5x+z=-3$

  $\boxed{2}\quad 4x-y=-1$

  $\boxed{3}\ \ 3y-7z=1$

$\boxed{2}\quad 4x-y=-1$

  $y=4x+1$

$\boxed{3}\qquad\quad 3y-7z=1$

  $3(4x+1)-7z=1$

  $12x+3-7z=1$

$\boxed{A}\qquad 12x-7z=-2$

$7\times\boxed{1}\ \ -35x+7z=-21$

$\boxed{A}\quad\underline{\ 12x-7z=-2\ }$

  $-23x=-23$

  $x=1$

$\boxed{2}\quad 4x-y=-1$

  $4(1)-y=-1$

  $4-y=-1$

  $-y=-5$

  $y=5$

$\boxed{3}\quad 3y-7z=1$

  $3(5)-7z=1$

  $15-7z=1$

  $-7z=-14$

  $z=2$

  $(1,5,2)$

27) $\boxed{1}$ $4a + 2b = -11$

$\boxed{2}$ $-8a - 3c = -7$

$\boxed{3}$ $b + 2c = 1$

$\boxed{3}$ $b + 2c = 1$

$b = -2c + 1$

$\boxed{1}$ $4a + 2b = -11$

$4a + 2(-2c + 1) = -11$

$4a - 4c + 2 = -11$

$\boxed{A}$ $4a - 4c = -13$

$\boxed{2}$ $-8a - 3c = -7$

$2 \times \boxed{A}$ $\underline{8a - 8c = -26}$

$-11c = -33$

$c = 3$

$\boxed{3}$ $b + 2c = 1$

$b + 2(3) = 1$

$b + 6 = 1$

$b = -5$

$\boxed{1}$ $4a + 2b = -11$

$4a + 2(-5) = -11$

$4a - 10 = -11$

$4a = -1$

$a = -\dfrac{1}{4}$

$\left(-\dfrac{1}{4}, -5, 3 \right)$

29) $\boxed{1}$ $6x + 3y - 3z = -1$

$\boxed{2}$ $10x + 5y - 5z = 4$

$\boxed{3}$ $x - 3y + 4z = 6$

$5 \times \boxed{1}$ $30x + 15y - 15z = -5$

$-3 \times \boxed{2}$ $\underline{-30x - 15y + 15z = -12}$

$\boxed{A}$ $0 \neq -17$ $\varnothing$, inconsistent

31) $\boxed{1}$ $7x + 8y - z = 16$

$\boxed{2}$ $-\dfrac{1}{2}x - 2y + \dfrac{3}{2}z = 1$

$\boxed{3}$ $\dfrac{4}{3}x + 4y - 3z = -\dfrac{2}{3}$

$\boxed{1}$ $7x + 8y - z = 16$

$14 \times \boxed{2}$ $\underline{-7x - 28y + 21z = 14}$

$\boxed{A}$ $-20y + 20z = 30$

$8 \times \boxed{2}$ $-4x - 16y + 12z = 8$

$3 \times \boxed{3}$ $\underline{4x + 12y - 9z = -2}$

$\boxed{B}$ $-4y + 3z = 6$

$\boxed{A}$ $-20y + 20z = 30$

$-5 \times \boxed{B}$ $\underline{20y - 15z = -30}$

$5z = 0, z = 0$

$\boxed{A}$ $-20y + 20z = 30$

$-20y + 20(0) = 30$

$-20y = 30$

$y = -\dfrac{30}{20}, y = -\dfrac{3}{2}$

$\boxed{1}$ $7x + 8y - z = 16$

$7x + 8\left(-\dfrac{3}{2} \right) - 0 = 16$

$7x - 12 = 16$

$7x = 28$

$x = 4$

$\left(4, -\dfrac{3}{2}, 0 \right)$

Section 5.2: Solving Systems of Linear Equations in Three Variables

33) $\boxed{1}$ $\quad 2a-3b=-4$

$\boxed{2}$ $\quad 3b-c=8$

$\boxed{3}$ $\quad -5a+4c=-4$

$\boxed{2}$ $\quad 3b-c=8$

$\quad\quad c=3b-8$

$\boxed{3}$ $\quad\quad -5a+4c=-4$

$\quad\quad -5a+4(3b-8)=-4$

$\quad\quad -5a+12b-32=-4$

$\boxed{A}$ $\quad\quad -5a+12b=28$

$4\times\boxed{1}$ $\quad 8a-12b=-16$

$\boxed{A}$ $\quad \underline{-5a+12b=28}$

$\quad\quad\quad 3a=12$

$\quad\quad\quad a=4$

$\boxed{1}$ $\quad 2a-3b=-4$

$\quad 2(4)-3b=-4$

$\quad\quad 8-3b=-4$

$\quad\quad -3b=-12, b=4$

$\boxed{2}$ $\quad 3b-c=8$

$\quad 3(4)-c=8$

$\quad\quad 12-c=8$

$\quad\quad -c=-4$

$\quad\quad c=4$

$\quad\quad\quad (4,4,4)$

35) $\boxed{1}$ $\quad -4x+6y+3z=3$

$\boxed{2}$ $\quad -\dfrac{2}{3}x+y+\dfrac{1}{2}z=\dfrac{1}{2}$

$\boxed{3}$ $\quad 12x-18y-9z=-9$

$3\times\boxed{1}$ $\quad -12x+18y+9z=9$

$\boxed{3}$ $\quad \underline{12x-18y-9z=-9}$

$\boxed{A}$ $\quad\quad\quad\quad\quad 0=0$

dependent, $\{(a,b,c)\mid -4x+6y+3z=3\}$

37) $\boxed{1}$ $\quad a+b+9c=-3$

$\boxed{2}$ $\quad -5a-2b+3c=10$

$\boxed{3}$ $\quad 4a+3b+6c=-15$

$5\times\boxed{1}$ $\quad 5a+5b+45c=-15$

$\boxed{2}$ $\quad \underline{-5a-2b+3c=10}$

$\boxed{A}$ $\quad\quad 3b+48c=-5$

$-4\times\boxed{1}$ $\quad -4a-4b-36c=12$

$\boxed{3}$ $\quad \underline{4a+\ 3b+\ 6c=-15}$

$\boxed{B}$ $\quad\quad\quad -b-30c=-3$

$\boxed{A}$ $\quad 3b+48c=-5$

$3\times\boxed{B}$ $\quad \underline{-3b-90c=-9}$

$\quad\quad\quad -42c=-14$

$\quad\quad\quad c=\dfrac{-14}{-42}=\dfrac{1}{3}$

$\boxed{B}-b-30c=-3$

$\quad -b-30\left(\dfrac{1}{3}\right)=-3$

$\quad\quad -b-10=-3$

$\quad\quad -b=7, b=-7$

$\boxed{1}$ $\quad a+b+9c=-3$

$\quad a+(-7)+9\left(\dfrac{1}{3}\right)=-3$

$\quad\quad a-7+3=-3 \quad \left(1,-7,\dfrac{1}{3}\right)$

$\quad\quad a-4=-3$

$\quad\quad a=1$

39) $\boxed{1}$ $\quad x+5z=10$

$\boxed{2}$ $\quad 4y+z=-2$

$\boxed{3}$ $3x-2y=2$

$\boxed{2}$ $4y+z=-2$

$\qquad z=-4y-2$

$\boxed{1}$ $\qquad x+5z=10$

$x+5(-4y-2)=10$

$x-20y-10=10$

$\boxed{A}$ $\qquad x-20y=20$

$\boxed{3}$ $3x-2y=2$

$-3\times\boxed{A}$ $\underline{-3x+60y=-60}$

$\qquad 58y=-58$

$\qquad y=-1$

$\boxed{2}$ $\quad 4y+z=-2$

$4(-1)+z=-2$

$\qquad -4+z=-2$

$\qquad z=2$

$\boxed{3}$ $\quad 3x-2y=2$

$3x-2(-1)=2$

$\qquad 3x+2=2$

$\qquad 3x=0$

$\qquad x=0$

$\qquad (0,-1,2)$

41) $\boxed{1}$ $2x-y+4z=-1$

$\boxed{2}$ $\quad x+3y+z=-5$

$\boxed{3}$ $\quad -3x+2y=7$

$\boxed{1}$ $\quad 2x-\;y\;+4z=-1$

$-4\times\boxed{2}$ $\underline{-4x-12y-4z=20}$

$\boxed{A}$ $\qquad -2x-13y=19$

$2\times\boxed{3}$ $\;-6x+4y=\;\;14$

$-3\times\boxed{A}$ $\underline{6x+39y=-57}$

$\qquad 43y=-43$

$\qquad y=-1$

$\boxed{3}$ $-3x+2y=7$

$-3x+2(-1)=7$

$\qquad -3x-2=7$

$\qquad -3x=9$

$\qquad x=-3$

$\boxed{2}$ $\qquad x+3y+z=-5$

$(-3)+3(-1)+z=-5$

$\qquad -3-3+z=-5$

$\qquad -6+z=-5$

$\qquad z=1$

$\qquad (3,-1,1)$

42) $\boxed{1}$ $-2a+3b=3$

$\boxed{2}$ $\quad a+5c=-1$

$\boxed{3}$ $\quad b-2c=-5$

$2\times\boxed{2}$ $\;2a+10c=-2$

$\boxed{A}$ $\underline{-2a+6c=18}$

$\qquad 16c=16$

$\qquad c=1$

Section 5.2: Solving Systems of Linear Equations in Three Variables

43) Answers may vary.

$\boxed{3}\ b-2c=-5$

$\qquad b=2c-5$

$\boxed{2}\ a+5c=-1$

$\quad a+5(1)=-1$

$\qquad a+5=-1$

$\qquad a=-6$

$\boxed{1}\qquad -2a+3b=3$

$\quad -2a+3(2c-5)=3$

$\qquad -2a+6c-15=3$

$\boxed{A}\qquad -2a+6c=18$

$\boxed{3}\ b-2c=-5$

$\quad b-2(1)=-5$

$\qquad b-2=-5$

$\qquad b=-3$

$\qquad (-6,-3,1)$

45) $\boxed{1}\quad a-2b-c+d=0$

$\boxed{2}\ -a+2b+3c+d=6$

$\boxed{3}\quad 2a+b+c-d=8$

$\boxed{4}\quad a-b+2c+3d=9$

Solution worked out below.

$(a,b,c,d)=(3,1,2,1)$

Add $\boxed{1}$ and $-1\times\boxed{4}$

$a-2b-c+d=0$

$\underline{-a+b-2c-3d=-9}$

$\boxed{A}\ -b-3c-2d=-9$

Add $\boxed{A}$ and $\boxed{C}$

Add $-2\times\boxed{1}$ and $\boxed{3}$

$-2a+4b+2c-2d=0$

$\underline{2a+b+c-d=8}$

$\boxed{B}\quad 5b+3c-3d=8$

Add $\boxed{B}$ and $-5\times\boxed{C}$

Add $\boxed{2}$ and $\boxed{4}$

$-a+2b+3c+d=6$

$\underline{a-b+2c+3d=9}$

$\boxed{C}\ b+5c+4d=15$

Add $11\times\boxed{D}$ and $\boxed{E}$

$$-b-3c-2d=-9$$
$$\underline{b+5c+4d=15}$$
$$\boxed{D}\quad 2c+2d=6$$

$$5b+3c-3d=8$$
$$\underline{-5b-25c-20d=-75}$$
$$\boxed{E}\quad -22c-23d=-67$$

$$22c+22d=66$$
$$\underline{-22c-23d=-67}$$
$$-d=-1$$
$$d=1$$

Substitute for d in $\boxed{D}$
$$2c+2(1)=6$$
$$2c+2=6$$
$$2c=4$$
$$c=2$$

Substitute for c and d in $\boxed{A}$
$$-b-3(2)-2(1)=-9$$
$$-b-6-2=-9$$
$$-b-8=-9$$
$$-b=-1,\ b=1$$

Substitute for b,c and d in $\boxed{1}$
$$a-2(1)-2+1=0$$
$$a-2-1=0$$
$$a-3=0$$
$$a=3$$

47)

$\boxed{1}\quad 3a+4b+c-d=-$
$\boxed{2}\quad -3a-2b-c+d=1$ Solution worked out below.
$\boxed{3}\quad a+2b+3c-2d=5$ $(a,b,c,d)=(0,-3,1,-4)$
$\boxed{4}\quad 2a+b+c-d=2$

Add $\boxed{1}$ and $-3\times\boxed{3}$
$$3a+4b+c-d=-7$$
$$\underline{-3a-6b-9c+6d=-15}$$
$$\boxed{A}\quad -2b-8c+5d=-22$$

Add $\boxed{2}$ and $3\times\boxed{3}$
$$-3a-2b-c+d=1$$
$$\underline{3a+6b+9c-6d=15}$$
$$\boxed{B}\quad 4b+8c-5d=16$$

Add $-2\times\boxed{3}$ and $\boxed{4}$
$$-2a-4b-6c+4d=-10$$
$$\underline{2a+b+c-d=2}$$
$$\boxed{C}\quad -3b-5c+3d=-8$$

Add $2\times\boxed{A}$ and $\boxed{B}$
$$-4b-16c+10d=-44$$
$$\underline{4b+8c-5d=16}$$
$$\boxed{D}\quad -8c+5d=-28$$

Add $-3\times\boxed{A}$ and $2\times\boxed{C}$
$$6b+24c-15d=66$$
$$\underline{-6b-10c+6d=-16}$$
$$\boxed{E}\quad 14c-9d=50$$

Add $7\times\boxed{D}$ and $4\times\boxed{E}$
$$-56c+35d=-196$$
$$\underline{56c-36d=200}$$
$$-d=4$$
$$d=-4$$

Substitute for d in $\boxed{D}$
$$-8c+5(-4)=-28$$
$$-8c-20=-28$$
$$-8c=-8$$
$$c=1$$

Substitute for c and d in $\boxed{A}$
$$-2b-8(1)+5(-4)=-22$$
$$-2b-8-20=-22$$
$$-2b-28=-22$$
$$-2b=6$$
$$b=-3$$

Substitute for b,c and d in $\boxed{1}$
$$3a+4(-3)+1-(-4)=-7$$
$$3a-12+1+4=-7$$
$$3a-7=-7$$
$$3a=0$$
$$a=0$$

Section 5.3

1) Let x = one number
   y = other number
   The sum of the numbers is 36
   $$x + y = 36$$
   One number is two more than the other number
   $$x = 2 + y$$
   The system of equations is $x + y = 36$
   $$x = 2 + y$$

   Substitute $x = 2 + y$ into $x + y = 36$ and solve.

   $$(2 + y) + y = 36$$
   $$2 + 2y = 36$$
   $$2y = 34$$
   $$y = 17$$
   Substitute $y = 17$ into $x = 2 + y$
   $$x = 2 + 17$$
   $$x = 19$$
   The numbers are 19 and 17.

3) Let x = number of Hugo awards, and
   let y = number of The Girl with the
   Dragon Tattoo awards.

   Hugo = 6 + The Girl with the Dragon Tattoo
   x   = 6 + y
   Hugo + The Girl with the Dragon Tattoo = 16
   x + y = 16

   Use substitution.
   $$(6 + y) + y = 16$$
   $$6 + 2y = 16 \qquad x = 6 + y; \; y = 5$$
   $$2y = 10 \qquad x = 6 + 5$$
   $$y = 5 \qquad x = 11$$
   Hugo, 11; The Girl with the Dragon Tattoo, 5

5) Let x = number of IHOPs, and
   let y = number of Waffle Houses.
   IHOP + Waffle = 2626
   $$x \; + \; y \; = 2626$$
   IHOP = Waffle -314
   $$x \; = \; y \; -314$$

   Use substitution.
   $$(y - 314) + y = 2626$$
   $$2y - 314 = 2626$$
   $$2y = 2940$$
   $$y = 1470$$
   $$x = y - 314; \; y = 1470$$
   $$x = 1470 - 314$$
   $$x = 1156$$
   IHOP: 1156; Waffle House: 1470

7) Let x = number of people is 1939, and let
   y = number of people in 2011
   people in 1939 + 64,876 = people in 2011
   x + 64,876 = y
   people in 1939 + people in 2011 = 75,876
   x + y = 75,876

   Use substitution.
   $$x + (x + 64,876) = 75,876$$
   $$2x + 64,876 = 75,876$$
   $$2x = 11,000$$
   $$x = 5500$$
   $$y = x + 64,876; \; x = 5500$$
   $$y = 5500 + 64,876$$
   $$y = 70,376$$
   1939: 5500; 2011: 70,376

9) Let x = pounds of chicken, and
let y = pounds of beef.
chicken = beef − 6.3
$x \quad = \quad y \quad − 6.3$
chicken + beef = 120.5
$x \quad + \quad y \quad = 120.5$

Use substitution.
$(y−6.3)+y=120.5$
$2y−6.3=120.5$
$2y=126.8$
$y=63.4$
$x=y−6.3;\ y=63.4$
$x=63.4−6.3$
$x=57.1$
beef: 63.4 lb; chicken: 57.1 lb

11) Let w = the width, and
let l = the length.
length = 2 · width
$l \quad = 2 \cdot \quad w$
$2l+2w=78$

Use substitution.
$2(2w)+2w=78$
$4w+2w=78$
$6w=78$
$w=13$
length: 26 in; width: 13 in

$l=2w;\ w=13$
$l=2(13)$
$l=26$

13) Let w = the width, and
let h = the height.
$2h+2w=220$
width = height − 50
$w \quad = \quad h \quad − 50$

Use substitution.
$2h+2(h−50)=220$
$2h+2h−100=220$
$4h−100=220$
$4h=320$
$h=80$
$w=h−50;\ h=80$
$w=80−50$
$w=30$
height: 80 in; width: 30 in

15) Let w = the width, and
let l = the length.
$2l+2w=28$
length = 4 + width
$l \quad = 4+ \quad w$

Use substitution.
$2(4+w)+2w=28$
$8+2w+2w=28$
$8+4w=28$
$4w=20$
$w=5$
$l=4+w;\ w=5$
$l=4+5$
$l=9$
length: 9 cm; width: 5 cm

Section 5.3: Applications of Systems of Linear Equations

17) $x° = \dfrac{2}{3}y°$

supplementary angles

$x° + y° = 180°$

Use substitution.

$\dfrac{2}{3}y + y = 180$

$\dfrac{5}{3}y = 180$

$y = 108$

$x = \dfrac{2}{3}y; \ y = 108$

$x = \dfrac{2}{3}(108)$

$x = 72$

$m\angle x = 72°; \ m\angle y = 108°$

19) Let x = cost of a *Marc Anthony* ticket
and let y = cost of a *Santana* ticket.

Jennifer's Purchase:

$5 \cdot Marc\ Anthony + 2 \cdot Santana = 563$

$\quad\quad 5x \quad\quad + \quad 2y \quad = 563$

Carlos's Purchase:

$3 \cdot Marc\ Anthony + 6 \cdot Santana = 657$

$\quad\quad 3x \quad\quad + \quad 6y \quad = 657$

$5x + 2y = 563$

$-3(5x + 2y) = -3(563)$

$-15x - 6y = -1689$

Add the equations.

$\quad -15x - 6y = -1689$

$\underline{+ \quad 3x + 6y = \quad 657}$

$\quad\quad\quad -12x = -1032$

$\quad\quad\quad\quad x = 86$

Substitute $x = 86$ into

$5x + 2y = 563$

$5(86) + 2y = 563$

$430 + 2y = 563$

$2y = 133$

$y = 66.5$

Marc Anthony: $86; *Santana*: $66.50

21) Let x = cost of a two-item meal
let y = cost of a three-item meal

$3 \cdot$ two-item $+ 1 \cdot$ three-item $= 21.96$

$\quad 3x \quad\quad + \quad\quad y \quad\quad = 21.96$

$2 \cdot$ two-item $+ 2 \cdot$ three-item $= 23.16$

$\quad 2x \quad\quad + \quad\quad 2y \quad\quad = 23.16$

$3x + y = 21.96$

$\quad y = 21.96 - 3x$

Use substitution.

$2x + 2(21.96 - 3x) = 23.16$

$2x + 43.92 - 6x = 23.16$

$-4x + 43.92 = 23.16$

$-4x = -20.76$

$x = 5.19$

$y = 21.96 - 3x; \ x = 5.19$

$y = 21.96 - 3(5.19)$

$y = 21.96 - 15.57$

$y = 6.39$

two-item: $5.19; three-item: $6.39

23) Let x = cost of a key chain, and

let y = cost of a postcard.

$3 \cdot$ key chain $+ 5 \cdot$ postcard $= 10.00$

$\quad 3x \quad + \quad 5y \quad = 10.00$

$2 \cdot$ key chain $+ 3 \cdot$ postcard $= 6.50$

$\quad 2x \quad + \quad 3y \quad = 6.50$

$3x + 5y = 10.00$

$-2(3x + 5y) = -2(10.00)$

$-6x - 10y = -20.00$

$2x + 3y = 6.50$

$3(2x + 3y) = 3(6.50)$

$6x + 9y = 19.50$

Add the equations.

$-6x - 10y = -20.00$

$\underline{+ \ 6x + \ 9y = \ 19.50}$

$\quad\quad\quad -y = -0.50$

$\quad\quad\quad\quad y = 0.50$

Substitute $y = 0.50$ into

$\quad 3x + 5y = 10.00$

$3x + 5(0.50) = 10.00$

$\quad 3x + 2.50 = 10.00$

$\quad\quad\quad 3x = 7.50$

$\quad\quad\quad\quad x = 2.50$

key chain : $2.50; postcard: $0.50

25) Let x = cost of a cantaloupe

let y = cost of a watermelon

$3 \cdot$ cantaloupe $+ 1 \cdot$ watermelon $= 7.50$

$\quad 3x \quad + \quad y \quad\quad = 7.50$

$2 \cdot$ cantaloupe $+ 2 \cdot$ watermelon $= 9.00$

$\quad 2x \quad + \quad 2y \quad\quad = 9.00$

$3x + y = 7.50$

$\quad y = 7.50 - 3x$

Use substitution.

$2x + 2(7.50 - 3x) = 9.00$

$\quad 2x + 15 - 6x = 9.00$

$\quad 15.00 - 4x = 9.00$

$\quad\quad\quad -4x = -6.00$

$\quad\quad\quad\quad x = 1.50$

$y = 7.50 - 3x; \ x = 1.50$

$y = 7.50 - 3(1.50)$

$y = 7.50 - 4.50$

$y = 3.00$

cantaloupe: $1.50;

watermelon: $3.00

Section 5.3: Applications of Systems of Linear Equations

27) Let $x =$ cost of a hamburger

let $y =$ cost of a small fry

$6 \cdot$ hamburger $+1 \cdot$ small fry $= 5.05$

$\qquad 6x \quad + \quad y \quad = 5.05$

$8 \cdot$ hamburger $+2 \cdot$ small fry $= 7.66$

$\qquad 8x \quad + \quad 2y \quad = 7.66$

$6x + y = 5.05$

$\quad y = 5.05 - 6x$

$8x + 2(5.05 - 6x) = 7.66$

$8x + 10.10 - 12x = 7.66$

Use substitution. $\quad -4x + 10.10 = 7.66$

$-4x = -2.44$

$x = 0.61$

hamburger: \$0.61; small fry: \$1.39

$y = 5.05 - 6x;\ x = 0.61$

$y = 5.05 - 6(0.61)$

$y = 5.05 - 3.66$

$y = 1.39$

29) $x =$ number of ounces of 9% solution

$y =$ number of ounces of 17% solution

Solution	Concentration	Number of ounces of solution	Number of ounces of alcohol in the solution
9%	0.09	x	$0.09x$
17%	0.17	y	$0.17y$
15%	0.15	12	$0.15(12)$

$x + y = 12 \qquad\qquad 0.09x + 0.17y = 0.15(12)$

$\quad y = 12 - x \qquad 100(0.09x + 0.17y) = 100[0.15(12)]$

$9x + 17y = 15(12)$

Use substitution.

112

$$9x + 17(12 - x) = 15(12)$$
$$9x + 204 - 17x = 180 \qquad y = 12 - x;\ x = 3$$
$$-8x + 204 = 180 \qquad y = 12 - 3$$
$$-8x = -24 \qquad y = 9$$
$$x = 3$$
$$9\%:3\,\text{oz};\ 17\%:9\,\text{oz}$$

31) x = number of pounds of peanuts

y = number of pounds of cashews

Nuts	Price per Pound	Number of lbs of Nuts	Value
peanuts	$1.80	x	$1.80x$
cashews	$4.50	y	$4.50y$
mixture	$2.61	10	$2.61(10)$

$$x + y = 10 \qquad\qquad 1.80x + 4.50y = 2.61(10)$$
$$y = 10 - x \qquad 100(1.80x + 4.50y) = 100\big[2.61(10)\big]$$
$$180x + 450y = 261(10)$$

Use substitution.

$$180x + 450(10 - x) = 261(10)$$
$$180x + 4500 - 450x = 2610 \qquad y = 10 - x;\ x = 7$$
$$-270x + 4500 = 2610 \qquad y = 10 - 7$$
$$-270x = -1890 \qquad y = 3$$
$$x = 7$$
peanuts : 7 lb; cashews : 3 lb

Section 5.3: Applications of Systems of Linear Equations

33) $x =$ amount Sally invested in the 3% account.

$y =$ amount Sally invested in the 5% account.

$x + y = 4000$

$\quad y = 4000 - x$

Total Interest Earned = Interest from 3% account + Interest from 5% account

$$144 \quad = \quad x(0.03)(1) \quad + \quad y(0.05)(1)$$

$$100(144) = 100\left[x(0.03)(1) + y(0.05)(1)\right]$$

$$14,400 = 3x + 5y$$

Use substitution.

$14,400 = 3x + 5(4000 - x)$

$14,400 = 3x + 20,000 - 5x$ $\qquad\qquad y = 4000 - x;\ x = 2800$

$14,400 = -2x + 20,000$ $\qquad\qquad\quad y = 4000 - 2800$

$-5600 = -2x$ $\qquad\qquad\qquad\qquad\quad y = 1200$

$\quad 2800 = x$

Sally invested $2800 in the 3% account and $1200 in the 5% account.

35) $q =$ number of quarters; $d =$ number of dimes

$q + d = 110$

$\quad q = 110 - d$

Value of Quarters + Value of Dimes = Total Value

$$0.25q \quad + \quad 0.10d \quad = \quad 18.80$$

$$100(0.25q + 0.10d) = 100(18.80)$$

$$25q + 10d = 1880$$

Use substituion.

$25(110 - d) + 10d = 1880$

$2750 - 25d + 10d = 1880$ $\qquad\qquad q = 110 - d;\ d = 58$

$2750 - 15d = 1880$ $\qquad\qquad\qquad q = 110 - 58$

$-15d = -870$ $\qquad\qquad\qquad\qquad q = 52$

$d = 58$

There are 52 quarters and 58 dimes.

37) $x =$ number of liters of 100% solution

$y =$ number of liters of 10% solution

Solution	Concentration	Number of liters of solution	Number of liters of acid in the solution
100%	1.00	x	$1.00x$
10%	0.10	y	$0.10y$
40%	0.40	12	$0.40(12)$

$x + y = 12$ $\qquad 1.00x + 0.10y = 0.40(12)$

$\quad y = 12 - x$ $\qquad 10(1.00x + 0.10y) = 10\left[0.40(12)\right]$

$10x + y = 4(12)$

Use substitution.

$10x + (12 - x) = 4(12)$ $\qquad\qquad y = 12 - x;\ x = 4$

$\qquad 9x + 12 = 48$ $\qquad\qquad\quad y = 12 - 4$

$\qquad\quad 9x = 36, x = 4$ $\qquad\qquad\quad y = 8$

4 liters of pure acid; 8 liters of 10% solution

39) Car's distance + Truck's distance $= 330$

	d	$=$	r	$\cdot$	t
Car	$3x$		x		3
Truck	$3y$		y		3

$y = x - 10$ $\qquad\qquad 3x + 3y = 330$

Use substitution.

$3x + 3(x - 10) = 330$

$\quad 3x + 3x - 30 = 330$ $\qquad\qquad y = x - 10;\ x = 60$

$\qquad\quad 6x - 30 = 330$ $\qquad\qquad\quad y = 60 - 10$

$\qquad\qquad\quad 6x = 360, x = 60$ $\qquad\quad y = 50$

Car: 60 mph; Truck: 50 mph

41) Walk time = Bike time

d	=	r	$\cdot$	t
Walk	8	x		t
Bike	22	y		t

$8 = xt$

$\dfrac{8}{x} = t$

$22 = yt$

Substitute.

$22 = y\left(\dfrac{8}{x}\right)$

$y = 7 + x$ $\qquad$ $22x = 8y$

Use substitution.

$22x = 8(7 + x)$

$22x = 56 + 8x$ $\qquad$ $y = 7 + x;\ x = 4$

$14x = 56$ $\qquad\qquad$ $y = 7 + 4$

$x = 4$ $\qquad\qquad\quad$ $y = 11$

walking: 4 mph; biking: 11 mph

43) Nick's distance + Scott's distance = 13

d	=	r	$\cdot$	t
Nick	$0.5x$	x		0.5
Scott	$0.5y$	y		0.5

$y = x - 2$ $\qquad$ $0.5x + 0.5y = 13$

Use substitution.

$0.5x + 0.5(x - 2) = 13$ $\qquad$ $y = x - 2;\ x = 14$

$0.5x + 0.5x - 1 = 13$ $\qquad\quad$ $y = 14 - 2$

$x = 14$ $\qquad\qquad\qquad$ $y = 12$

Nick: 14 mph; Scott: 12 mph

45) $x =$ cost of a hot dog

$y =$ cost of an order of fries

$z =$ cost of a large soda

Use the information from the problem to get the following system of equations.

M $2x + 2y + z = 9.00$

L $2x + y + 2z = 9.50$

C $3x + 2y + z = 11.00$

Add M and L $\cdot (-2)$

$$2x + 2y + z = 9.00$$
$$+ \quad -4x - 2y - 4z = -19.00$$

A $\quad\quad -2x - 3z = -10.00$

Add L $\cdot (-2)$ and C

$$-4x - 2y - 4z = -19.00$$
$$+ \quad\quad 3x + 2y + z = 11.00$$

B $\quad\quad -x - 3z = -8.00$

Add A and B $\cdot (-1)$

$$-2x - 3z = -10.00$$
$$+ \quad x + 3z = \quad 8.00$$

$\quad\quad -x = -2.00 \quad\quad x = 2.00$

Substitute $x = 2.00$ into A

$-2(2.00) - 3z = -10.00$

$\quad -4.00 - 3z = -10.00$

$\quad\quad -3z = -6.00 \quad\quad z = 2.00$

Substitute $x = 2.00$ and $z = 2.00$ into L

$2(2.00) + y + 2(2.00) = 9.50$

$4.00 + y + 4.00 = 9.50$

$y + 8.00 = 9.50$

$y = 1.50$

hot dog: $2.00, fries: $1.50, soda: $2.00

47) $c =$ grams of protein in a Clif Bar.
$b =$ grams of protein in a Balance Bar.
$p =$ grams of protein in a PowerBar.

Use the information from the problem to get the following system of equations.
I $\quad c = b - 4$
II $\quad p = b + 9$
III $c + b + p = 50$

Substitute I and II into III
$(b - 4) + b + (b + 9) = 50$
$$3b + 5 = 50$$
$$3b = 45$$
$$b = 15$$

Substitute $b = 15$ into I
$c = 15 - 4$
$c = 11$

Substitute $b = 15$ into II
$p = 15 + 9$
$p = 24$
Clif Bar: 11g, Balance Bar: 15g,
PowerBar: 24g

49) $k =$ revenue collected by the Knicks.
$l =$ revenue collected by the Lakers.
$b =$ revenue collected by the Bulls.

Use the information from the problem to get the following system of equations.
I $k + l + b = 428$
II $\quad l = b + 30$
III $\quad k = l + 11$

Solve III for l
$l = k - 11$
Substitute $l = k - 11$ into II
$k - 11 = b + 30$
A $\quad k = b + 41$

51) $b =$ cost of a bronze date
$s =$ cost of a silver date
$g =$ cost of a gold date

Use the information from the problem to get the following system of equations.
I $4b + 4s + 3g = 441$
II $4b + 3s + 2g = 341$
III $\quad 3b + 3s + g = 262$

Substitute II and A into I
$(b + 41) + (b + 30) + b = 428$
$$3b + 71 = 428$$
$$3b = 357$$
$$b = 119$$

Substitute $b = 119$ into II
$l = 119 + 30$
$l = 149$

Substitute $l = 149$ into III
$k = 149 + 11$
$k = 160$
Knicks: $160 million,
Lakers: $149 million,
Bulls: $119 million

Add I and III·(−3) Add II and III·(−2)

Add A·(2) and B·(−5)

$$4b + 4s + 3g = 441$$
$$+ \quad -9b - 9s - 3g = -786$$

A $\quad -5b - 5s = -345$

$$4b + 3s + 2g = 341$$
$$+ \quad -6b - 6s - 2g = -524$$

B $\quad -2b - 3s = -183$

$$-10b - 10s = -690$$
$$\underline{10b + 15s = 915}$$
$$5s = 225$$
$$s = 45$$

Substitute s = 45 into B

$$-2b - 3(45) = -183$$
$$-2b - 135 = -183$$
$$-2b = -28$$
$$b = 24$$

Substitute b = 24 and s = 45 into III

$$3(24) + 3(45) + g = 262$$
$$72 + 135 + g = 262$$
$$p + 207 = 55$$
$$g = 55$$

Bronze: \$24; Silver: \$45; Gold: \$55

53) l = measure of the largest angle.

m = measure of the middle angle.

s = measure of the smallest angle.

Use the information from the problem to get the following system of equations.

I $\qquad l = 2m$

II $\qquad s = m - 28$

III $l + m + s = 180$

Substitute I and II into III

$$(2m) + m + (m - 28) = 180$$
$$4m - 28 = 180$$
$$4m = 208$$
$$m = 52$$

Substitute $m = 52$ into I

$$l = 2(52) = 104$$

Substitute $m = 52$ into II

$$s = 52 - 28 = 24$$
$$104°, \ 52°, \ 24°$$

55) l = measure of the largest angle.

m = measure of the middle angle.

s = measure of the smallest angle.

Use the information from the problem to get the following system of equations.

I $\qquad s = l - 44$

II $\qquad s + m = l + 20$

III $l + m + s = 180$

Add II·(−1) and III

$$l - m - s = -20$$
$$+ \quad \underline{l + m + s = 180}$$
$$2l = 160$$
$$l = 80$$

Substitute $l = 80$ into I

$$s = 80 - 44 = 36$$

Substitute $l = 80$ and $s = 36$ into III

$$80 + m + 36 = 180$$
$$m + 116 = 180$$
$$m = 64$$
$$80°, \ 64°, \ 36°$$

Section 5.4: Solving Systems of Linear Equations Using Matrices

57) l = length of the longest side.

m = length of the medium side.

s = length of the shortest side.

Use the information from the problem to get the following system of equations.

I $\quad l + m + s = 29$

II $\qquad l = s + 5$

III $\quad m + s = l + 5$

Add I and III $\cdot (-1)$

$l + m + s = 29$

$+ \quad \underline{l - m - s = -5}$

$\qquad 2l = 24$

$\qquad l = 12$

Substitute $l = 12$ into II

$12 = s + 5$

$\quad 7 = s$

Substitute $l = 12$ and $s = 7$ into I

$12 + m + 7 = 29$

$\quad m + 19 = 29$

$\qquad m = 10$

12 cm, 10 cm, 7 cm

Section 5.4

1) $\begin{bmatrix} 1 & -7 & | & 15 \\ 4 & 3 & | & -1 \end{bmatrix}$

3) $\begin{bmatrix} 1 & 6 & -1 & | & -2 \\ 3 & 1 & 4 & | & 7 \\ -1 & -2 & 3 & | & 8 \end{bmatrix}$

5) $\quad 3x + 10y = -4$

$\qquad x - 2y = 5$

7) $\quad x - 6y = 8$

$\qquad y = -2$

9) $\qquad x - 3y + 2z = 7$

$\qquad 4x - y + 3z = 0$

$\qquad -2x + 2y - 3z = -9$

11) $\quad x + 5y + 2z = 14$

$\qquad y - 8z = 2$

$\qquad z = -3$

13) $\begin{bmatrix} 1 & 4 & | & -1 \\ 3 & 5 & | & 4 \end{bmatrix} \xrightarrow{-3R_1 + R_2 \to R_2}$

$\begin{bmatrix} 1 & 4 & | & -1 \\ 0 & -7 & | & 7 \end{bmatrix} \xrightarrow{-\frac{1}{7}R_2 \to R_2} \begin{bmatrix} 1 & 4 & | & -1 \\ 0 & 1 & | & -1 \end{bmatrix}$

$y = -1 \quad x + 4y = -1$

$\qquad x + 4(-1) = -1$

$\qquad x - 4 = -1$

$\qquad x = 3 \quad (3, -1)$

15) $\begin{bmatrix} 1 & -3 & | & 9 \\ -6 & 5 & | & 11 \end{bmatrix} \xrightarrow{6R_1 + R_2 \to R_2}$

$\begin{bmatrix} 1 & -3 & | & 9 \\ 0 & -13 & | & 65 \end{bmatrix} \xrightarrow{-\frac{1}{13}R_2 \to R_2} \begin{bmatrix} 1 & -3 & | & 9 \\ 0 & 1 & | & -5 \end{bmatrix}$

$y = -5 \quad x - 3y = 9$

$\qquad x - 3(-5) = 9$

$\qquad x + 15 = 9$

$\qquad x = -6 \; (-6, -5)$

17) $\begin{bmatrix} 4 & -3 & | & 6 \\ 1 & 1 & | & -2 \end{bmatrix} \xrightarrow{R_1 \leftrightarrow R_2} \begin{bmatrix} 1 & 1 & | & -2 \\ 4 & -3 & | & 6 \end{bmatrix} \xrightarrow{-4R_1 + R_2 \to R_2} \begin{bmatrix} 1 & 1 & | & -2 \\ 0 & -7 & | & 14 \end{bmatrix} \xrightarrow{-\frac{1}{7}R_2 \to R_2} \begin{bmatrix} 1 & 1 & | & -2 \\ 0 & 1 & | & -2 \end{bmatrix}$

$y = -2 \qquad x + y = -2$

$\qquad\qquad x + (-2) = -2$

$\qquad\qquad\qquad x = 0 \quad (0, -2)$

19) $\begin{bmatrix} 1 & 1 & -1 & | & -5 \\ 4 & 5 & -2 & | & 0 \\ 8 & -3 & 2 & | & -4 \end{bmatrix} \xrightarrow[-8R_1 + R_3 \to R_3]{-4R_1 + R_2 \to R_2} \begin{bmatrix} 1 & 1 & -1 & | & -5 \\ 0 & 1 & 2 & | & 20 \\ 0 & -11 & 10 & | & 36 \end{bmatrix}$

$\xrightarrow{11R_2 + R_3 \to R_3} \begin{bmatrix} 1 & 1 & -1 & | & -5 \\ 0 & 1 & 2 & | & 20 \\ 0 & 0 & 32 & | & 256 \end{bmatrix} \xrightarrow{\frac{1}{32}R_3 \to R_3} \begin{bmatrix} 1 & 1 & -1 & | & -5 \\ 0 & 1 & 2 & | & 20 \\ 0 & 0 & 1 & | & 8 \end{bmatrix}$

$x + y - z = -5 \quad y + 2z = 20 \quad x + y - z = -5$

$\quad y + 2z = 20 \quad y + 2(8) = 20 \quad x + 4 - 8 = -5$

$\qquad\quad z = 8 \qquad y + 16 = 20 \qquad x - 4 = -5$

$\qquad\qquad\qquad\qquad\quad y = 4 \qquad\qquad x = -1 \ (-1, 4, 8)$

21) $\begin{bmatrix} 1 & -3 & 2 & | & -1 \\ 3 & -8 & 4 & | & 6 \\ -2 & -3 & -6 & | & 1 \end{bmatrix} \xrightarrow[2R_1 + R_3 \to R_3]{-3R_1 + R_2 \to R_2} \begin{bmatrix} 1 & -3 & 2 & | & -1 \\ 0 & 1 & -2 & | & 9 \\ 0 & -9 & -2 & | & -1 \end{bmatrix} \xrightarrow{9R_2 + R_3 \to R_3}$

$\begin{bmatrix} 1 & -3 & 2 & | & -1 \\ 0 & 1 & -2 & | & 9 \\ 0 & 0 & -20 & | & 80 \end{bmatrix} \xrightarrow{-\frac{1}{20}R_3 \to R_3} \begin{bmatrix} 1 & -3 & 2 & | & -1 \\ 0 & 1 & -2 & | & 9 \\ 0 & 0 & 1 & | & -4 \end{bmatrix}$

$x - 3y + 2z = -1 \qquad y - 2z = 9 \qquad x - 3y + 2z = -1$

$\quad y - 2z = 9 \quad y - 2(-4) = 9 \quad x - 3(1) + 2(-4) = -1$

$\qquad\quad z = -4 \qquad y + 8 = 9 \qquad x - 3 - 8 = -1$

$\qquad\qquad\qquad\qquad\quad y = 1 \qquad\qquad x - 11 = -1$

$\qquad\qquad\qquad\qquad\qquad\qquad\qquad x = 10 \ (10, 1, -4)$

23) $\begin{bmatrix} -4 & -3 & 1 & | & 5 \\ 1 & 1 & -1 & | & -7 \\ 6 & 4 & 1 & | & 12 \end{bmatrix} \xrightarrow{R_1 \leftrightarrow R_2} \begin{bmatrix} 1 & 1 & -1 & | & -7 \\ -4 & -3 & 1 & | & 5 \\ 6 & 4 & 1 & | & 12 \end{bmatrix}$

$\xrightarrow[-6R_1+R_3 \to R_3]{4R_1+R_2 \to R_2} \begin{bmatrix} 1 & 1 & -1 & | & -7 \\ 0 & 1 & -3 & | & -23 \\ 0 & -2 & 7 & | & 54 \end{bmatrix} \xrightarrow{2R_2+R_3 \to R_3} \begin{bmatrix} 1 & 1 & -1 & | & -7 \\ 0 & 1 & -3 & | & -23 \\ 0 & 0 & 1 & | & 8 \end{bmatrix}$

$x+y-z=-7 \qquad y-3z=-23 \quad x+y-z=-7$

$y-3z=-23 \quad y-3(8)=-23 \quad x+1-8=-7$

$z=8 \qquad y-24=-23 \qquad x-7=-7$

$\qquad\qquad\qquad y=1 \qquad\qquad x=0 \quad (0,1,8)$

25) $\begin{bmatrix} 1 & -3 & 1 & | & -4 \\ 4 & 5 & -1 & | & 0 \\ 2 & -6 & 2 & | & 1 \end{bmatrix} \xrightarrow[-2R_1+R_3 \to R_3]{-4R_1+R_2 \to R_2} \begin{bmatrix} 1 & -3 & 1 & | & -4 \\ 0 & 17 & -5 & | & 16 \\ 0 & 0 & 0 & | & 9 \end{bmatrix}$

$\varnothing$; inconsistent

27) $\begin{bmatrix} 1 & 1 & 3 & 1 & | & -1 \\ -1 & 0 & 1 & -1 & | & 7 \\ 2 & 3 & 9 & -2 & | & 7 \\ 1 & -2 & 1 & 3 & | & -11 \end{bmatrix} \xrightarrow[\substack{-2R_1+R_3 \to R_3 \\ -R_1+R_4 \to R_4}]{R_1+R_2 \to R_2} \begin{bmatrix} 1 & 1 & 3 & 1 & | & -1 \\ 0 & 1 & 4 & 0 & | & 6 \\ 0 & 1 & 3 & -4 & | & 9 \\ 0 & -3 & -2 & 2 & | & -10 \end{bmatrix}$

$\xrightarrow[3R_2+R_4 \to R_4]{-R_2+R_3 \to R_3} \begin{bmatrix} 1 & 1 & 3 & 1 & | & -1 \\ 0 & 1 & 4 & 0 & | & 6 \\ 0 & 0 & -1 & -4 & | & 3 \\ 0 & 0 & 10 & 2 & | & 8 \end{bmatrix} \xrightarrow{-R_3 \to R_3} \begin{bmatrix} 1 & 1 & 3 & 1 & | & -1 \\ 0 & 1 & 4 & 0 & | & 6 \\ 0 & 0 & 1 & 4 & | & -3 \\ 0 & 0 & 10 & 2 & | & 8 \end{bmatrix}$

$\xrightarrow{-10R_3+R_4 \to R_4} \begin{bmatrix} 1 & 1 & 3 & 1 & | & -1 \\ 0 & 1 & 4 & 0 & | & 6 \\ 0 & 0 & 1 & 4 & | & -3 \\ 0 & 0 & 0 & -38 & | & -8 \end{bmatrix} \xrightarrow{-\frac{1}{38}R_4 \to R_4} \begin{bmatrix} 1 & 1 & 3 & 1 & | & -1 \\ 0 & 1 & 4 & 0 & | & 6 \\ 0 & 0 & 1 & 4 & | & -3 \\ 0 & 0 & 0 & 1 & | & -1 \end{bmatrix}$

$$a+b+3c+d=-1 \qquad c+4d=-3 \qquad b+4c=6 \qquad a+b+3c+d=-1$$

$$b+4c=6 \qquad c+4(-1)=-3 \qquad b+4(1)=6 \qquad a+2+3(1)+(-1)=-1$$

$$c+4d=-3 \qquad c=1 \qquad b=2 \qquad a+4=-1$$

$$d=-1 \qquad\qquad\qquad\qquad\qquad\qquad a=-5$$

$$(-5,2,1,-1)$$

29) $\begin{bmatrix} 1 & -3 & 2 & -1 & | & -2 \\ -3 & 8 & -5 & 1 & | & 2 \\ 2 & -1 & 1 & 3 & | & 7 \\ 1 & -2 & 1 & 2 & | & 3 \end{bmatrix} \xrightarrow[\substack{3R_1+R_2\to R_2 \\ -2R_1+R_3\to R_3 \\ -R_1+R_4\to R_4}]{} \begin{bmatrix} 1 & -3 & 2 & -1 & | & -2 \\ 0 & -1 & 1 & -2 & | & -4 \\ 0 & 5 & -3 & 5 & | & 11 \\ 0 & 1 & -1 & 3 & | & 5 \end{bmatrix}$

$\xrightarrow[-R_2\to R_2]{} \begin{bmatrix} 1 & -3 & 2 & -1 & | & -2 \\ 0 & 1 & -1 & 2 & | & 4 \\ 0 & 5 & -3 & 5 & | & 11 \\ 0 & 1 & -1 & 3 & | & 5 \end{bmatrix} \xrightarrow[\substack{-5R_2+R_3\to R_3 \\ -R_2+R_4\to R_4}]{} \begin{bmatrix} 1 & -3 & 2 & -1 & | & -2 \\ 0 & 1 & -1 & 2 & | & -8 \\ 0 & 0 & 2 & -5 & | & -9 \\ 0 & 0 & 0 & 1 & | & 1 \end{bmatrix}$

$\xrightarrow[\frac{1}{2}R_3+R_4\to R_4]{} \begin{bmatrix} 1 & -3 & 2 & -1 & | & -2 \\ 0 & 1 & -1 & 2 & | & 4 \\ 0 & 0 & 1 & -\frac{5}{2} & | & -\frac{9}{2} \\ 0 & 0 & 0 & 1 & | & 1 \end{bmatrix}$

$$a-3b+2c-d=-2 \qquad c-\frac{5}{2}d=-\frac{9}{2} \qquad b-c+2d=4 \qquad a-3b+2c-d=-2$$

$$b-c+2d=4 \qquad c-\frac{5}{2}(1)=-\frac{9}{2} \qquad b-(-2)+2(1)=4 \qquad a-3(0)+2(-2)-(1)=-2$$

$$c-\frac{5}{2}d=-\frac{9}{2} \qquad c=-2 \qquad b=0 \qquad a-5=-2$$

$$d=1 \qquad\qquad\qquad\qquad\qquad\qquad\qquad a=3$$

$$(3,0,-2,1)$$

Chapter 5 Review

1) No

$$-2x + y = 3$$
$$-2(-4) + (-5) = 3$$
$$8 - 5 = 3$$
$$3 = 3$$

$$3x - y = -17$$
$$3(-4) - (-5) = -17$$
$$-12 + 5 = -17$$
$$-7 \neq -17$$

3) $(-2, -3)$

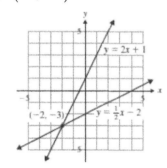

5) $\varnothing$, inconsistent

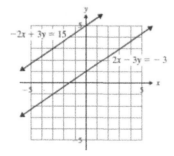

7) infinite number of solutions of the form $\{(x, y) \mid 4x + y = -4\}$; dependent system

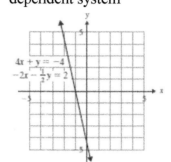

9) $y = -\dfrac{2}{3}x - 3$

$$4x + 6y = 5$$
$$6y = -4x + 5$$
$$y = -\dfrac{2}{3}x + \dfrac{5}{6}$$

no solution because the lines have
the same slope but different y-intercepts

11) $15x - 10y = 4$
$$-10y = -15x + 4$$
$$y = \dfrac{3}{2}x - \dfrac{2}{5}$$

$$-9x + 6y = 1$$
$$6y = 9x + 1$$
$$y = \dfrac{3}{2}x + \dfrac{1}{6}$$

no solution because the lines have
the same slope but different y-intercepts

13) $x + 8y = -2$

$x = -8y - 2$

Substitute $x = -8y - 2$ into

$2x + 11y = -9$

$2(-8y - 2) + 11y = -9$

$-16y - 4 + 11y = -9$

$-5y - 4 = -9$

$-5y = -5$

$y = 1$

Substitute $y = 1$ into

$x = -8y - 2$

$x = -8(1) - 2$

$x = -8 - 2$

$x = -10 \qquad (-10,\ 1)$

15) $-2x + y = -18$

$y = 2x - 18$

Substitute $y = 2x - 18$ into

$x + 7y = 9$

$x + 7(2x - 18) = 9$

$x + 14x - 126 = 9$

$15x - 126 = 9$

$15x = 135$

$x = 9$

Substitute $x = 9$ into

$y = 2x - 18$

$y = 2(9) - 18$

$y = 18 - 18$

$y = 0 \qquad (9,\ 0)$

17) $\dfrac{5}{2}x + \dfrac{9}{2}y = \dfrac{1}{2}$

$2\left(\dfrac{5}{2}x + \dfrac{9}{2}y\right) = 2\left(\dfrac{1}{2}\right)$

$5x + 9y = 1$

$\dfrac{1}{6}x + \dfrac{2}{3}y = -\dfrac{1}{3}$

$6\left(\dfrac{1}{6}x + \dfrac{2}{3}y\right) = 6\left(-\dfrac{1}{3}\right)$

$x + 4y = -2$

$x = -4y - 2$

Substitute $x = -4y - 2$ into

$5x + 9y = 1$

$5(-4y - 2) + 9y = 1$

$-20y - 10 + 9y = 1$

$-11y - 10 = 1$

$-11y = 11$

$y = -1$

Substitute $y = -1$ into

$x = -4y - 2$

$x = -4(-1) - 2$

$x = 4 - 2$

$x = 2 \qquad (2,\ -1)$

19) Add the equations.

$x - y = -8$

$+\ 3x + y = -12$

$\overline{4x = -20}$

$x = -5$

Substitute $x = -5$ into

$x - y = -8$

$(-5) - y = -8$

$-y = -3$

$y = 3 \qquad (-5,\ 3)$

21) $4x - 5y = -16$

$3(4x - 5y) = 3(-16)$

$12x - 15y = -48$

$-3x + 4y = 13$

$4(-3x + 4y) = 4(13)$

$-12x + 16y = 52$

Add the equations.

$12x - 15y = -48$

$+ \ -12x + 16y = \ \ 52$

$\overline{\hspace{1.5cm} y = 4}$

Substitute $y = 4$ into

$4x - 5y = -16$

$4x - 5(4) = -16$

$4x - 20 = -16$

$4x = 4$

$x = 1 \qquad (1, \ 4)$

23) $0.12x + 0.01y = 0.06$

$100(0.12x + 0.01y) = 100(0.06)$

$12x + y = 6$

$-2(12x + y) = -2(6)$

$-24x - 2y = -12$

$0.5x + 0.2y = -0.7$

$10(0.5x + 0.2y) = 10(-0.7)$

$5x + 2y = -7$

Add the equations.

$-24x - 2y = -12$

$+ \ \ \ \ 5x + 2y = -7$

$\overline{\hspace{0.8cm} -19x = -19}$

$x = 1$

Substitute $x = 1$ into

$12x + y = 6$

$12(1) + y = 6$

$12 + y = 6$

$y = -6 \qquad (1, \ -6)$

25) $2x - 3y = 3$

$2x = 3y + 3$

$x = \dfrac{3}{2}y + \dfrac{3}{2}$

Substitute $x = \dfrac{3}{2}y + \dfrac{3}{2}$ into

$3x + 4y = -21$

$3\left(\dfrac{3}{2}y + \dfrac{3}{2}\right) + 4y = -21$

$\dfrac{9}{2}y + \dfrac{9}{2} + 4y = -21$

$2\left(\dfrac{9}{2}y + \dfrac{9}{2} + 4y\right) = 2(-21)$

$9y + 9 + 8y = -42$

$17y + 9 = -42$

$17y = -51$

$y = -3$

Substitute $y = -3$ into

$2x - 3y = 3$

$2x - 3(-3) = 3$

$2x + 9 = 3$

$2x = -6$

$x = -3 \qquad (-3, \ -3)$

27) $6x - 4y = 12$

$\frac{1}{2}(6x - 4y) = \frac{1}{2}(12)$

$3x - 2y = 6$

$15x - 10y = 30$

$-\frac{1}{5}(15x - 10y) = -\frac{1}{5}(30)$

$-3x + 2y = -6$

Add the equations.

$\begin{array}{r} 3x - 2y = 6 \\ + \ -3x + 2y = -6 \\ \hline 0 = 0 \end{array}$

infinite number of solutions of the

form $\{(x,\ y)\ |\ 6x - 4y = 12\}$;

dependent equations

29) $15 - y = y + 2(4x + 5)$

$15 - y = y + 8x + 10$

$-2y = 8x - 5$

$y = -4x + \frac{5}{2}$

$2(x + 1) + 3 = 2(y + 4) + 10x$

$2x + 2 + 3 = 2y + 8 + 10x$

$2x + 5 = 2y + 8 + 10x$

$-8x - 2y = 3$

Substitute $y = -4x + \frac{5}{2}$ into

$-8x - 2y = 3$

$-8x - 2\left(-4x + \frac{5}{2}\right) = 3$

$-8x + 8x - 5 = 3$

$-5 \neq 3$

$\varnothing$; inconsistent

31) Eliminate x.

$2x + 7y = -8$

$3(2x + 7y) = 3(-8)$

$6x + 21y = -24$

$3x - y = 13$

$-2(3x - y) = -2(13)$

$-6x + 2y = -26$

Add the equations.

$\begin{array}{r} 6x + 21y = -24 \\ + \ -6x + 2y = -26 \\ \hline 23y = -50 \end{array}$

$y = -\frac{50}{23}$

$3x - \left(-\frac{50}{23}\right) = 13$

$3x + \frac{50}{23} = 13$

$23\left(3x + \frac{50}{23}\right) = 23(13)$

$69x + 50 = 299$

$69x = 249$

$x = \frac{249}{69} = \frac{83}{23}$

$\left(\frac{83}{23},\ -\frac{50}{23}\right)$

33)
$$x - 6y + 4z = 13$$
$$(-3) - 6(-2) + 4(1) = 13$$
$$-3 + 12 + 4 = 13$$
$$13 = 13$$
$$5x + y + 7z = 8$$
$$5(-3) + (-2) + 7(1) = 8$$
$$-15 - 2 + 7 = 8$$
$$-10 \neq 8$$
$$2x + 3y - z = -5$$
$$2(-3) + 3(-2) - (1) = -5$$
$$-6 - 6 - 1 = -5$$
$$-13 \neq -5$$

The ordered triple is not a solution of the system.

35) I $2x - 5y - 2z = 3$
   II $x + 2y + z = 5$
   III $-3x - y + 2z = 0$

Add I and II $\cdot (-2)$
$$2x - 5y - 2z = 3$$
$$+ \; \underline{-2x - 4y - 2z = -10}$$
A $-9y - 4z = -7$

Add II $\cdot (3)$ and III
$$3x + 6y + 3z = 15$$
$$+ \; \underline{-3x - y + 2z = 0}$$
B $5y + 5z = 15$

Add A and B $\cdot (9/5)$
$$-9y - 4z = -7$$
$$\underline{9y + 9z = 27}$$
$$5z = 20$$
$$z = 4$$

Substitute $z = 4$ into B

$$5y + 5(4) = 15$$
$$5y + 20 = 15$$
$$5y = -5$$
$$y = -1$$

Substitute $y = -1$ and $z = 4$ into II
$$x + 2(-1) + (4) = 5$$
$$x - 2 + 4 = 5$$
$$x = 3$$

$$(3, -1, 4)$$

37) I $5a - b + 2c = -6$
   II $-2a - 3b + 4c = -2$
   III $a + 6b - 2c = 10$

Add I $\cdot (-2)$ and II
$$-10a + 2b - 4c = 12$$
$$+ \; \underline{-2a - 3b + 4c = -2}$$
A $-12a - b = 10$

Add I and III
$$5a - b + 2c = -6$$
$$+ \; \underline{a + 6b - 2c = 10}$$
B $6a + 5b = 4$

Add A $\cdot (5)$ and B
$$-60a - 5b = 50$$
$$+ \; \underline{6a + 5b = 4}$$
$$-54a = 54$$
$$a = -1$$

Substitute $a = -1$ into A
$$-12(-1) - b = 10$$
$$12 - b = 10$$
$$-b = -2$$
$$b = 2$$

Substitute $a = -1$ and $b = 2$ into I

$$5(-1)-(2)+2c=-6$$
$$-5-2+2c=-6$$
$$-7+2c=-6$$
$$2c=1$$
$$c=\frac{1}{2}$$
$$\left(-1,\ 2,\ \frac{1}{2}\right)$$

Substitute $y=\dfrac{2}{3}$ into II

$$x+3\left(\frac{2}{3}\right)=5$$
$$x+2=5$$
$$x=3$$
$$\left(3,\ \frac{2}{3},\ -\frac{1}{2}\right)$$

39) I $4x-9y+8z=2$

    II $x+3y=5$

    III $6y+10z=-1$

Add I and II $\cdot(-4)$

$$4x-9y+8z=2$$
$$+\ \ \underline{-4x-12y\ \ \ \ \ =-20}$$
$$A\ \ \ \ \ -21y+8z=-18$$

Add A $\cdot(-5)$ and III

$$105y-40z=90$$
$$+\ \ \underline{24y+40z=-4}$$
$$129y=86$$
$$y=\frac{86}{129}=\frac{2}{3}$$

Substitute $y=\dfrac{2}{3}$ into III

$$6\left(\frac{2}{3}\right)+10z=-1$$
$$4+10z=-1$$
$$10z=-5$$
$$z=-\frac{5}{10}=-\frac{1}{2}$$

41) I $x+3y-z=0$

    II $11x-4y+3z=8$

    III $5x+15y-5z=1$

Add I $\cdot(-5)$ and III

$$-5x-15y+5z=0$$
$$+\ \ \underline{5x+15y-5z=1}$$
$$0\neq1$$

$\varnothing$; inconsistent

43)   I $12a-8b+4c=8$

       II $3a-2b+c=2$

      III $-6a+4b-2c=-4$

Equation I $=$ II $\cdot4$

Equation I $=$ III $\cdot(-2)$

$\{(a,b,c)\,|\,3a-2b+c=2\}$;

dependent equations

45) I $5y+2z=6$

    II $-x+2y=-1$

    III $4x-z=1$

Add II $\cdot(4)$ and III

$-4x + 8y \quad = -4$

$\underline{\quad 4x \quad - z = 1}$

A $\quad 8y - z = -3$

Add I and A $\cdot (2)$

$5y + 2z = 6$

$\underline{+ \; 16y - 2z = -6}$

$21y = 0$

$y = 0$

Substitute $y = 0$ into I

$5(0) + 2z = 6$

$2z = 6$

$z = 3$

Substitute $y = 0$ into II

$-x + 2(0) = -1$

$-x = -1$

$x = 1$

$(1, 0, 3)$

47) I $\quad 8x + z = 7$

II $\; 3y + 2z = -4$

III $\; 4x - y = 5$

Add I $\cdot (-2)$ and II

$-16x \quad - 2z = -14$

$\underline{+ \qquad 3y + 2z = -4}$

A $\quad -16x + 3y = -18$

Add III $\cdot (3)$ and A

$12x - 3y = 15$

$\underline{+ \; -16x + 3y = -18}$

$-4x = -3$

$x = \dfrac{3}{4}$

Substitute $x = \dfrac{3}{4}$ into I

$8\left(\dfrac{3}{4}\right) + z = 7$

$6 + z = 7$

$z = 1$

Substitute $x = \dfrac{3}{4}$ into III

$4\left(\dfrac{3}{4}\right) - y = 5$

$3 - y = 5$

$-y = 2$

$y = -2$

$\left(\dfrac{3}{4}, -2, 1\right)$

49) $\quad$ Let $x =$ number of dogs, and let $y =$ number of cats.

dogs $= 2 \cdot$ cats

$x \quad = 2y$

dogs + cats $= 51$

$x + y = 51$

Use substitution.

$(2y) + y = 51 \qquad x = 2y; \; y = 17$

$3y = 51 \qquad\qquad x = 2(17)$

$y = 17 \qquad\qquad\; x = 34$

34 dogs; 17 cats

51) Let x = the cost of a hot dog,
and let y = the cost of a soda.
$4 \cdot \text{hot dog} + 2 \cdot \text{soda} = 26.50$
$$4x \quad + \quad 2y \quad = 26.50$$
$3 \cdot \text{hot dog} + 4 \cdot \text{soda} = 28.00$
$$3x \quad + \quad 4y \quad = 28.00$$

$$4x + 2y = 26.50$$
$$-2(4x + 2y) = -2(26.50)$$
$$-8x - 4y = -53.00$$
Add the equations.
$$-8x - 4y = -53.00$$
$$+ \quad 3x + 4y = \quad 28.00$$
$$\overline{\quad\quad -5x = -25.00}$$
$$x = 5.00$$
Substitute $x = 5.00$ into
$$4x + 2y = 26.50$$
$$4(5.00) + 2y = 26.50$$
$$20.00 + 2y = 26.50$$
$$2y = 6.50$$
$$y = 3.25$$
hot dog : $5.00; soda: $3.25

53) $x = 2y$
sum of the angles of a triangle $= 180$
$x + y + 123 = 180$
Use substitution.
$$2y + y + 123 = 180$$
$$3y + 123 = 180$$
$$3y = 57$$
$$y = 19$$
$x = 2y; \ y = 19$
$$x = 2(19)$$
$$x = 38$$
$m\angle x = 38°; \ m\angle y = 19°$

55) $x =$ number of milliliters of pure alcohol

$y =$ number of milliliters of 4% solution

Solution	Concentration	Number of milliliters of solution	Number of milliliters of alcohol in the solution
100%	1.00	x	$1.00x$
4%	0.04	y	$0.04y$
8%	0.08	480	$0.08(480)$

$x+y=480$ $\qquad$ $1.00x+0.04y=0.08(480)$

$\quad y=480-x$ $\qquad$ $100(1.00x+0.04y)=100\left[0.08(480)\right]$

$100x+4y=8(480)$

Use substitution.

$100x+4(480-x)=8(480)$

$\quad 100x+1920-4x=3840$ $\qquad$ $y=480-x;\ x=20$

$\qquad 96x+1920=3840$ $\qquad\quad$ $y=480-20$

$\qquad\qquad 96x=1920$ $\qquad\qquad$ $y=460$

$\qquad\qquad\quad x=20$

Pure alcohol: 20 ml; 4% solution: 460 ml

57) $p =$ mg of sodium in a Powerade.

$r =$ mg of sodium in a Propel.

$g =$ mg of sodium in a Gatorade.

Use the information from the problem to get the following system of equations.

I $\qquad p=r+17$

II $\qquad g=p+58$

III $p+r+g=197$

Subtract r in I to get equation A:

A $\quad p-r=17$

Add A and III

$$p - r \quad = 17$$
$$\underline{p + r + g = 197}$$
B $\quad 2p + g = 214$

Substitute II into B

$$2p + (p + 58) = 214$$
$$3p = 156$$
$$p = 52$$

Substitute $p = 52$ into II

$$g = 52 + 58 = 110$$

Substitute $p = 52$ into I

$$52 = r + 17$$
$$35 = r$$

Propel: 35 mg, Powerade: 52 mg,

Gatorade Extreme: 110 mg

59) $\ s = $ Serena's texts

$b = $ Blair's texts

$c = $ Chuck's texts

Use the information from the

problem to get the following

system of equations.

I $\ s + b + c = 140$

II $\qquad b = s + 15$

III $\qquad c = 0.5s$

Substitute II and III into I

$$s + (s + 15) + 0.5s = 140$$
$$2.5s + 15 = 140$$
$$2.5s = 125$$
$$s = 50$$

Substitute $s = 50$ into II

$$b = 50 + 15$$
$$b = 65$$

Substitute $s = 50$ into III

$$c = 0.5(50)$$
$$c = 25$$

Serena: 50 texts,

Blair: 65 texts,

Chuck: 25 texts

61) $\ x = $ cost of a cone.

$y = $ cost of a shake.

$z = $ cost of a sundae.

Use the information from the

problem to get the following

system of equations.

I $\ 2x + 3y + z = 13.50$

II $\ 3x + y + 2z = 13.00$

III $\ 4x + y + z = 11.50$

Add I $\cdot (-2)$ and II

$$-4x - 6y - 2z = -27.00$$
$$+ \quad \underline{3x + 1y + 2z = 13.00}$$
A $\qquad -x - 5y = -14.00$

Add I and III $\cdot (-1)$

$$2x + 3y + 1z = 13.50$$
$$+ \quad \underline{-4x - 1y - 1z = -11.50}$$
B $\qquad -2x + 2y = 2.00$

Add A $\cdot (-2)$ and B

$$2x + 10y = 28.00$$
$$+ \quad \underline{-2x + 2y = 2.00}$$
$$12y = 30.00$$
$$y = 2.50$$

Substitute $y = 2.50$ into A

$$-x - 5(2.50) = -14.00$$
$$-x - 12.50 = -14.00$$
$$-x = -1.50$$
$$x = 1.50$$

Substitute $x = 1.50$ and $y = 2.50$ into I

$$2(1.50)+3(2.50)+1z=13.50$$
$$3.00+7.50+z=13.50$$
$$z=3.00$$

ice cream cone: \$1.50, shake: \$2.50, sundae: \$3.00

63) l = measure of the largest angle.

m = measure of the middle angle.

s = measure of the smallest angle.

Use the information from the problem to get the following system of equations.

I $\quad s=\dfrac{1}{3}m$

II $\quad l=s+70$

III $\quad l+m+s=180$

Solve I for m

$$s=\frac{1}{3}m$$

A $\quad 3s=m$

Substitute A and II into III

$$(s+70)+(3s)+s=180$$
$$5s+70=180$$
$$5s=110$$
$$s=22$$

Substitute $s=22$ into II

$$l=(22)+70$$
$$l=92$$

Substitute $s=22$ into A

$$3(22)=m$$
$$66=m$$

$92°$, $66°$, $22°$

65) $\begin{bmatrix}1 & -1 & | & -11\\ 2 & 9 & | & 0\end{bmatrix} \xrightarrow{-2R_1+R_2\to R_2} \begin{bmatrix}1 & -1 & | & -11\\ 0 & 11 & | & 22\end{bmatrix} \xrightarrow{\frac{1}{11}R_2\to R_2} \begin{bmatrix}1 & -1 & | & -11\\ 0 & 1 & | & 2\end{bmatrix}$

$y=2$ $\quad x-y=-11$
$$x-2=-11$$
$$x=-9 \quad (-9,2)$$

67) $\begin{bmatrix}5 & 3 & | & 5\\ -1 & 8 & | & -1\end{bmatrix} \xrightarrow{R_1\leftrightarrow R_2} \begin{bmatrix}-1 & 8 & | & -1\\ 5 & 3 & | & 5\end{bmatrix} \xrightarrow{-R_1\to R_1} \begin{bmatrix}1 & -8 & | & 1\\ 5 & 3 & | & 5\end{bmatrix}$

$\xrightarrow{-5R_1+R_2\to R_2} \begin{bmatrix}1 & -8 & | & 1\\ 0 & 43 & | & 0\end{bmatrix} \xrightarrow{\frac{1}{43}R_2\to R_2} \begin{bmatrix}1 & -8 & | & 1\\ 0 & 1 & | & 0\end{bmatrix}$

$y=0$ $\quad x-8y=1$
$$x-8(0)=1$$
$$x=1 \quad (1,0)$$

69)

$$\begin{bmatrix} 1 & -3 & -3 & | & -7 \\ 2 & -5 & -3 & | & 2 \\ -3 & 5 & 4 & | & -1 \end{bmatrix} \xrightarrow[\substack{-2R_1+R_2 \to R_2 \\ 3R_1+R_3 \to R_3}]{} \begin{bmatrix} 1 & -3 & -3 & | & -7 \\ 0 & 1 & 3 & | & 16 \\ 0 & -4 & -5 & | & -22 \end{bmatrix} \xrightarrow{4R_2+R_3 \to R_3}$$

$$\begin{bmatrix} 1 & -3 & -3 & | & -7 \\ 0 & 1 & 3 & | & 16 \\ 0 & 0 & 7 & | & 42 \end{bmatrix} \xrightarrow{\frac{1}{7}R_3 \to R_3} \begin{bmatrix} 1 & -3 & -3 & | & -7 \\ 0 & 1 & 3 & | & 16 \\ 0 & 0 & 1 & | & 6 \end{bmatrix}$$

$$x - 3y - 3z = -7 \quad y + 3z = 16 \qquad x - 3y - 3z = -7$$
$$y + 3z = 16 \quad y + 3(6) = 16 \quad x - 3(-2) - 3(6) = -7$$
$$z = 6 \qquad y + 18 = 16 \qquad x + 6 - 18 = -7$$
$$y = -2 \qquad x - 12 = -7$$
$$x = 5 \quad (5, -2, 6)$$

Chapter 5 Test

1) Yes

$$8x + y = 1$$
$$8\left(\frac{3}{4}\right) + (-5) = 1$$
$$6 - 5 = 1$$
$$1 = 1$$

$$-12x - 4y = 11$$
$$-12\left(\frac{3}{4}\right) - 4(-5) = 11$$
$$-9 + 20 = 11$$
$$11 = 11$$

3) $\varnothing$, inconsistent

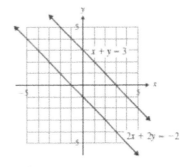

5) Substitute $y = \frac{3}{4}x + \frac{7}{4}$ into

$$-9x + 12y = 21$$
$$-9x + 12\left(\frac{3}{4}x + \frac{7}{4}\right) = 21$$
$$-9x + 9x + 21 = 21$$
$$21 = 21$$

infinite number of solutions of the form $\{(x, y) \mid -9x + 12y = 21\}$; dependent

Chapter 5: Test

7) $7x + 8y = 28$

$-3(7x + 8y) = -3(28)$

$-21x - 24y = -84$

$-5x + 6y = -20$

$4(-5x + 6y) = 4(-20)$

$-20x + 24y = -80$

Add the equations.

$-21x - 24y = -84$

$+ \ -20x + 24y = -80$

$\overline{ -41x = -164}$

$x = 4$

Substitute $x = 4$ into

$-5x + 6y = -20$

$-5(4) + 6y = -20$

$-20 + 6y = -20$

$6y = 0$

$y = 0 \qquad (4, 0)$

9) $x - 8y = 1$

$x = 8y + 1$

Substitute $x = 8y + 1$ into

$-2x + 9y = -9$

$-2(8y + 1) + 9y = -9$

$-16y - 2 + 9y = -9$

$-7y - 2 = -9$

$-7y = -7$

$y = 1$

Substitute $y = 1$ into

$x = 8y + 1$

$x = 8(1) + 1$

$x = 8 + 1$

$x = 9 \qquad (9, 1)$

11) $5(y + 4) - 9 = -3(x - 4) + 4y$

$5y + 20 - 9 = -3x + 12 + 4y$

$5y + 11 = -3x + 12 + 4y$

$3x + y = 1$

$-3(3x + y) = -3(1)$

$-9x - 3y = -3$

$13x - 2(3x + 2) = 3(1 - y)$

$13x - 6x - 4 = 3 - 3y$

$7x - 4 = 3 - 3y$

$7x + 3y = 7$

Add the equations.

$-9x - 3y = -3$

$+ \quad 7x + 3y = \ \ 7$

$\overline{ -2x = 4}$

$x = -2$

Substitute $x = -2$ into

$3x + y = 1$

$3(-2) + y = 1$

$-6 + y = 1$

$y = 7 \qquad (-2, 7)$

13) Let x = the cost of a box of screws

Let y = the cost of a box of nails

$3 \cdot \text{screws} + 2 \cdot \text{nails} = 18$

$\quad 3x \quad + \quad 2y \quad = 18$

$1 \cdot \text{screws} + 4 \cdot \text{nails} = 16$

$\quad x \quad + \quad 4y \quad = 16$

$x + 4y = 16$

$\quad x = 16 - 4y$

Use substitution.

$3(16 - 4y) + 2y = 18$

$48 - 12y + 2y = 18$

$48 - 10y = 18$

$-10y = -30$

$y = 3$

$x = 16 - 4y; \ y = 3$

$x = 16 - 4(3)$

$x = 16 - 12$

$x = 4$

screws: \$4; nails: \$3

15) $\begin{bmatrix} 1 & 5 & | & -4 \\ 3 & 2 & | & 14 \end{bmatrix} \xrightarrow{-3R_1 + R_2 \rightarrow R_2}$

$\begin{bmatrix} 1 & 5 & | & -4 \\ 0 & -13 & | & 26 \end{bmatrix} \xrightarrow{-\frac{1}{13}R_2 \rightarrow R_2} \begin{bmatrix} 1 & 5 & | & -4 \\ 0 & 1 & | & -2 \end{bmatrix}$

$y = -2 \quad x + 5y = -4$

$\quad x + 5(-2) = -4$

$\quad x - 10 = -4$

$\quad x = 6 \quad (6, -2)$

Cumulative Review: Chapters 1–5

1) $\dfrac{3}{10} - \dfrac{7}{15} = \dfrac{9}{30} - \dfrac{14}{30} = -\dfrac{5}{30} = -\dfrac{1}{6}$

3) $(5-8)^3 + 40 \div 10 - 6$

$= (-3)^3 + 40 \div 10 - 6$

$= -27 + 40 \div 10 - 6$

$= -27 + 4 - 6$

$= -29$

5) $-8(3x^2 - x - 7) = -24x^2 + 8x + 56$

7) $0.04(3p - 2) - 0.02p = 0.1(p + 3)$

$100(0.04(3p-2) - 0.02p) = 100(0.1(p+3))$

$4(3p - 2) - 2p = 10(p + 3)$

$12p - 8 - 2p = 10p + 30$

$10p - 8 = 10p + 30$

$-8 \neq 30$

$\varnothing$

9) Let x = Google's second quarter revenue in 2011

second quarter revenue in 2012 = second quarter revenue in 2011 + amount of increase

\$12.2 billion = x + x(0.35)

The second quarter revenue in 2011 was about \$9.0 billion.

Chapter 5: Cumulative Review Chapters 1–5

11) a)　　$A = \dfrac{1}{2}h(b_1 + b_2)$

　　　$2A = h(b_1 + b_2)$

　　　$\dfrac{2A}{b_1 + b_2} = h$

　　b)　$h = \dfrac{2A}{b_1 + b_2} = \dfrac{2(39 \text{ cm}^2)}{8 \text{ cm} + 5 \text{ cm}}$

　　　$= \dfrac{78 \text{ cm}^2}{13 \text{ cm}}$

　　　$= 6 \text{ cm}$

13)

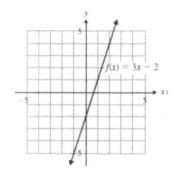

15)　$m = \dfrac{-3-4}{1-(-7)} = \dfrac{-7}{8} = -\dfrac{7}{8}$　$(x_1,\ y_1) = (-7,\ 4)$

$y - y_1 = m(x - x_1)$

$y - 4 = -\dfrac{7}{8}\left(x - (-7)\right)$

$y - 4 = -\dfrac{7}{8}(x + 7)$

$y - 4 = -\dfrac{7}{8}x - \dfrac{49}{8}$

$y = -\dfrac{7}{8}x - \dfrac{17}{8}$

17)　$(-\infty, \infty)$

19)　$f(x) = 4x + 9$

　　a) $f(-5) = 4(-5) + 9$
　　　　　$= -20 + 9$
　　　　　$= -11$
　　b) $f(p) = 4p + 9$

　　c) $f(n+2) = 4(n+2) + 9$
　　　　　　$= 4n + 8 + 9$
　　　　　　$= 4n + 17$

21)　$3(2x-1) - (y+10) = 2(2x-3) - 2y$

　　　$6x - 3 - y - 10 = 4x - 6 - 2y$

　　　$6x - y - 13 = 4x - 6 - 2y$

　　　　　$2x + y = 7$

　　　$3x + 13 = 4x - 5(y-3)$

　　　$3x + 13 = 4x - 5y + 15$

　　　　　$-x + 5y = 2$

　　　$2(-x + 5y) = 2(2)$

　　　　$-2x + 10y = 4$

Add the equations.

$$2x + \quad y = 7$$
$$+ \quad -2x + 10y = 4$$
$$\overline{\qquad 11y = 11}$$
$$y = 1$$

Substitute $y = 1$ into

$$-x + 5y = 2$$
$$-x + 5(1) = 2$$
$$-x + 5 = 2$$
$$-x = -3$$
$$x = 3 \qquad\qquad (3, 1)$$

23) $\boxed{1}\ 4a - 3b = -5$

$\boxed{2}\ -a + 5c = 2$

$\boxed{3}\ 2b + c = -2$

$\boxed{3}\ 2b + c = -2$
$$c = -2b - 2$$

$\boxed{2}\qquad -a + 5c = 2$
$$-a + 5(-2b - 2) = 2$$
$$-a - 10b - 10 = 2$$

$\boxed{A}\qquad -a - 10b = 12$

$\boxed{1}\qquad\qquad 4a - 3b = -5$

$4 \times \boxed{A}\ \underline{-4a - 40b = 48}$
$$-43b = 43$$
$$b = -1$$

$$c = -2b - 2$$
$$c = -2(-1) - 2$$
$$c = 0$$

$$-a + 5c = 2$$
$$-a + 5(0) = 2$$
$$-a = 2$$
$$a = -2$$
$$(-2, -1, 0)$$

25) Let p = puppy video views, let g = pig video views and let k = kitty video views

p = 2g

k = 1.6 + p or k = 1.6 + 2g

p + g + k = 7.1

Use substitution.

$$2g + g + (1.6 + 2g) = 7.1$$
$$5g + 1.6 = 7.1$$
$$5g = 5.5$$
$$g = 1.1$$

$$p = 2g;\ g = 1.1$$
$$p = 2(1.1)$$
$$p = 2.2$$

$$k = 1.6 + 2.2$$
$$k = 3.8$$

Puppy: 2.2 million; pig: 1.1 million; kitty: 3.8 million

Chapter 6: Polynomials and Polynomial Functions

Section 6.1

1) $(-7r)^5$

   Base: -7r   Exponent: 5

3) 9^6

5) $(3+4)^2 \quad 3^2+4^2$

   $\quad (7)^2 \quad 9+16$

   $\quad 49 \quad\quad 25$

No, they are not equivalent.

7) No;

   $-3^4 = -1\cdot 3^4 = -1\cdot 81 = -81$

   $(-3)^4 = (-3)\cdot(-3)\cdot(-3)\cdot(-3) = 81$

9) $2^2\cdot 2^4 = 2^{2+4} = 2^6 = 64$

11) $3^3\cdot 5^2 = 27\cdot 25 = 675$

13) $(-4)^2\cdot(-4)^3\cdot(-4)^2 = (-4)^{2+3+2} = (-4)^7$

15) $a^2\cdot a^3 = a^{2+3} = a^5$

17) $k\cdot k^2\cdot k^3 = k^{1+2+3} = k^6$

19) $(7h^3)(8h^{12}) = 56h^{3+12} = 56h^{15}$

21) $\left(5n^3\right)\left(-6n^7\right)\left(2n^2\right) = -60n^{3+7+2} = -60n^{12}$

23) $\left(\dfrac{49}{24}t^5\right)\left(-4t^7\right)\left(-\dfrac{12}{7}t\right)$

$= \left(\dfrac{\overset{7}{\cancel{49}}}{\underset{2}{\cancel{24}}}t^5\right)\left(-\overset{2}{\cancel{4}}\,t^7\right)\left(-\dfrac{\overset{1}{\cancel{12}}}{\overset{}{\cancel{7}}}t\right)$

$= 14t^{5+7+1}$

$= 14t^{13}$

25) $\left(a^9\right)^4 = a^{9\cdot 4} = a^{36}$

27) $\left(2^3\right)^2 = 2^{3\cdot 2} = 2^6 = 64$

29) $\left(\dfrac{1}{2}\right)^5 = \dfrac{1^5}{2^5} = \dfrac{1}{32}$

31) $\left(\dfrac{4}{y}\right)^3 = \dfrac{4^3}{y^3} = \dfrac{64}{y^3}$

33) $(-10r)^4$

   $= (-10)^4 r^4$

   $= (-10)(-10)(-10)(-10)r^4$

   $= 10,000r^4$

35) $(-4ab)^3$

   $= (-4)^3 a^3 b^3$

   $= (-4)(-4)(-4)a^3 b^3$

   $= -64a^3 b^3$

37) $\left(k^9\right)^2\left(k^3\right)^2$

$= k^{9\cdot 2}k^{3\cdot 2}$

$= k^{18}k^6$

$= k^{18+6}$

$= k^{24}$

39) $(5+3)^2 = 8^2 = 64$

41) $8\left(6k^7l^2\right)^2$

$= 8\left(6\right)^2 k^{7\cdot 2}l^{2\cdot 2}$

$= 8\cdot 36k^{14}l^4$

$= 288k^{14}l^4$

43) $-m^4\left(-5m^2\right)^3\left(-m^6\right)^2$

$= -m^4\left(-5\right)^3 m^{2\cdot 3}\left(-1\right)^2 m^{6\cdot 2}$

$= -m^4 - 125m^6\cdot 1m^{12}$

$= 125m^{4+6+12}$

$= 125m^{22}$

45.

$$\frac{\left(4d^9\right)^2}{\left(-2c^5\right)^6} = \frac{4^2 d^{9\cdot 2}}{\left(-2\right)^6 c^{5\cdot 6}} = \frac{\left(4d^9\right)^2}{\left(-2c^5\right)^6} = \frac{16d^{18}}{64c^{30}} = \frac{d^{18}}{4c^{30}}$$

47. $\dfrac{6\left(a^8 b^3\right)^5}{(2c)^3} = \dfrac{6a^{8\cdot 5}b^{3\cdot 5}}{(2)^3 c^3} = \dfrac{\overset{3}{\cancel{6}}\, a^{40}b^{15}}{\underset{4}{\cancel{8}}\, c^3} = \dfrac{3a^{40}b^{15}}{4c^3}$

49. $\left(8u^3 v^8\right)^2\left(-\dfrac{13}{4}uv^5\right)^2$

$= 8^2 u^{3\cdot 2}v^{8\cdot 2}\left(-\dfrac{13}{4}\right)^2 u^2 v^{5\cdot 2}$

$= \overset{4}{\cancel{64}}\, u^6 v^{16}\dfrac{169}{\underset{1}{\cancel{16}}}u^2 v^{10}$

$= 676u^{6+2}v^{16+10}$

$= 676u^8 v^{26}$

51. $\left(\dfrac{4r^6 s^2}{t^3}\right)^3 = \dfrac{4^3 r^{6\cdot 3}s^{2\cdot 3}}{t^{3\cdot 3}} = \dfrac{64r^{18}s^6}{t^9}$

53. $\left(\dfrac{36n^4}{4m^8 p^3}\right)^2 = \left(\dfrac{\overset{9}{\cancel{36}}\, n^4}{\underset{1}{\cancel{4}}\, m^8 p^3}\right)^2 = \dfrac{81n^{4\cdot 2}}{m^{8\cdot 2}p^{3\cdot 2}} = \dfrac{81n^8}{m^{16}p^6}$

55. False

57. True

59. $6^0 = 1$

61. $-\left(-7\right)^{0}=-1\cdot 1=-1$

63. $\left(11\right)^{0}+\left(-11\right)^{0}=1+1=2$

65. $-6y^{0}=-6\cdot y^{0}-6\cdot 1=-6$

67. $6^{-2}=\dfrac{1}{6^{2}}=\dfrac{1}{36}$

69. $\left(\dfrac{1}{7}\right)^{-2}=\left(\dfrac{7}{1}\right)^{2}=\dfrac{49}{1}=49$

71. $\left(\dfrac{4}{3}\right)^{-3}=\left(\dfrac{3}{4}\right)^{3}=\dfrac{3^{3}}{4^{3}}=\dfrac{27}{64}$

73. $\left(-\dfrac{1}{2}\right)^{-5}=\left(-\dfrac{2}{1}\right)^{5}=\dfrac{(-2)^{5}}{1^{5}}=\dfrac{-32}{1}=-32$

75. $-2^{-6}=\dfrac{-1}{2^{6}}=-\dfrac{1}{64}$

77. $2^{-3}-4^{-2}=\dfrac{1}{2^{3}}-\dfrac{1}{4^{2}}$

$=\dfrac{1}{8}\cdot\dfrac{2}{2}-\dfrac{1}{16}$

$=\dfrac{2}{16}-\dfrac{1}{16}$

$=\dfrac{1}{16}$

79. $-9^{-2}+3^{-3}+\left(-7\right)^{0}=-\dfrac{1}{9^{2}}+\dfrac{1}{3^{3}}+1$

$=-\dfrac{1}{81}+\dfrac{1}{27}\cdot\dfrac{3}{3}+\dfrac{1}{1}\cdot\dfrac{81}{81}$

$=-\dfrac{1}{81}+\dfrac{3}{81}+\dfrac{81}{81}$

$=\dfrac{83}{81}$

81. $y^{-4}=\dfrac{1}{y^{4}}$

83. $\dfrac{a^{-10}}{b^{-3}}=\dfrac{b^{3}}{a^{10}}$

85. $\dfrac{x^{4}}{10y^{-5}}=\dfrac{x^{4}y^{5}}{10}$

87. $8x^{3}y^{-7}=\dfrac{8x^{3}}{y^{7}}$

89. $\dfrac{8a^{6}b^{-1}}{5c^{-10}d}=\dfrac{8a^{6}c^{10}}{5bd}$

91. $\left(\dfrac{a}{6}\right)^{-2}=\left(\dfrac{6}{a}\right)^{2}=\dfrac{6^{2}}{a^{2}}=\dfrac{36}{a^{2}}$

93. $\left(\dfrac{12b}{cd}\right)^{-2}=\left(\dfrac{cd}{12b}\right)^{2}=\dfrac{c^{2}d^{2}}{12^{2}b^{2}}=\dfrac{c^{2}d^{2}}{144b^{2}}$

95. $-6r^{-2}=-\dfrac{6}{r^{2}}$

97. $-p^{-8}=-\dfrac{1}{p^{8}}$

99. $\left(\dfrac{1}{x}\right)^{-1} = \left(\dfrac{x}{1}\right)^{1} = x$

101. $\dfrac{n^9}{n^4} = n^{9-4} = n^5$

103. $\dfrac{8^{11}}{8^9} = 8^{11-9} = 8^2 = 64$

105. $\dfrac{5^6}{5^9} = 5^{6-9} = 5^{-3} = \dfrac{1}{5^3} = \dfrac{1}{125}$

107. $\dfrac{9d^4}{d} = 9d^{4-1} = 9d^3$

109. $\dfrac{t^3}{t^5} = t^{3-5} = t^{-2} = \dfrac{1}{t^2}$

111. $\dfrac{x^{-3}}{x^6} = \dfrac{1}{x^3 x^6} = \dfrac{1}{x^{3+6}} = \dfrac{1}{x^9}$

113. $\dfrac{-6k}{k^4} = -6k^{1-4} = -6k^{-3} = -\dfrac{6}{k^3}$

115. $\dfrac{6v^{-1}w}{54v^2 w^{-5}} = \dfrac{\overset{1}{\cancel{6}}\, w \cdot w^5}{\underset{9}{\cancel{54}}\, v \cdot v^2} = \dfrac{w^{1+5}}{9v^{1+2}} = \dfrac{w^6}{9v^3}$

117. $\dfrac{20m^{-3}n^4}{4m^8 n^6} = \dfrac{20n^{4-6}}{4m^8 m^3} = \dfrac{5n^{-2}}{m^{11}} = \dfrac{5}{m^{11}n^2}$

119. $\dfrac{(x+y)^8}{(x+y)^3} = (x+y)^{8-3} = (x+y)^5$

Section 6.2

1) $-10\left(-3g^4\right)^3 = -10\cdot(-3)^3 g^{4\cdot 3}$

$\qquad = -10\cdot -27g^{12}$

$\qquad = 270g^{12}$

3) $\dfrac{23t}{t^{11}} = 23t^{1-11} = 23t^{-10} = \dfrac{23}{t^{10}}$

5) $\left(\dfrac{2xy^4}{3x^{-9}y^{-2}}\right)^3 = \left(\dfrac{2x^1 x^9 y^4 y^2}{3}\right)^3$

$\qquad = \left(\dfrac{2x^{10}y^6}{3}\right)^3$

$\qquad = \dfrac{2^3 x^{10\cdot 3} y^{6\cdot 3}}{3}$

$\qquad = \dfrac{8}{3}x^{30}y^{18}$

7) $\left(\dfrac{7r^3}{s^8}\right)^{-2} = \left(\dfrac{s^8}{7r^3}\right)^2$

$\qquad = \dfrac{s^{8\cdot 2}}{7^2 r^{3\cdot 2}}$

$\qquad = \dfrac{s^{16}}{49r^6}$

9) $\left(-k^4\right)^3 = (-1)^3 k^{4\cdot 3} = -k^{12}$

11) $\left(-2m^5 n^2\right)^5 = (-2)^5 m^{5\cdot 5} n^{2\cdot 5} = -32m^{25}n^{10}$

13) $\left(-\dfrac{9}{4}z^5\right)\left(\dfrac{2}{3}z^{-1}\right)=\left(-\dfrac{\cancel{9}^{\,3}}{\cancel{4}_{\,2}}z^5\right)\left(\dfrac{\cancel{2}^{\,1}}{\cancel{3}_{\,1}}z^{-1}\right)$

$=-\dfrac{3}{2}z^{5+(-1)}=-\dfrac{3}{2}z^4$

15) $\left(\dfrac{a^5}{b^4}\right)^{-6}=\left(\dfrac{b^4}{a^5}\right)^6=\dfrac{b^{4\cdot6}}{a^{5\cdot6}}=\dfrac{b^{24}}{a^{30}}$

17) $\left(-ab^3c^5\right)^2\left(\dfrac{a^4}{bc}\right)^3=(-1)^2a^2b^{3\cdot2}c^{5\cdot2}\cdot\dfrac{a^{4\cdot3}}{b^3c^3}$

$=1a^2b^6c^{10}\cdot\dfrac{a^{12}}{b^3c^3}$

$=a^{2+12}b^{6-3}c^{10-3}$

$=a^{14}b^3c^7$

19) $\left(\dfrac{48u^{-7}v^2}{36u^3v^{-5}}\right)^{-3}=\left(\dfrac{36u^3v^{-5}}{48u^{-7}v^2}\right)^3$

$=\left(\dfrac{3u^3u^7}{4v^2v^5}\right)^3$

$=\left(\dfrac{3u^{3+7}}{4v^{2+5}}\right)^3$

$=\left(\dfrac{3u^{10}}{4v^7}\right)^3$

$=\dfrac{3^3u^{10\cdot3}}{4^3v^{7\cdot3}}$

$=\dfrac{27u^{30}}{64v^{21}}$

21) $\left(\dfrac{-3t^4u}{t^2u^{-4}}\right)^4=\left(\dfrac{-3t^4uu^4}{t^2}\right)^4$

$=\left(-3t^{4-2}u^{1+4}\right)^4$

$=\left(-3t^2u^5\right)^4$

$=(-3)^4t^{2\cdot4}u^{5\cdot4}$

$=81t^8u^{20}$

23) $\left(h^{-3}\right)^7=\left(\dfrac{1}{h^3}\right)^7=\dfrac{(1)^7}{h^{3\cdot7}}=\dfrac{1}{h^{21}}$

24) $\left(-n^4\right)^{-5}=\dfrac{1}{\left(-n^4\right)^5}=\dfrac{1}{-n^{4\cdot5}}=-\dfrac{1}{n^{20}}$

25) $\left(\dfrac{h}{2}\right)^5=\dfrac{h^5}{2^5}=\dfrac{h^5}{32}$

27) $-7c^8\left(-2c^2\right)^3=-7c^8\cdot(-2)^3c^{2\cdot3}$

$=-7c^8\cdot-8c^6$

$=56c^{8+6}$

$=56c^{14}$

29) $\left(12a^7\right)^{-1}(6a)^2=\dfrac{(6a)^2}{12a^7}=\dfrac{6^2a^2}{12a^7}$

$=\dfrac{36a^{2-7}}{12}$

$=4a^{-5}$

$=\dfrac{4}{a^5}$

31) $\left(\dfrac{9}{20}d^5\right)\left(2d^{-3}\right)\left(\dfrac{4}{33}d^9\right) = \dfrac{\cancel{72}^{\,6}}{\cancel{660}_{\,55}}d^{5+(-3)+9}$

$\qquad = \dfrac{6}{55}d^{11}$

33) $\left(\dfrac{56m^4n^8}{21m^4n^5}\right)^{-2} = \left(\dfrac{21\cancel{m^4}n^5}{56\cancel{m^4}n^8}\right)^2$

$\qquad = \left(\dfrac{3n^{5-8}}{8}\right)^2$

$\qquad = \left(\dfrac{3n^{-3}}{8}\right)^2$

$\qquad = \left(\dfrac{3}{8n^3}\right)^2$

$\qquad = \dfrac{3^2}{8^2 n^{3\cdot2}}$

$\qquad = \dfrac{9}{64n^6}$

35) $\dfrac{\left(-r^{-1}t^2\right)^{-5}}{\left(2r^7t^{-1}\right)^{-5}} = \dfrac{\left(2r^7t^{-1}\right)^5}{\left(-r^{-1}t^2\right)^5}$

$\qquad = \left(-2r^{7-(-1)}t^{-1-2}\right)^5$

$\qquad = \left(-2r^8t^{-3}\right)^5$

$\qquad = \left(\dfrac{-2r^8}{t^3}\right)^5$

$\qquad = \dfrac{(-2)^5 r^{8\cdot5}}{t^{3\cdot5}}$

$\qquad = -\dfrac{32r^{40}}{t^{15}}$

37) $\left(\dfrac{c^5}{6a^4b^{-1}}\right)^{-2}\left(\dfrac{a^{-8}c}{2b^3}\right)^5\left(\dfrac{3b^5c^{-2}}{a}\right)^{-3}$

$\qquad = \left(\dfrac{c^5b}{6a^4}\right)^{-2}\left(\dfrac{c}{2a^8b^3}\right)^5\left(\dfrac{3b^5}{ac^2}\right)^{-3}$

$\qquad = \left(\dfrac{6a^4}{c^5b}\right)^2\left(\dfrac{c}{2a^8b^3}\right)^5\left(\dfrac{ac^2}{3b^5}\right)^3$

$\qquad = \left(\dfrac{6^2 a^{4\cdot2}}{c^{5\cdot2}b^2}\right)\left(\dfrac{c^5}{2^5 a^{8\cdot5}b^{3\cdot5}}\right)\left(\dfrac{a^3c^{2\cdot3}}{3^3 b^{5\cdot3}}\right)$

$\qquad = \left(\dfrac{\cancel{36}^{\,1} a^8}{c^{10}b^2}\right)\left(\dfrac{c^5}{\cancel{32}_{\,8} a^{40}b^{15}}\right)\left(\dfrac{a^3c^6}{\cancel{27}_{\,3} b^{15}}\right)$

$\qquad = \dfrac{a^{8-40+3}c^{5+6-10}}{24b^{2+15+15}}$

$\qquad = \dfrac{a^{-29}c}{24b^{32}}$

$\qquad = \dfrac{c}{24a^{29}b^{32}}$

39)

$\qquad \dfrac{(x+y)^4}{(x+y)^9} = (x+y)^{4-9} = (x+y)^{-5} = \dfrac{1}{(x+y)^5}$

41) $\left(p^{2n}\right)^5 = p^{2n\cdot5} = p^{10n}$

43) $y^m \cdot y^{10m} = y^{m+10m} = y^{11m}$

45) $x^{5a} \cdot x^{-8a} = x^{5a+(-8a)} = x^{-3a} = \dfrac{1}{x^{3a}}$

47) $\dfrac{21c^{2x}}{35c^{8x}} = \dfrac{3c^{2x-8x}}{5} = \dfrac{3c^{-6x}}{5} = \dfrac{3}{5c^{6x}}$

Section 6.2: More on Exponents and Scientific Notation

49) $A = \dfrac{1}{2}bh$

$A = \dfrac{1}{2} \cdot x \cdot \dfrac{3}{4}x$

$A = \dfrac{3}{8}x^2$ sq units

51) yes

53) No, decimal is not in the correct place, there should be one number to the left of the decimal.

55) yes

57) no

59) Answers may vary.

61) $-6.8 \times 10^{-5} = -0.000068$

63) $3.45029 \times 10^4 = 34,502.9$

65) $-5 \times 10^{-5} = -0.00005$

67) $-8.1 \times 10^{-4} = -0.00081$

69) $3 \times 10^6 = 3,000,000$

71) $-3.921 \times 10^{-2} = -0.03921$

73) $2110.5 = 2.1105 \times 10^3$

75) $0.0048 = 4.8 \times 10^{-3}$

77) $-400,000 = -4 \times 10^5$

79) $11000 = 1.1 \times 10^4$

81) $0.0008 = 8 \times 10^{-4}$

83) $-0.054 = -5.4 \times 10^{-2}$

85) $6500 = 6.5 \times 10^3$

87) $\$15,000,000 = \$1.5 \times 10^7$

89) $0.00000001 = 1 \times 10^{-8}$ cm

91) $\dfrac{8 \times 10^6}{4 \times 10^2} = \dfrac{8}{4} \times \dfrac{10^6}{10^2}$

$= 2 \times 10^{6-2}$

$= 2 \times 10^4$

$= 20,000$

93) $(2.3 \times 10^3)(3 \times 10^2) = (2.3 \times 3)(10^3 \times 10^2)$

$= 6.9 \times 10^{3+2}$

$= 6.9 \times 10^5$

$= 690,000$

95) $\dfrac{9.6 \times 10^{11}}{-4 \times 10^4} = -\dfrac{9.6}{4} \times \dfrac{10^{11}}{10^4}$

$= -2.4 \times 10^{11-4}$

$= -2.4 \times 10^7$

$= -24,000,000$

97) $\left(-1.5\times10^{-6}\right)\left(6\times10^{2}\right)$

$= \left(-1.5\times6\right)\left(10^{-6}\times10^{2}\right)$

$= -9\times10^{-6+2} = -9\times10^{-4}$

$= -0.0009$

99) $\dfrac{\left(0.004\right)\left(600,000\right)}{0.0003} = 18,000,000$

101) $\left(\dfrac{60\,\text{sec}}{1\,\text{min}}\right)\left(\dfrac{60\,\text{min}}{1\text{hr}}\right)\left(\dfrac{24\text{hr}}{}\right) = 86,400$

$= 8.64\times10^{4}$

$7.8\times10^{3}\left(8.64\times10^{4}\right)$

$= \left(7.8\times1.44\right)\left(10^{3}\times10^{4}\right) = 67.392\times10^{7}$

$= 6.7392\times10^{8}\,\text{m}$

103) $\dfrac{6\times10^{9}}{3\times10^{8}} = \dfrac{6}{3}\times\dfrac{10^{9}}{10^{8}}$

$= 2\times10^{9-8}$

$= 2\times10^{1}$

20 metric tons

105) $4\times10^{9}\left(6\right) = 24\times10^{9}$

24,000,000,000 kW-hr

Section 6.3

1) It is a polynomial because the variables have whole number exponents and the coefficients are real numbers.

2) It is a polynomial because the variables have whole number exponents and the coefficients are real numbers.

3) It is not a polynomial because one of the variables doesn't have a whole number exponent.

5) It is not a polynomial because the variables do not have whole number exponents.

7) binomial

9) trinomial

11) monomial

13) Look at the variable with the highest degree. That is the degree of the polynomial.

15) Add the exponents on the variables.

17)

Term	Coefficient	Degree
$7y^3$	7	3
$10y^2$	10	2
$-y$	-1	1
2	2	0

The degree of this polynomial is 3.

19)

Term	Coefficient	Degree
$-9r^3s^2$	-9	5
$-r^2s^2$	-1	4
$\dfrac{1}{2}rs$	$\dfrac{1}{2}$	2
6s	6	1

The degree of this polynomial is 5.

21) $(11w^2 + 2w - 13) + (-6w^2 + 5w + 7)$

$= (11w^2 - 6w^2) + (2w + 5w) + (-13 + 7)$

$= 5w^2 + 7w - 6$

23) $(-p + 16) + (-7p - 9)$

$= (-p - 7p) + (16 - 9)$

$= -8p + 7$

25) $\left(-7a^4 - \dfrac{3}{2}a + 1\right) + \left(2a^4 + 9a^3 - a^2 - \dfrac{3}{8}\right)$

$= \left(-7a^4 + 2a^4\right) + 9a^3 - \dfrac{3}{2}a - a^2 + \left(1 - \dfrac{3}{8}\right)$

$= -5a^2 + 9a^3 - a^2 - \dfrac{3}{2}a + \left(\dfrac{8}{8} - \dfrac{3}{8}\right)$

$= -5a^2 + 9a^3 - a^2 - \dfrac{3}{2}a + \dfrac{5}{8}$

27) $\left(\dfrac{11}{4}x^3 - \dfrac{5}{6}\right) + \left(\dfrac{3}{8}x^3 + \dfrac{11}{12}\right)$

$= \left(\dfrac{11}{4} \cdot \dfrac{2}{2}x^3 + \dfrac{3}{8}x^3\right) + \left(-\dfrac{5}{6} \cdot \dfrac{2}{2} + \dfrac{11}{12}\right)$

$= \left(\dfrac{22}{8}x^3 + \dfrac{3}{8}x^3\right) + \left(-\dfrac{10}{12} + \dfrac{11}{12}\right)$

$= \dfrac{25}{8}x^3 + \dfrac{1}{12}$

29) $\left(6.8k^3 + 3.5k^2 - 10k - 3.3\right)$

$+ \left(-4.2k^3 + 5.2k^2 + 2.7k - 1.1\right)$

$= \left(6.8k^3 - 4.2k^3\right) + \left(3.5k^2 + 5.2k^2\right)$

$\left(-10k + 2.7k\right) + \left(-3.3 - 1.1\right)$

$= 2.6k^3 + 8.7k^2 - 7.3k - 4.4$

31) $12x - 11$

$\underline{+5x + 3}$

$17x - 8$

33) $9r^2 + 16r + 2$

$\underline{+3r^2 - 10r + 9}$

$12r^2 + 6r + 11$

35) $-2.6q^3 - q^2 + 6.9q - 1$

$\underline{+ \quad 4.1q^3 \qquad -2.3q + 16}$

$\quad\;\; 1.5q^3 - q^2 + 4.6q + 15$

37) $\left(8a^4 - 9a^2 + 17\right) - \left(15a^4 + 3a^2 + 3\right)$

$= \left(8a^4 - 9a^2 + 17\right) + \left(-15a^4 - 3a^2 - 3\right)$

$= -7a^4 - 12a^2 + 14$

39) $\left(j^2 + 18j + 2\right) - \left(-7j^2 + 6j + 2\right)$

$= \left(j^2 + 18j + 2\right) + \left(7j^2 - 6j - 2\right)$

$= 8j^2 + 12j$

41) $\left(19s^5 - 11s^2\right) - \left(10s^5 + 3s^4 - 8s^2 - 2\right)$

$= \left(19s^5 - 11s^2\right) + \left(-10s^5 - 3s^4 + 8s^2 + 2\right)$

$= 9s^5 - 3s^4 - 3s^2 + 2$

43)

$\left(-3b^4 - 5b^2 + b + 2\right) - \left(-2b^4 + 10b^3 - 5b^2 - 18\right)$

$= \left(-3b^4 - 5b^2 + b + 2\right) + \left(2b^4 - 10b^3 + 5b^2 + 18\right)$

$= -b^4 - 10b^3 + b + 20$

45) $\left(-\dfrac{5}{7}r^2 + \dfrac{4}{9}r + \dfrac{2}{3}\right) - \left(-\dfrac{5}{14}r^2 - \dfrac{5}{9}r + \dfrac{11}{6}\right)$

$= \left(-\dfrac{5}{7}r^2 + \dfrac{4}{9}r + \dfrac{2}{3}\right) + \left(\dfrac{5}{14}r^2 + \dfrac{5}{9}r - \dfrac{11}{6}\right)$

$= \left(-\dfrac{5}{7}\cdot\dfrac{2}{2}r^2 + \dfrac{5}{14}r^2\right) + \left(\dfrac{4}{9}r + \dfrac{5}{9}r\right) + \left(\dfrac{2}{3}\cdot\dfrac{2}{2} - \dfrac{11}{6}\right)$

$= \left(-\dfrac{10}{14}r^2 + \dfrac{5}{14}r^2\right) + \left(\dfrac{4}{9}r + \dfrac{5}{9}r\right) + \left(\dfrac{4}{6} - \dfrac{11}{6}\right)$

$= -\dfrac{5}{14}r^2 + r - \dfrac{7}{6}$

47) $\qquad 17v + 3 \qquad\quad 17v + 3$

$\quad\; \underline{- (2v + 9)} + \underline{(-2v - 9)}$

$\qquad\qquad\qquad\qquad 15v - 6$

49) $\qquad 2b^2 - 7b + 4 \qquad\quad 2b^2 - 7b + 4$

$\quad\; \underline{- \left(3b^2 + 5b - 3\right)} + \underline{(-3b^2 - 5b + 3)}$

$\qquad\qquad\qquad\qquad\qquad -b^2 - 12b + 7$

51) $\qquad a^4 - 2a^3 + 6a^2 - 7a + 11$

$\;\; \underline{- \;-2a^4 + 9a^3 - \;a^2 \qquad\quad + 3}$

$\qquad\qquad a^4 \;-\; 2a^3 + 6a^2 - 7a + 11$

$\;\; + \;\; \underline{(2a^4 \; -9a^3 + \;a^2 \qquad\quad -3)}$

$\qquad\qquad 3a^4 - 11a^3 + 7a^2 - 7a \;+9$

53) Answers may vary.

55) No. $= \left(j^2 + 18j + 2\right) + \left(7j^2 - 6j - 2\right)$

$\qquad\qquad = 8j^2 + 12j$

57) $\left(-3b^4 + 4b^2 - 6\right) + \left(2b^4 - 18b^2 + 4\right)$

$+\left(b^4 + 5b^2 - 2\right)$

$= \left(-3b^4 + 2b^4 + b^4\right) + \left(4b^2 - 18b^2 + 5b^2\right)$

$+\left(-6 + 4 - 2\right)$

$= -9b^2 - 4$

59) $\left(n^3 - \dfrac{1}{2}n^2 - 4n + \dfrac{5}{8}\right) + \left(\dfrac{1}{4}n^3 - n^2 + 7n - \dfrac{3}{4}\right)$

$= \left(\dfrac{4}{4}n^3 + \dfrac{1}{4}n^3\right) + \left(-\dfrac{1}{2}n^2 - \dfrac{2}{2}n^2\right) + \left(-4n + 7n\right)$

$+\left(\dfrac{5}{8} - \dfrac{3}{4} \cdot \dfrac{2}{2}\right)$

$= \dfrac{5}{4}n^3 - \dfrac{3}{2}n^2 + 3n + \left(\dfrac{5}{8} - \dfrac{6}{8}\right)$

$= \dfrac{5}{4}n^3 - \dfrac{3}{2}n^2 + 3n - \dfrac{1}{8}$

61) $\left(u^3 + 2u^2 + 1\right) - \left(4u^3 - 7u^2 + u + 9\right)$

$+\left(8u^3 - 19u^2 + 2\right)$

$= \left(u^3 + 2u^2 + 1\right) + \left(-4u^3 + 7u^2 - u - 9\right)$

$+\left(8u^3 - 19u^2 + 2\right)$

$= 5u^3 - 10u^2 - u - 6$

63) $\left(\dfrac{3}{8}k^2 + k - \dfrac{1}{5}\right) - \left(2k^2 + k - \dfrac{7}{10}\right) + \left(k^2 - 9k\right)$

$= \left(\dfrac{3}{8}k^2 + \cancel{k} - \dfrac{1}{5}\right) + \left(-2k^2 \cancel{-k} + \dfrac{7}{10}\right)$

$+\left(k^2 - 9k\right)$

$= \left(\dfrac{3}{8}k^2 - 2 \cdot \dfrac{8}{8}k^2 + \dfrac{8}{8}k^2\right) - 9k$

$+\left(-\dfrac{1}{5} \cdot \dfrac{2}{2} + \dfrac{7}{10}\right)$

$= \left(\dfrac{3}{8}k^2 - \dfrac{16}{8}k^2 + \dfrac{8}{8}k^2\right) - 9k$

$+\left(-\dfrac{2}{10} + \dfrac{7}{10}\right)$

$= \dfrac{-5}{8}k^2 - 9k + \dfrac{1}{2}$

65)

$\left(2t^3 - 8t^2 + t + 10\right)$

$-\left[\left(5t^3 + 3t^2 - t + 8\right) + \left(-6t^3 - 4t^2 + 3t + 5\right)\right]$

$= \left(2t^3 - 8t^2 + t + 10\right) - \left(5t^3 + 3t^2 - t + 8\right)$

$-\left(-6t^3 - 4t^2 + 3t + 5\right)$

$= \left(2t^3 - 8t^2 + t + 10\right) + \left(-5t^3 - 3t^2 + t - 8\right)$

$+\left(6t^3 + 4t^2 - 3t - 5\right)$

$= 3t^3 - 7t^2 - t - 3$

67) $\left(-12a^2+9\right)-\left(-9a^3+7a+6\right)$

$\quad+\left(12a^2-a+10\right)$

$\quad=\left(-12a^2+9\right)+\left(9a^3-7a-6\right)$

$\quad+\left(12a^2-a+10\right)$

$\quad=9a^3-8a+13$

69) $\left(4a+13b\right)-\left(a+5b\right)$

$\quad=\left(4a+13b\right)+\left(-a-5b\right)$

$\quad=3a+8b$

71) $\left(5m+\dfrac{5}{6}n+\dfrac{1}{2}\right)+\left(-6m+n-\dfrac{3}{4}\right)$

$\quad=-m+\left(\dfrac{5}{6}n+\dfrac{6}{6}n\right)+\left(\dfrac{1}{2}\cdot\dfrac{2}{2}-\dfrac{3}{4}\right)$

$\quad=-m+\dfrac{11}{6}n+\left(\dfrac{2}{4}-\dfrac{3}{4}\right)$

$\quad=-m+\dfrac{11}{6}n-\dfrac{1}{4}$

73) $\left(-12y^2z^2+5y^2z-25yz^2+16\right)$

$\quad+\left(17y^2z^2+2y^2z-15\right)$

$\quad=5y^2z^2+7y^2z-25yz^2+1$

75) $\quad 8x^3y^2-7x^2y^2+7x^2y-3$

$\quad+\ 2x^3y^2\qquad\qquad+x^2y-1$

$\quad-(\qquad\ \ 4x^2y^2\ +2x^2y+8)$

$\quad 8x^3y^2-7x^2y^2+7x^2y-3$

$+\ 2x^3y^2\qquad\qquad+\ x^2y-1$

$+\qquad\ \ -4x^2y^2\ -2x^2y-8$

$\overline{10x^3y^2-11x^2y^2+6x^2y-12}$

77) $\left(v^2-9\right)+\left(4v^2+3v+1\right)$

$\quad=5v^2+3v-8$

79) $\left(5g^2+3g+6\right)-\left(g^2-7g+16\right)$

$\quad=\left(5g^2+3g+6\right)+\left(-g^2+7g-16\right)$

$\quad=4g^2+10g-10$

81) $\left(2n^2+n+4\right)$

$\quad-\left[\left(4n^2+1\right)+\left(6n^2-10n+3\right)\right]$

$\quad=\left(2n^2+n+4\right)-\left[10n^2-10n+4\right]$

$\quad=\left(2n^2+n+4\right)+\left[-10n^2+10n-4\right]$

$\quad=-8n^2+11n$

83) $p=2l+2w$

$\quad p=2(3x+8)+2(x-1)$

$\quad p=6x+16+2x-2$

$\quad p=8x+14$ units

85) $\quad p=2l+2w$

$\quad p=2(3w^2-2w+4)+2(w-7)$

$\quad p=6w^2-4w+8+2w-14$

$\quad p=6w^2-2w-6$ units

Section 6.3: Addition and Subtraction of Polynomials and Polynomial Functions

87) $f(x) = 5x^2 + 7x - 8$

  a) $f(-3) = 5(-3)^2 + 7(-3) - 8$
  $f(-3) = 5 \cdot 9 - 21 - 8$
  $f(-3) = 45 - 21 - 8$
  $f(-3) = 16$

  b) $f(1) = 5(1)^2 + 7(1)(-8) = 5 + 7 - 8 = 4$

89) $P(t) = t^3 - 3t^2 + 2t + 5$

  a) $P(4) = 4^3 - 3(4)^2 + 2 \cdot 4 + 5$
  $= 64 - 48 + 8 + 5$
  $= 29$

  b) $P(0) = t^3 - 3t^2 + 2t + 5$
  $= 0^3 - 3 \cdot 0 + 2 \cdot 0 + 5$
  $= 5$

91) $H(z) = -4z + 9$
  $11 = -4z + 9$
  $11 - 9 = -4z$
  $2 = -4z$
  $-\dfrac{1}{2} = z$

93) $r(k) = \dfrac{2}{5}k - 3$ $13 = \dfrac{2}{5}k - 3$
  $13 + 3 = \dfrac{2}{5}k$
  $16 = \dfrac{2}{5}k$
  $80 = 2k$
  $40 = k$

95) $f(x) = -3x + 1$ $g(x) = 2x - 11$

  a) $(f + g)(x) = -3x + 1 + 2x - 11 = -x - 10$

  b) $(f + g)(5) = -x - 10 = -5 - 10 = -15$

  c) $(f - g)(x) = -3x + 1 - (2x - 11)$
  $= -3x + 1 - 2x + 11 = -5x + 12$

  d) $(f - g)(2) = -5 \cdot 2 + 12 = -10 + 12 = 2$

97) $f(x) = 4x^2 - 7x - 1$ $g(x) = x^2 + 3x - 6$

   a) $(f + g)(x) = 4x^2 - 7x - 1 + x^2 + 3x - 6$

               $= 5x^2 - 4x - 7$

   b)

      $(f + g)(5) = 5x^2 - 4x - 7 = 5(5)^2 - 4 \cdot 5 - 7$

                $= 125 - 20 - 7 = 98$

   c) $(f - g)(x) = 4x^2 - 7x - 1 - (x^2 + 3x - 6)$

                $= 4x^2 - 7x - 1 - x^2 - 3x + 6$

                $= 3x^2 - 10x + 5$

   d) $(f - g)(2) = 3x^2 - 10x + 5$

                $= 3(2)^2 - 10(2) + 5$

                $= 12 - 20 + 5$

                $= -3$

99) $(g - h)(t) = -t^2 + 6 - (3t^2 - 4t)$

               $= -t^2 + 6 - 3t^2 + 4t$

               $= -4t^2 + 4t + 6$

101) $(g - h)(5) = -4(5)^2 + 4 \cdot 5 + 6$

                $= -100 + 20 + 6$

                $= -74$

103) $(f + g)(-1) = 4t - 1 + -t^2 + 6$

                $= -t^2 + 4t + 5$

                $= -(-1)^2 + 4 \cdot (-1) + 5$

                $= -1 - 4 + 5$

                $= 0$

105) $(h - f)\left(\dfrac{1}{2}\right) = 3t^2 - 4t - (4t - 1)$

                 $= 3t^2 - 4t - 4t + 1$

                 $= 3t^2 - 8t + 1$

                 $= 3\left(\dfrac{1}{2}\right)^2 - \overset{4}{\cancel{8}} \cdot \dfrac{1}{\underset{1}{\cancel{2}}} + 1$

                 $= 3 \cdot \dfrac{1}{4} - 4 + 1$

                 $= \dfrac{3}{4} - 3 \cdot \dfrac{4}{4} = \dfrac{3}{4} - \dfrac{12}{4} = -\dfrac{9}{4}$

107)   Answers may vary.

109) $R(x) = 12x$ $C(x) = 8x + 2000$

   a) $P(x) = R(x) - C(x)$

              $= 12x - (8x + 2000)$

              $= 12x - 8x - 2000$

              $= 4x - 2000$

   b) $P(x) = 4x - 2000$

              $= 4 \cdot 1500 - 2000$

              $= \$4000$

111) $R(x) = 18x$ $C(x) = 15x + 2400$

a) $P(x) = R(x) - C(x)$

$= 18x - (15x + 2400)$

$= 18x - 15x - 2400$

$= 3x - 2400$

1) Answers may vary.

3) $\left(7k^4\right)\left(2k^2\right) = 14k^{4+2} = 14k^6$

b) $P(x) = 3x - 2400$

$= 3 \cdot 800 - 2400$

$= \$0$

5) $\left(\dfrac{7}{\cancel{10}_{\,2}} d^9\right)\left(\dfrac{\cancel{5}^{\,1}}{2} d^2\right) = \dfrac{7}{4} d^{11}$

113) $R(x) = -0.2x^2 + 23x$ $C(x) = 4x + 9$

7) $7y(4y - 9) = 28y^2 - 63y$

a) $P(x) = R(x) - C(x)$

$= -0.2x^2 + 23x - (4x + 9)$

$= -0.2x^2 + 23x - 4x - 9$

$= -0.2x^2 + 19x - 9$

9) $6v^3\left(v^2 - 4v - 2\right)$

$= \left(6v^3\right)\left(v^2\right) + \left(6v^3\right)(-4v) + \left(6v^3\right)(-2)$

$= 6v^5 - 24v^4 - 12v^3$

b) $P(x) = -0.2x^2 + 19x - 9$

$= -0.2(20)^2 + 19 \cdot 20 - 9$

$= -80 + 380 - 9$

$= 291$

Profit: $291,000

11) $3t^2\left(9t^3 - 6t^2 - 4t - 7\right)$

$= \left(3t^2\right)\left(9t^3\right) + \left(3t^2\right)\left(-6t^2\right)$

$+ \left(3t^2\right)(-4t) + \left(3t^2\right)(-7)$

$= 27t^5 - 18t^4 - 12t^3 - 21t^2$

13) $2x^3y\left(xy^2+8xy-11y+2\right)=\left(2x^3y\right)\left(xy^2\right)+\left(2x^3y\right)\left(8xy\right)+\left(2x^3y\right)\left(-11y\right)+\left(2x^3y\right)\left(2\right)$
$$=2x^4y^3+16x^4y^2-22x^3y^2+4x^3y$$

15) $-\dfrac{3}{4}t^4\left(20t^3+8t^2-5t\right)=-\dfrac{60}{4}t^7-\dfrac{24}{4}t^6+\dfrac{15}{4}t^5=-15t^7-6t^6+\dfrac{15}{4}t^5$

17) $2\left(10g^3+5g^2+4\right)-\left(2g^3-14g-20\right)=\left(20g^3+10g^2+8\right)+\left(-2g^3+14g+20\right)$
$$=\left(20g^3-2g^3\right)+10g^2+14g+\left(8+20\right)$$
$$=18g^3+10g^2+14g+28$$

19) $7\left(a^3b^3+6a^3b^2+3\right)-9\left(6a^3b^3+9a^3b^2-12a^2b+8\right)$
$$=\left(7a^3b^3+42a^3b^2+21\right)+\left(-54a^3b^3-81a^3b^2+108a^2b-72\right)$$
$$=\left(7a^3b^3-54a^3b^3\right)+\left(42a^3b^2-81a^3b^2\right)+108a^2b+\left(21-72\right)$$
$$=-47a^3b^3-39a^3b^2+108a^2b-51$$

21) $(q+3)\left(5q^2-15q+9\right)=q\left(5q^2-15q+9\right)+3\left(5q^2-15q+9\right)$
$$=5q^3-15q^2+9q+15q^2-45q+27$$
$$=5q^3-36q+27$$

23) $(p-6)\left(2p^2+3p-5\right)=p\left(2p^2+3p-5\right)-6\left(2p^2+3p-5\right)$
$$=2p^3+3p^2-5p-12p^2-18p+30$$
$$=2p^3-9p^2-23p+30$$

25) $\left(5y^3-y^2+8y+1\right)(3y-4)=\left(5y^3\right)(3y-4)-y^2(3y-4)+8y(3y-4)+(3y-4)$
$$=15y^4-20y^3-3y^3+4y^2+24y^2-32y+3y-4$$
$$=15y^4-23y^3+28y^2-29y-4$$

27)

$$12k^2 + 5k - 10$$
$$\times \quad \frac{1}{2}k^2 + 3$$
$$\overline{36k^2 + 15k - 30}$$
$$6k^4 + \frac{5}{2}k^3 - 5k^2$$
$$\overline{6k^4 + \frac{5}{2}k^3 + 31k^2 + 15k - 30}$$

29)

$$a^2 - a + 3$$
$$a^2 + 4a - 2$$
$$-2a^2 + 2a - 6$$
$$4a^3 - 4a^2 + 12a$$
$$a^4 - a^3 + 3a^2$$
$$\overline{a^4 + 3a^3 - 3a^2 + 14a - 6}$$

31)

$$3v^2 - v + 2$$
$$-8v^3 + 6v^2 + 5$$
$$15v^2 - 5v + 10$$
$$18v^4 - 6v^3 + 12v^2$$
$$-24v^5 + 8v^4 \qquad -16v^2$$
$$\overline{-24v^5 + 26v^4 - 6v^3 + 11v^2 - 5v + 10}$$

33) Horizontally

$$(2x-3)(4x^2 - 5x + 2)$$
$$= 2x(4x^2 - 5x + 2) - 3(4x^2 - 5x + 2)$$
$$= 8x^3 - 10x^2 + 4x - 12x^2 + 15x - 6$$
$$= 8x^3 - 22x^2 + 19x - 6$$

Vertically

$$4x^2 - 5x + 2$$
$$\times \quad 2x - 3$$
$$-12x^2 + 15x - 6$$
$$8x^3 - 10x^2 + 4x$$
$$\overline{8x^3 - 22x^2 + 19x - 6}$$

35) First, Outer, Inner, Last

37) $(w+8)(w+7) = w^2 + 7w + 8w + 56$
$$= w^2 + 15w + 56$$

39) $(n-11)(n-4) = n^2 - 4n - 11n + 44$
$$= n^2 - 15n + 44$$

41) $(4p+5)(p-3) = 4p^2 - 12p + 5p - 15$
$$= 4p^2 - 7p - 15$$

43) $(8n+3)(3n+4) = 24n^2 + 32n + 9n + 12$
$$= 24n^2 + 41n + 12$$

45) $(0.4g - 0.9)(0.1g + 1.1)$
$$= 0.04g^2 + 0.44g - 0.09g - 0.99$$
$$= 0.04g^2 + 0.35g - 0.99$$

47)

$$(4a-5b)(3a+4b) = 12a^2 + 16ab - 15ab - 20b^2$$
$$= 12a^2 + ab - 20b^2$$

49) $(6p+5q)(10p+3q)$

$= 60p^2 + 18pq + 50pq + 15q^2$

$= 60p^2 + 68pq + 15q^2$

51) $\left(2a - \dfrac{1}{4}b\right)(2b+a)$

$= 4ab + 2a^2 - \dfrac{2}{4}b^2 - \dfrac{1}{4}ab$

$= 2a^2 + 4\cdot\dfrac{4}{4}ab - \dfrac{1}{4}ab - \dfrac{1}{2}b^2$

$= 2a^2 + \dfrac{16}{4}ab - \dfrac{1}{4}ab - \dfrac{1}{2}b^2$

$= 2a^2 + \dfrac{15}{4}ab - \dfrac{1}{2}b^2$

53) Perimeter

a) $P = 2l + 2w = 2(y+6) + 2(y-2)$

$= 2y + 12 + 2y - 4$

$= 4y + 8$ units

Area

b) $A = lw = (y+6)(y-2)$

$= y^2 - 2y + 6y - 12$

$= y^2 + 4y - 12$ sq units

55) Perimeter

a) $P = 2l + 2w = 2(a^2 - a + 8) + 2(3a)$

$= 2a^2 - 2a + 16 + 6a$

$= 2a^2 + 4a + 16$ units

Area

b) $A = lw = 3a(a^2 - a + 8)$

$= 3a^3 - 3a^2 + 24a$ sq units

57) Both are correct.

59) $3(y+4)(5y-2) = 3(5y^2 - 2y + 20y - 8)$

$= 3(5y^2 + 18y - 8)$

$= 15y^2 + 54y - 24$

61)

$-7r^2(r-9)(r-2) = -7r^2(r^2 - 2r - 9r + 18)$

$= -7r^2(r^2 - 11r + 18)$

$= -7r^4 + 77r^3 - 126r^2$

63) $(c+3)(c+4)(c-1)$

$= (c^2 + 4c + 3c + 12)(c-1)$

$= (c^2 + 7c + 12)(c-1)$

$= c^3 + 7c^2 + 12c - c^2 - 7c - 12$

$= c^3 + 6c^2 + 5c - 12$

65) $\overset{5}{\cancel{10}}\, n\left(\dfrac{1}{\underset{1}{\cancel{2}}}n^2 + 3\right)(n^2 + 5)$

$= (5n^3 + 30n)(n^2 + 5)$

$= 5n^5 + 25n^3 + 30n^3 + 150n$

$= 5n^5 + 55n^3 + 150n$

67) $(r+t)(r-2t)(2r-t)$

$= (r+t)(2r^2 - rt - 4rt + 2t^2)$

$= (r+t)(2r^2 - 5rt + 2t^2)$

$= 2r^3 - 5r^2t + 2rt^2 + 2r^2t - 5rt^2 + 2t^3$

$= 2r^3 - 3r^2t - 3rt^2 + 2t^3$

69) $(3m+2)(3m-2) = (3m)^2 - 2^2 = 9m^2 - 4$

71) $(7a-8)(7a+8) = (7a)^2 - 8^2 = 49a^2 - 64$

73) $(6a-b)(6a+b) = (6a)^2 - b^2 = 36a^2 - b^2$

75) $\left(n+\dfrac{1}{2}\right)\left(n-\dfrac{1}{2}\right) = n^2 - \left(\dfrac{1}{2}\right)^2 = n^2 - \dfrac{1}{4}$

77) $\left(\dfrac{2}{3}-k\right)\left(\dfrac{2}{3}+k\right) = \left(\dfrac{2}{3}\right)^2 - k^2 = \dfrac{4}{9} - k^2$

79)

$(0.3x - 0.4y)(0.3x + 0.4y) = (0.3x)^2 - (0.4y)^2$

$= 0.09x^2 - 0.16y^2$

81) $(5x^2 + 4)(5x^2 - 4) = (5x^2)^2 - 4^2$

$= 25x^4 - 16$

83) $(y+8)^2 = y^2 + 2\cdot 8\cdot y + 8^2 = y^2 + 16y + 64$

85) $(t-11)^2 = t^2 - 2\cdot 11\cdot t + 11^2 = t^2 - 22t + 121$

87)

$(4w+1)^2 = (4w)^2 + 2\cdot 4\cdot w + 1^2 = 16w^2 + 8w + 1$

89) $(2d-5)^2 = (2d)^2 + 2\cdot 2d\cdot(-5) + (-5)^2$

$= 4d^2 - 20d + 25$

91) $(6a-5b)^2 = (6a)^2 + 2\cdot 6a\cdot(-5b) + (-5b)^2$

$= 36a^2 - 60ab + 25b^2$

93) No, they are not the same.

$4(t+3)^2$

$= 4(t^2 + 2\cdot 3t + 3^2)$

$= 4t^2 + 24t + 36$

$(4t+12)^2$

$= (4t)^2 + 2\cdot 4t\cdot 12 + 12^2$

$= 16t^2 + 96t + 144$

95) $6(x+1)^2 = 6(x^2+2\cdot x\cdot 1+1^2)$

$= 6(x^2+2x+1)$

$= 6x^2+12x+6$

97) $2a(a+3)^2 = 2a\left[a^2+2\cdot a\cdot 3+3^2\right]$

$= 2a\left[a^2+6a+9\right]$

$= 2a^3+12a^2+18a$

99) $\left[(3m+n)+2\right]^2$

$= (3m+n)^2+2(3m+n)(2)+2^2$

$= (3m)^2+2\cdot 3m\cdot 2n+n^2+12m+4n+4$

$= 9m^2+6mn+n^2+12m+4n+4$

$= 9m^2+6mn+n^2+12m+4n+4$

101) $\left[(x-4)-y\right]^2 = (x-4)^2+2(x-y)(-y)+(-y)^2$

$= (x)^2+2\cdot x\cdot(-4)+(-4)^2-2xy+2y^2+y^2$

$= x^2-8x+16-2xy+2y^2+y^2$

$= x^2-8x+16-2xy+3y^2$

103) $(r+5)^3 = (r+3)(r+3)^2 = (r+3)\left[r^2+2\cdot r\cdot 3+3^2\right] = (r+3)\left[r^2+6r+9\right]$

$= r\left[r^2+6r+9\right]+3\left[r^2+6r+9\right] = r^3+6r^2+9r+3r^2+18r+27$

$= r^3+9r^2+27r+27$

105) $(x-2)^3 = (x-2)(x-2)^2 = (x-2)\left[x^2+2\cdot x\cdot -2+(-2)^2\right] = (x-2)\left[x^2-4x+4\right]$

$= x\left[x^2-4x+4\right]-2\left[x^2-4x+4\right] = x^3-4x^2+4x-2x^2+8x-8$

$= x^3-6x^2+12x-8$

107) $(c^2-9)^2 = (c^2)^2-2\cdot c^2\cdot(-9)+(-9)^2 = c^4+18c^2+81$

109) $(y+2)^4 = (y+2)^2(y+2)^2 = y^2+4y+24+y^2+4y+4 = 2y^2+8y+8$

111) $\left[(v-5w)+4\right]\left[(v-5w)-4\right]=(v-5w)^2-16$

$$=v^2+2v(-5w)+(-5w)^2$$
$$=v^2-10vw+25w^2-16$$

113) $\left[(2a+b)+c\right]\left[(2a+b)-c\right]=(2a+b)^2-c^2$

$$=(2a)^2+2\cdot 2a\cdot b+b^2-c^2$$
$$=4a^2+4ab+b^2-c^2$$

115) No; $(x+5)^2=(x+5)(x+5)=x^2+10x+25$

117) $(h+2)^3=(h+2)(h+2)^2=(h+2)\left[(h)^2+2\cdot h\cdot 2+2^2\right]=(h+2)\left[h^2+4h+4\right]$

$$=h\left[h^2+4h+4\right]+2\left[h^2+4h+4\right]=h^3+4h^2+4h+2h^2+8h+8$$
$$=h^3+6h^2+12h+8 \text{ cubic units}$$

119) $\left[(5x+3)(2x+5)\right]-\left[(x-1)(x-1)\right]=10x^2+25x+6x+15-\left[x^2-x-x+1\right]$

$$=10x^2+25x+6x+15-x^2+x+x-1$$
$$=9x^2+33x+14 \text{ sq units}$$

121) $(fg)(x)=(3x-5)(x^2)=3x^3-5x^2$

123) $(fg)(1)=3x^3-5x^2$

$$=3(1)^3-5(1)^2$$
$$=3-5$$
$$=-2$$

125) $(fh)(x) = (3x-5)(2x^2-3x-1) = 3x(2x^2-3x-1) - 5(2x^2-3x-1)$

$= 6x^3 - 9x^2 - 3x - 10x^2 + 15x + 5 = 6x^3 - 19x^2 + 12x + 5$

127) $(fh)\left(-\dfrac{1}{3}\right) = 6\left(-\dfrac{1}{3}\right)^3 - 19\left(-\dfrac{1}{3}\right)^2 + 12\left(-\dfrac{1}{3}\right) + 5 = \overset{2}{\cancel{6}} \cdot -\dfrac{1}{\underset{9}{\cancel{27}}} - 19 \cdot \dfrac{1}{9} - 4 + 5$

$= -\dfrac{6}{27} - \dfrac{19}{9} \cdot \dfrac{3}{3} + \dfrac{27}{27} = -\dfrac{2}{9} - \dfrac{19}{9} + \dfrac{9}{9} = -\dfrac{36}{27} = -\dfrac{4}{3}$

129) $f(1) = 3x - 5 = 3 \cdot 1 - 5 = -2 \qquad g(1) = x^2 = 1^2 = 1 \qquad f(1) \cdot g(1) = -2 \cdot 1 = -2$

and

$(fg)(1) = 3x^3 - 5x^2 = 3(1)^3 - 5(1)^2 = 3 - 5 = -2$

They are the same.

Section 6.5

1) $12c^3 + 20c^2 - 4c$ is the *Dividend*.

   4c is the *Divisor*.

   $3c^2 + 5c - 1$ is the *Quotient*.

3) Divide each term in the polynomial by the monomial and simplify.

5) $\dfrac{4a^5 - 10a^4 + 6a^3}{2a^3}$

$= \dfrac{4a^5}{2a^3} - \dfrac{10a^4}{2a^3} + \dfrac{6\cancel{a^3}}{2\cancel{a^3}}$

$= 2a^2 - 5a + 3$

7) $\dfrac{18u^7 + 18u^5 + 45u^4 - 72u^2}{9u^2}$

$= \dfrac{18u^7}{9u^2} + \dfrac{18u^5}{9u^2} + \dfrac{45u^4}{9u^2} - \dfrac{72u^2}{9u^2}$

$= 2u^5 + 2u^3 + 5u^2 - 8$

9) $\dfrac{35d^5 - 7d^2}{-7d^2} = \dfrac{35d^5}{-7d^2} - \dfrac{7d^2}{-7d^2} = -5d^{5-2} + 1$

$= -5d^3 + 1$

11) $\dfrac{9w^5 + 42w^4 - 6w^3 + 3w^2}{6w^3}$

$= \dfrac{9w^5}{6w^3} + \dfrac{42w^4}{6w^3} - \dfrac{6w^3}{6w^3} + \dfrac{3w^2}{6w^3}$

$= \dfrac{3}{2}w^2 + 7w - 1 + \dfrac{1}{2}w^{-1}$

$= \dfrac{3}{2}w^2 + 7w - 1 + \dfrac{1}{2w}$

13) $\dfrac{10v^7 - 36v^5 - 22v^4 - 5v^2 + 1}{4v^4}$

$= \dfrac{10v^7}{4v^4} - \dfrac{36v^5}{4v^4} - \dfrac{22v^4}{4v^4} - \dfrac{5v^2}{4v^4} + \dfrac{1}{4v^4}$

$= \dfrac{5}{2}v^3 - 9v - \dfrac{11}{2} - \dfrac{5}{4v^2} + \dfrac{1}{4v^4}$

15) $\dfrac{90a^4b^3 + 60a^3b^3 - 40a^3b^2 + 100a^2b^2}{10ab^2}$

$= \dfrac{90a^4b^3}{10ab^2} + \dfrac{60a^3b^3}{10ab^2} - \dfrac{40a^3b^2}{10ab^2} + \dfrac{100a^2b^2}{10ab^2}$

$= 9a^3b + 6a^2b - 4a^2 + 10a$

17) $\dfrac{9t^5u^4 - 63t^4u^4 - 108t^3u^4 + t^3u^2}{-9tu^2}$

$= \dfrac{9t^5u^4}{-9tu^2} - \dfrac{63t^4u^4}{-9tu^2} - \dfrac{108t^3u^4}{-9tu^2} + \dfrac{t^3u^2}{-9tu^2}$

$= t^4u^2 + 7t^3u^2 + 12t^2u^2 - \dfrac{1}{9}t^2$

19) No; $\dfrac{16t^3 - 36t^2 + 4t}{4t} = 4t^2 - 9t + 1$

Irene canceled the last terms to zero

instead of 1.

21)

$$\begin{array}{r} g+4 \\ g+5\overline{)g^2 + 9g + 20} \\ \underline{-(g^2 + 5g)} \\ 4g + 20 \\ \underline{-(4g + 20)} \\ 0 \end{array}$$

23)

$$\begin{array}{r} p+6 \\ p+2\overline{)p^2 + 8p + 12} \\ \underline{-(p^2 + 2p)} \\ 6p + 12 \\ \underline{-(6p + 12)} \\ 0 \end{array}$$

25)

$$\begin{array}{r} k-5 \\ k+9\overline{)k^2 + 4k - 45} \\ \underline{-(k^2 + 9k)} \\ -5k - 45 \\ \underline{-(-5k - 45)} \\ 0 \end{array}$$

27)

$$\begin{array}{r} h+8 \\ h-3\overline{)h^2 + 5h - 24} \\ \underline{-(h^2 - 3h)} \\ 8h - 24 \\ \underline{-(8h - 24)} \\ 0 \end{array}$$

29)

$$\begin{array}{r} 2a^2 - 7a - 3 \\ 2a-5\overline{)4a^3 - 24a^2 + 29a + 15} \\ \underline{-(4a^3 - 10a^2)} \\ -14a^2 + 29a \\ \underline{-(-14a^2 + 35a)} \\ -6a + 15 \\ \underline{-(-6a + 15)} \\ 0 \end{array}$$

31)

$$\begin{array}{r} 3p^2 + 7p - 1 \\ 6p+1\overline{)18p^3 + 45p^2 + p - 1} \\ \underline{-\left(18p^3 + 3p^2\right)} \\ 42p^2 + p \\ \underline{-\left(42p^2 + 7p\right)} \\ -6p - 1 \\ \underline{-\left(-6p - 1\right)} \\ 0 \end{array}$$

33)

$$\begin{array}{r} 6t + 23 \\ t-5\overline{)6t^2 - 7t 4} \\ \underline{-\left(6t^2 - 30t\right)} \\ 23t + 4 \\ \underline{-\left(23t - 115\right)} \\ 119 \end{array}$$

35)

$$\begin{array}{r} 4z^2 + 8z + 7 \\ 3z+5\overline{)12z^3 + 44z^2 + 61z - 37} \\ \underline{-\left(12z^3 + 20z^2\right)} \\ 24z^2 + 61z \\ \underline{-\left(24z^2 + 40z\right)} \\ 21z - 37 \\ \underline{-\left(21z + 35\right)} \\ -72 \end{array}$$

37)

$$\begin{array}{r} w^2 - 4w + 16 \\ w+4\overline{)w^3 + 0w^2 + 0w + 64} \\ \underline{-\left(w^3 + 4w^2\right)} \\ -4w^2 + 0w \\ \underline{-\left(-4w^2 - 16w\right)} \\ 16w + 64 \\ \underline{-\left(16w + 64\right)} \\ 0 \end{array}$$

39)

$$\begin{array}{r} 2r^2 + 8r + 3 \\ 8r-3\overline{)16r^3 + 58r^2 + 0r - 9} \\ \underline{-\left(16r^3 - 6r^2\right)} \\ 64r^2 + 0r \\ \underline{-\left(64r^2 - 24r\right)} \\ 24r - 9 \\ \underline{-\left(24r - 9\right)} \\ 0 \end{array}$$

41) $\dfrac{6x^4 y^4 + 30x^4 y^3 - x^2 y^2 + 3xy}{6x^2 y^2}$

$$= \frac{6x^4 y^4}{6x^2 y^2} + \frac{30x^4 y^3}{6x^2 y^2} - \frac{x^2 y^2}{6x^2 y^2} + \frac{3xy}{6x^2 y^2}$$

$$= x^2 y^2 + 5x^2 y - \frac{1}{6} + \frac{1}{2xy}$$

43)

$$\begin{array}{r} -2g^3 - 5g^2 + g - 4 \\ 4g - 9\overline{)-8g^4 - 2g^3 + 49g^2 - 25g + 36} \end{array}$$

$$\underline{-\left(-8g^4 + 18g^3\right)}$$

$$-20g^3 + 49g^2$$

$$\underline{-\left(-20g^3 + 45g^2\right)}$$

$$4g^2 - 25g$$

$$\underline{-\left(4g^2 - 9g\right)}$$

$$-16g + 36$$

$$\underline{-\left(-16g + 36\right)}$$

$$0$$

45) $$\begin{array}{r} 6t + 5 \\ t - 8\overline{)6t^2 - 43t - 20} \end{array}$$

$$\underline{-\left(6t^2 - 48t\right)}$$

$$5t - 20$$

$$\underline{-\left(5t - 40\right)}$$

$$20$$

47)

$$\begin{array}{r} 4n^2 + 10n + 25 \\ 2n - 5\overline{)8n^3 + 0n^2 + 0n - 125} \end{array}$$

$$\underline{-\left(8n^3 - 20n^2\right)}$$

$$20n^2 + 0n$$

$$\underline{-\left(20n^2 - 50n\right)}$$

$$50n - 125$$

$$\underline{-\left(50n - 125\right)}$$

$$0$$

49)

$$\begin{array}{r} 5x^2 - 7x + 3 \\ x^2 + 2\overline{)5x^4 - 7x^3 + 13x^2 - 14x + 6} \end{array}$$

$$\underline{-\left(5x^4 \qquad + 10x^2\right)}$$

$$-7x^3 + 3x^2 \quad -14x$$

$$\underline{-\left(-7x^3 \qquad -14x\right)}$$

$$3x^2 \qquad +6$$

$$\underline{-\left(3x^2 \qquad +6\right)}$$

$$0$$

51) $$\dfrac{-12a^3 + 9a^2 - 21a}{-3a}$$

$$= \dfrac{-12a^3}{-3a} + \dfrac{9a^2}{-3a} + \dfrac{-21a}{-3a}$$

$$= 4a^2 - 3a + 7$$

53)

$$\begin{array}{r} 5h^2 - 3h - 2 + \dfrac{1}{2h^2 - 9} \\ 2h^2 - 9\overline{)10h^4 - 6h^3 - 49h^2 + 27h + 19} \end{array}$$

$$\underline{-\left(10h^4 \qquad - 45h^2\right)}$$

$$-6h^3 - 4h^2 + 27h$$

$$\underline{-\left(-6h^3 \qquad + 27h\right)}$$

$$-4h^2 \qquad +19$$

$$\underline{-\left(-4h^2 \qquad +18\right)}$$

$$1$$

55)

$$\begin{array}{r} 3d^2-d-8 \\ 2d^2+7d+5\overline{)6d^4+19d^3-8d^2-61d-40} \\ -(6d^4+21d^3+15d^2) \\ \hline -2d^3-23d^2-61d \\ -(-2d^3-7d^2-5d) \\ \hline -16d^2-56d-40 \\ -(-16d^2-56d-40) \\ \hline 0 \end{array}$$

57)

$$\begin{array}{r} 9c^2+8c+3 \\ c^2-10c+4\overline{)9c^4-82c^3-41c^2+9c+16} \\ -(9c^4-90c^3+36c^2) \\ \hline 8c^3-77c^2+9c \\ -(8c^3-80c^2+32c) \\ \hline 3c^2-23c+16 \\ -(3c^2-30c+12) \\ \hline 7c+4 \end{array}$$

59)

$$\begin{array}{r} k^2-9 \\ k^2+9\overline{)k^4+0k^3+0k^2+0k-81} \\ -(k^4+9k^2) \\ \hline -9k^2-81 \\ -(-9k^2-81) \\ \hline 0 \end{array}$$

61) $\dfrac{5a^6+49a^4-14a^3-15a^2}{-7a^3}$

$\quad = \dfrac{5a^6}{-7a^3}+\dfrac{49a^4}{-7a^3}-\dfrac{14a^3}{-7a^3}-\dfrac{15a^2}{-7a^3}$

$\quad = -\dfrac{5}{7}a^3-7a+2+\dfrac{15}{7a}$

63) $\begin{array}{r} \frac{1}{2}x+3 \\ 2x+1\overline{)x^2+\frac{13}{2}x+3} \\ -\left(x^2+\frac{1}{2}x\right) \\ \hline \frac{12}{2}x+3 \\ -(6x+3) \\ \hline 0 \end{array}$

65) $\begin{array}{r} \frac{2}{3}w+2 \\ 3w-4\overline{)2w^2+\frac{10}{3}w-8} \\ -\left(2w^2-\frac{8}{3}w\right) \\ \hline \frac{18}{3}w-8 \\ -(6w-8) \\ \hline 0 \end{array}$

67) $\begin{array}{r} 4y+1 \\ y-6\overline{)4y^2-23y-6} \\ -(4y^2-24y) \\ \hline y-6 \\ -(y-6) \\ \hline 0 \end{array}$

69) $\dfrac{18a^5 - 45a^4 + 9a^3}{9a^3}$

$= \dfrac{18a^5}{9a^3} - \dfrac{45a^4}{9a^3} + \dfrac{9a^3}{9a^3}$

$= 2a^2 - 5a + 1$

71) $A = \dfrac{1}{2}bh \qquad b = \dfrac{2A}{h}$

$b = \dfrac{2\left(6h^3 + 3h^2 + h\right)}{h}$

$b = \dfrac{12h^3}{h} + \dfrac{6h^2}{h} + \dfrac{2h}{h}$

$b = 12h^2 + 6h + 2$

73) $f(x) = 5x^2 + 6x - 27 \qquad g(x) = x + 3$

$\left(\dfrac{f}{g}\right)(x) = \dfrac{5x^2 + 6x - 27}{x + 3}; \quad x \neq -3$

-3 is not in the domain of $\left(\dfrac{f}{g}\right)(x)$

75) $f(x) = 12x^3 - 18x^2 + 2x \qquad g(x) = 2x$

$\left(\dfrac{f}{g}\right)(x) = \dfrac{12x^3 - 18x^2 + 2x}{2x}; \quad x \neq 0$

0 is not in the domain of $\left(\dfrac{f}{g}\right)(x)$

77)

$f(x) = 3x^4 - 10x^3 + 9x^2 + 2x - 4 \qquad g(x) = x - 1$

$\left(\dfrac{f}{g}\right)(x) = \dfrac{3x^4 - 10x^3 + 9x^2 + 2x - 4}{x - 1}; \quad x \neq 1$

1 is not in the domain of $\left(\dfrac{f}{g}\right)(x)$

79)

$\left(\dfrac{f}{g}\right)(x) = \dfrac{4x^2 - 1}{2x + 1} = 2x - 1; \quad \text{where } x \neq -\dfrac{1}{2}$

$$
\begin{array}{r}
2x-1 \\
2x+1\overline{\smash{\big)}\,4x^2 - 0x - 1} \\
\underline{-\left(4x^2 + 2x\right)} \\
-2x - 1 \\
\underline{-\left(-2x - 1\right)} \\
0
\end{array}
$$

Since $\left(\dfrac{f}{g}\right)(x) = 2x - 1$

81)

then $\left(\dfrac{f}{g}\right)(5) = 2 \cdot 5 - 1 = 10 - 1 = 9$

83) $\left(\dfrac{g}{h}\right)(x) = \dfrac{2x + 1}{3x}; \quad \text{where } x \neq 0$

85) $\left(\dfrac{g}{h}\right)\left(-\dfrac{2}{3}\right) = \dfrac{2 \cdot \left(-\dfrac{2}{3}\right) + 1}{\overset{1}{\cancel{3}} \cdot \left(-\dfrac{2}{\underset{1}{\cancel{3}}}\right)} = \dfrac{-\dfrac{4}{3} + \dfrac{3}{3}}{-2} = \dfrac{-\dfrac{1}{3}}{-2}$

$= -\dfrac{1}{3} \div -2 = -\dfrac{1}{3} \cdot -\dfrac{1}{2} = \dfrac{1}{6}$

Chapter 6 Review

1) $\dfrac{3^{10}}{3^6} = 3^4 = 81$

3) $\left(\dfrac{5}{4}\right)^{-3} = \left(\dfrac{4}{5}\right)^3 = \dfrac{64}{125}$

5) $\left(z^6\right)^3 = z^{6\cdot3} = z^{18}$

7) $\dfrac{70r^9}{10r^4} = 7r^{9-4} = 7r^5$

9) $\left(-9t\right)\left(6t^6\right) = -54t^{1+6} = -54t^7$

11) $\dfrac{k^3}{k^{11}} = k^{3-11} = k^{-8} = \dfrac{1}{k^8}$

13) $\left(-2a^2b\right)^3\left(5a^{-12}b\right) = \left(-8a^6b^3\right)\left(5a^{-12}b\right)$
$= -40a^{6+(-12)}b^4$
$= -40a^{-6}b^4$
$= \dfrac{-40b^4}{a^6}$

15) $\left(\dfrac{3pq^{-10}}{2p^{-2}q^5}\right)^{-2} = \left(\dfrac{2p^{-2}q^5}{3pq^{-10}}\right)^2 = \left(\dfrac{2q^5q^{10}}{3pp^2}\right)^2$
$= \left(\dfrac{2q^{15}}{3p^3}\right)^2$
$= \dfrac{4q^{30}}{9p^6}$

17) $\left(\dfrac{\cancel{40}^{2}}{\cancel{21}_{3}}x^{10}\right)\left(3x^{-12}\right)\left(\dfrac{\cancel{49}^{7}}{\cancel{20}_{1}}x^2\right)$
$= \dfrac{42}{3}x^{10+(-12)+2} = 14x^0 = 14$

19) $x^{5t}\cdot x^{3t} = x^{5t+3t} = x^{8t}$

21) $\left(y^{2p}\right)^3 = y^{6p}$

23) No, they are not the same. $-x^2 = -1\cdot x^2$

25) $9.38\times10^5 = 938,000$

27) $1.05\times10^{-6} = 0.00000105$

29) $0.0000575 = 5.75\times10^{-5}$

31) $32,000,000 = 3.2\times10^7$

33) $\dfrac{8\times10^6}{2\times10^{13}} = \dfrac{8}{2}\times10^{6-13} = 4\times10^{-7} = 0.0000004$

35)

$\dfrac{-3\times10^{10}}{-4\times10^6} = \dfrac{3}{4}\times10^{10-6} = 0.75\times10^4 = 7500$

37) $\dfrac{2.4 \times 10^5}{8} = \dfrac{240,000}{8} = 30,000 \, \text{quills}$

39) $4r^3 - 7r^2 + r + 5$

Term	Coefficient	Degree
$4r^3$	4	3
$-7r^2$	-7	2
r	1	1
5	5	0

The degree of the polynomial is 3.

41) $-x^2 y^2 - 7xy + 2x + 5$

$= -(-3)^2 (2)^2 - 7(-3)(2) + 2 \cdot (-3) + 5$

$= -9 \cdot 4 + 42 - 6 + 5$

$= -36 + 42 - 6 + 5$

$= 5$

43) $\quad 5.8p^3 - 1.2p^2 + \quad p - 7.5$

$\underline{+ \quad 2.1p^3 + 6.3p^2 + 3.8p + 3.9}$

$\quad 7.9p^3 + 5.1p^2 + 4.8p - 3.6$

45) $\left(a^2 b^2 + 7a^2 b - 3ab + 11 \right)$

$\qquad - \left(3a^2 b^2 - 10a^2 b + ab + 6 \right)$

$= \left(a^2 b^2 + 7a^2 b - 3ab + 11 \right)$

$\qquad + \left(-3a^2 b^2 + 10a^2 b - ab - 6 \right)$

$= -2a^2 b^2 + 17a^2 b - 4ab + 5$

47) Answers may vary. For example:

$\quad x^5 + 3x^2 - 7$

49) $f(x) = -2x^2 + 5x - 8$

a) $f(-3) = -2(-3)^2 + 5 \cdot -3 - 8$

$\qquad = -18 - 15 - 8 = -41$

b) $f\left(\dfrac{1}{2}\right) = -2\left(\dfrac{1}{2}\right)^2 + 5 \cdot \dfrac{1}{2} - 8$

$= -2\left(\dfrac{1}{4}\right) + \dfrac{5}{2} \cdot \dfrac{2}{2} - 8 \cdot \dfrac{4}{4}$

$= -\dfrac{1}{2} + \dfrac{10}{4} - \dfrac{32}{4}$

$= -\dfrac{24}{4}$

$= -6$

51. a) $N(x) = 0.5x^3 - 4.357x^2 + 16.429x + 122.686$

$\quad N(0) = 0.5 \cdot 0^3 - 4.357 \cdot 0^2 + 16.429 \cdot 0 + 122.686 = 122.686$

$\quad$ 123 ships

b) $N(x) = 0.5x^3 - 4.357x^2 + 16.429x + 122.686$

$\quad N(4) = 0.5 \cdot 4^3 - 4.357 \cdot 4^2 + 16.429 \cdot 4 + 122.686 = 32 - 69.712 + 65.716 + 122.686 = 150.69$

$\quad$ 150 ships

c) $N(4) = 0.5x^3 - 4.357x^2 + 16.429x + 122.686$

$N(4) = 0.5 \cdot 1^3 - 4.357 \cdot 1^2 + 16.429 \cdot 1 + 122.686 = 0.5 - 4.357 + 16.429 + 122.686 = 135.258$

135 ships. In 2003 there were approximately 135 cruise ships operating in North America.

53) $-6m^3\left(9m^2 - 3m + 7\right) = -54m^{3+2} + 18m^{3+1} - 42m^3 = -54m^5 + 18m^4 - 42m^3$

55) $(2w+5)\left(-12w^3 + 6w^2 - 2w + 3\right) = 2w\left(-12w^3 + 6w^2 - 2w + 3\right) + 5\left(-12w^3 + 6w^2 - 2w + 3\right)$

$\qquad = -24w^4 + 12w^3 - 4w^2 + 6w - 60w^3 + 30w^2 - 10w + 15$

$\qquad = -24w^4 - 48w^3 + 26w^2 - 4w + 15$

57) $(y-7)(y+8) = y^2 + 8y - 7y + 56 = y^2 + y + 56$

59) $(ab+5)(ab+6) = a^2b^2 + 6ab + 5ab + 30 = a^2b^2 + 11ab + 30$

61) $-(4d+3)(6d+7) = -\left(24d^2 + 28d + 18d + 21\right)$

$\qquad = -\left(24d^2 + 46d + 21\right)$

$\qquad = -24d^2 - 46d - 21$

63) $(p+3)(p-6)(p+2)$

$= (p+3)\left(p^2 + 2p - 6p - 12\right)$

$= (p+3)\left(p^2 - 4p - 12\right)$

$= p\left(p^2 - 4p - 12\right) + 3\left(p^2 - 4p - 12\right)$

$= p^3 - 4p^2 - 12p + 3p^2 - 12p - 36$

$= p^3 - p^2 - 24p - 36$

65) $\left(\dfrac{3}{5}m+2\right)\left(\dfrac{1}{3}m-4\right)$

$= \dfrac{3}{15}m^2 - \dfrac{12}{5}m + \dfrac{2}{3}m - 8$

$= \dfrac{1}{5}m^2 - \dfrac{12}{5}\cdot\dfrac{3}{3}m + \dfrac{2}{3}\cdot\dfrac{5}{5}m - 8$

$= \dfrac{1}{5}m^2 - \dfrac{36}{15}m + \dfrac{10}{15}m - 8$

$= \dfrac{1}{35}m^2 - \dfrac{26}{15}m - 8$

67) $(z+9)(z-9) = z^2 - 81$

69) $\left(\dfrac{7}{8}-r^2\right)\left(\dfrac{7}{8}+r^2\right) = \dfrac{49}{64} - r^4$

71) $(b+7)^2 = b^2 + 2\cdot 7b + 49 = b^2 + 14b + 49$

73) $(5q-2)^2 = 25q^2 - 2\cdot 5q\cdot 2 + 4$

$\qquad = 25q^2 - 20q + 4$

75) $(x-2)^3 = (x-2)(x-2)^2$

$\qquad = (x-2)(x^2 - 2\cdot 2x + 4)$

$\qquad = (x-2)(x^2 - 4x + 4)$

$\qquad = x(x^2 - 4x + 4) - 2(x^2 - 4x + 4)$

$\qquad = x^3 - 4x^2 + 4x - 2x^2 + 8x - 8$

$\qquad = x^3 - 6x^2 + 12x - 8$

77) $-2(3c-4)^2 = -2(9c^2 - 2\cdot 4\cdot 3c + 16)$

$\qquad = -2(9c^2 - 24c + 16)$

$\qquad = -18c^2 + 48c - 32$

79) $\left[(m-5)+n\right]^2$

$\qquad = (m-5)^2 + 2\cdot(m-5)\cdot n + n^2$

$\qquad = m^2 + 2\cdot(-5)m + 5^2 + 2mn - 10n + n^2$

$\qquad = m^2 - 10m + 25 + 2mn - 10n + n^2$

81) $f(x) = 2x - 9 \qquad g(x) = -5x + 4$

a) $(fg)(x) = (2x-9)(-5x+4)$

$\qquad = -10x^2 + 8x + 45x - 36$

$\qquad = -10x^2 + 53x - 36$

b) $(fg)(2) = -10\cdot 2^2 + 53\cdot 2 - 36$

$\qquad = -40 + 106 - 36$

$\qquad = 30$

83) $A = lw \qquad P = 2l + 2w$

a) $A = (4m+5)(m-3)$

$\qquad A = 4m^2 - 12m + 5m - 15$

$\qquad A = 4m^2 - 7m - 15$ sq units

b) $P = 2(4m+5) + 2(m-3)$

$\qquad P = 8m + 10 + 2m - 6$

$\qquad P = 10m + 4$ units

85) $\dfrac{8t^5 - 14t^4 - 20t^3}{2t^3}$

$= \dfrac{8t^5}{2t^3} - \dfrac{14t^4}{2t^3} - \dfrac{20t^3}{2t^3}$

$= 4t^2 - 7t - 10$

87)
$$\begin{array}{r} c+10 \\ c-2\overline{)c^2+8c-20} \end{array}$$
$$-\left(c^2-2c\right)$$
$$10c-20$$
$$-(10c-20)$$
$$0$$

95)
$$\begin{array}{r} 3a+7 \\ 5a-4\overline{)15a^2+23a-7} \end{array}$$
$$-\left(15a^2-12a\right)$$
$$35a-7$$
$$-(35a-28)$$
$$21$$

89)
$$\begin{array}{r} 4r^2-7r+3 \\ 3r+2\overline{)12r^3-13r^2-5r+6} \end{array}$$
$$-\left(12r^3+8r^2\right)$$
$$-21r^2-5r$$
$$-\left(-21r^2-14r\right)$$
$$9r+6$$
$$-(9r+6)$$
$$0$$

97)
$$\begin{array}{r} t^2+4t-4 \\ 8t^2-11\overline{)8t^4+32t^3-43t^2-44t+48} \end{array}$$
$$-\left(8t^4 \quad -11t^2\right)$$
$$32t^3-32m^2-44t$$
$$-\left(32t^3 \quad -44t\right)$$
$$-32t^2 \quad +48$$
$$-\left(-32t^2 \quad +44\right)$$
$$4$$

91)
$$\frac{15x^4y^4-42x^3y^4-6x^2y+10y}{-6x^2y}$$
$$=\frac{15x^4y^4}{-6x^2y}-\frac{42x^3y^4}{-6x^2y}-\frac{6x^2y}{-6x^2y}+\frac{10y}{-6x^2y}$$
$$=-\frac{5}{2}x^2y^3+7xy^2+1-\frac{5}{3x^2}$$

93)
$$\begin{array}{r} 2q-4 \\ 3q+7\overline{)6q^2+2q-35} \end{array}$$
$$-\left(6q^2+14q\right)$$
$$-12q-35$$
$$-(-12q-28)$$
$$7$$

99)
$$\begin{array}{r} f^2-5f+25 \\ f+5\overline{)f^3+0f^2+0f+125} \end{array}$$
$$-\left(f^3+5f^2\right)$$
$$-5f^2+0f$$
$$-\left(-5f^2-25f\right)$$
$$25f+125$$
$$-(25f+125)$$
$$0$$

101)

$$\begin{array}{r} 5k^2 + 6k + 4 \\ 3k-2\overline{\smash)15k^3 + 8k^2 - 0k - 8} \\ \underline{-\left(15k^3 - 10k^2\right)} \\ 18k^2 - 0k \\ \underline{-\left(18k^2 - 12k\right)} \\ 12k - 8 \\ \underline{-\left(12k-8\right)} \\ 0 \end{array}$$

103)

$$\begin{array}{r} 3c^2 - c - 2 \\ 2c^2 + 5c - 4\overline{\smash)6c^4 + 13c^3 - 21c^2 - 9c + 10} \\ \underline{-\left(6c^4 + 15c^3 - 12c^2\right)} \\ -2c^3 - 9c^2 \quad -9c \\ \underline{-\left(-2c^3 - 5c^2 + 4c\right)} \\ -4c^2 - 13c + 10 \\ \underline{-\left(-4c^2 - 10c + 8\right)} \\ -3c + 2 \end{array}$$

105) $A = lw$

$$\left(6x^3 - x^2 + 13x - 10\right) = \left(3x-2\right)l$$

$$\begin{array}{r} 2x^2 + x + 5 \\ 3x-2\overline{\smash)6x^3 - x^2 + 13x - 10} \\ \underline{-\left(6x^3 - 4x^2\right)} \\ 3x^2 + 13x \\ \underline{-\left(3x^2 - 2x\right)} \\ 15x - 10 \\ \underline{-\left(15x - 10\right)} \\ 0 \end{array}$$

Chapter 6 Test

1) Evaluate

a) $-3^4 = -1 \cdot 3^4 = -81$

b) $2^{-5} = \dfrac{1}{2^5} = \dfrac{1}{32}$

c) $-6^0 - 9^0 = -1 - 1 = -2$

d) $\left(\dfrac{3}{10}\right)^{-3} = \left(\dfrac{10}{3}\right)^3 = \dfrac{1000}{27}$

e) $\dfrac{2^{11}}{2^{17}} = 2^{11-17} = 2^{-6} = \dfrac{1}{2^6} = \dfrac{1}{64}$

3) $\dfrac{a^5 b}{a^9 b^7} = a^{-4}b^{-6} = \dfrac{1}{a^4 b^6}$

5) $\left(\dfrac{36xy^8}{54x^3y^{-1}}\right)^{-2} = \left(\dfrac{54x^3y^{-1}}{36xy^8}\right)^2 = \left(\dfrac{3x^2}{2y^8y}\right)^2$

$= \left(\dfrac{3x^2}{2y^9}\right)^2 = \dfrac{9x^4}{4y^{18}}$

7) $7.283\times10^5 = 728,300$

9)

$\dfrac{-7.5\times10^{12}}{1.5\times10^8} = \dfrac{-7.5}{1.5}\times10^{12-8} = -5\times10^4 = -50,000$

11) Given: $5p^3 - p^2 + 12p + 9$

    1) Coefficient is -1   b) Degree is 3

13) $10r^3s^2 + 7r^2s^2 - 11rs + 5$

     $+ \; 4r^3s^2 - \; 9r^2s^2 + \; 6rs + 3$

     $\overline{14r^3s^2 - \; 2r^2s^2 \; -5rs \; +8}$

15) $(c-8)(c-7) = c^2 - 7c - 8c + 56$

           $= c^2 - 15c + 56$

17) $\left(u+\dfrac{3}{4}\right)\left(u-\dfrac{3}{4}\right) = u^2 - \dfrac{9}{16} = u^2 - \dfrac{9}{16}$

19) $(7m-5)^2 = 49m^2 + 2\cdot7m\cdot(-5) + 25$

           $= 49m^2 - 70m + 25$

21) $(3-8m)(2m^2 + 4m - 7)$

     $= 3(2m^2 + 4m - 7) - 8m(2m^2 + 4m - 7)$

     $= 6m^2 + 12m - 21 - 16m^3 - 32m^2 + 56m$

     $= -16m^3 - 26m^2 + 68m - 21$

23) $\left[(5a-b)-3\right]^2$

     $= (5a-b)^2 + 2\cdot(5a-b)\cdot(-3) + 9$

     $= 25a^2 + 2\cdot(-5ab) + b^2 - 30a + 6b + 9$

     $= 25a^2 - 10ab + b^2 - 30a + 6b + 9$

25) $r+7\overline{\smash{\big)}\,r^2 + 10r + 21}$ quotient $r+3$

$\underline{-(r^2+7r)}$

$3r+21$

$\underline{-(3r+21)}$

0

27)

$5v-6\overline{\smash{\big)}\,30v^3 - 51v^2 + 38v - 31}$ quotient $6v^2 - 3v + 4$

$\underline{-(30v^3 - 36v^2)}$

$-15v^2 + 38v$

$\underline{-(-15v^2 + 18v)}$

$20v - 31$

$\underline{-(20v - 24)}$

-7

29) $R(x) = 5x + 9 \quad C(x) = 3x + 8$

1) $P(x) = R(x) - C(x) = 5x + 9 - (3x + 8)$

$\qquad = 5x + 9 - 3x - 8 = 2x + 1$

2) $P(8000) = 2 \cdot 8000 + 1000 = \$17,000$

Cumulative Review: Chapters 1–6

1)

$$\left\{ \frac{6}{11}, -14, 2.7, \sqrt{19}, 43, 0.\overline{65}, 0, 8.21079314... \right\}$$

a) whole numbers: $43, 0$

b) integers: $-14, 43, 0$

c) rational numbers:

$\qquad \dfrac{6}{11}, -14, 2.7, 43, 0.\overline{65}, 0$

3) $2\dfrac{6}{7} \div 1\dfrac{4}{21} = \dfrac{20}{7} \div \dfrac{25}{21} = \dfrac{\overset{4}{\cancel{20}}}{\underset{1}{\cancel{7}}} \cdot \dfrac{\overset{3}{\cancel{21}}}{\underset{5}{\cancel{25}}} = \dfrac{12}{5}$

5) $6(w + 4) + 2w = 1 + 8(w - 1)$

$\qquad 6w + 24 + 2w = 1 + 8w - 8$

$\qquad 8w + 24 = 8w - 7$

$\qquad 8w - 8w = -7 - 24$

$\qquad 0 = -31$

No solution

7)

$$A = \frac{1}{2} h(b_1 + b_2)$$

$$A \cdot 2 = \frac{1}{\cancel{2}} \cdot \cancel{2} h(b_1 + b_2)$$

$$2A = h(b_1 + b_2)$$

$$\frac{2A}{h} = b_1 + b_2$$

$$\frac{2A}{h} - b_1 = b_2$$

9)

$12\% = 0.12$	x	$0.12x$
$4\% = 0.04$	$60 - x$	$0.04(60 - x)$
$10\% = 0.10$	60	$0.10(60)$

$0.12x + 0.04(60 - x) = 0.10 \cdot 60$

$\qquad 0.12x + 2.4 - 0.04x = 6$

$\qquad 0.08x = 6 - 2.4$

$\qquad 0.08x = 3.6$

$\qquad x = 45$ ml

45ml of 12% solution, 15 ml of 4% solution.

11)

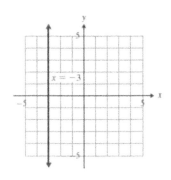

13) $(-\infty, \infty)$

15)

equation 1	equation 2
$P = 2l + 2w$	$l = 2w - 7$
$76 = 2l + 2w$	

substitute

$76 = 2(2w - 7) + 2w$ $l = 2 \cdot 15 - 7$

$76 = 4w - 14 + 2w$ $l = 30 - 7$

$76 + 14 = 6w$ $l = 23 \text{ cm}$

$90 = 6w$

$15 \text{ cm} = w$

width: 15cm; length: 23 cm

17) $\left(\dfrac{2n^{-10}}{n^{-4}}\right)^3 = \left(2n^{-10-(-4)}\right)^3 = \left(2n^{-6}\right)^3 = 8n^{-6 \cdot 3}$

$= 8n^{-18} = \dfrac{8}{n^{18}}$

19)

$\left(4q^2 + 11q - 2\right) - 3\left(6q^2 - 5q + 4\right) + 2\left(-10q - 3\right)$

$= \left(4q^2 + 11q - 2\right) + \left(-18q^2 + 15q - 12\right)$

$\qquad + \left(-20q - 6\right)$

$= \left(4q^2 - 18q^2\right) + \left(11q + 15q - 20q\right)$

$\qquad + \left(-2 - 12 - 6\right)$

$= -14q^2 + 6q - 20$

21) $\dfrac{8a^4b^4 - 20a^3b^2 + 56ab + 8b}{8a^3b^3}$

$= \dfrac{8a^4b^4}{8a^3b^3} - \dfrac{20a^3b^2}{8a^3b^3} + \dfrac{56ab}{8a^3b^3} + \dfrac{8b}{8a^3b^3}$

$= ab - \dfrac{5}{2b} + \dfrac{7}{a^2b^2} + \dfrac{1}{a^3b^2}$

23.

$$
\begin{array}{r}
2p^2 - 3p + 1 \\
p+4\overline{\smash{\big)}\,2p^3 + 5p^2 - 11p + 9} \\
\underline{-\left(2p^3 + 8p^2\right)} \\
-3p^2 - 11p \\
\underline{-\left(-3p^2 - 12p\right)} \\
p + 9 \\
\underline{-\left(p + 4\right)} \\
5
\end{array}
$$

25. $f(x) = x^2 - 6$ $g(x) = x^2 + 5x + 4$

a)

$$(fg)(x) = (x^2 - 6)(x^2 + 5x + 4)$$
$$= x^2(x^2 + 5x + 4) - 6(x^2 + 5x + 4)$$
$$= x^4 + 5x^3 + 4x^2 - 6x^2 - 30x - 24$$
$$= x^4 + 5x^3 - 2x^2 - 30x - 24$$

b) $(fg)(2) = 2^4 + 5 \cdot 2^3 - 2 \cdot 2^2 - 30 \cdot 2 - 24$
$$= 16 + 40 - 8 - 60 - 24$$
$$= -36$$

Chapter 7: Factoring Polynomials

Section 7.1

1) $45m^3 = 3 \cdot 3 \cdot 5 \cdot m \cdot m \cdot m,$
$20m^2 = 2 \cdot 2 \cdot 5 \cdot m \cdot m$
GCF of $45m^3$ and $20m^2$
$= 5 \cdot m \cdot m$
$= 5m^2$

3) $42k^5 = 2 \cdot 3 \cdot 7 \cdot k \cdot k \cdot k \cdot k \cdot k,$
$54k^7 = 2 \cdot 3 \cdot 3 \cdot 3 \cdot k \cdot k \cdot k \cdot k \cdot k \cdot k \cdot k$
$72k^9 = 2 \cdot 2 \cdot 2 \cdot 3 \cdot 3 \cdot k \cdot k \cdot k \cdot k$
$\qquad\qquad\quad \cdot k \cdot k \cdot k \cdot k \cdot k$
GCF of $42k^5$, $54k^7$ and $72k^9$
$= 2 \cdot 3 \cdot k \cdot k \cdot k \cdot k \cdot k$
$= 6k^5$

5) $9x^2 y$

7) $4uv^3$

9) $s^2 t$

11) $(n-7)$

13) Answers may vary.

15) $\text{GCF} = 6$
$30s + 18 = (6)(5s) + (6)(3)$
$\qquad\qquad = 6(5s+3)$

17) $\text{GCF} = 4$
$24z - 4 = (4)(6z) - (4)(1)$
$\qquad\quad = 4(6z-1)$

19) $\text{GCF} = 3d$
$3d^2 - 6d = (3d)(d) - (3d)(2)$
$\qquad\qquad = 3d(d-2)$

21) $\text{GCF} = 7y^2$
$42y^2 + 35y^3 = (7y^2)(6) + (7y^2)(5y)$
$\qquad\qquad\quad = 7y^2(6+5y)$

23) $\text{GCF} = t^4$
$t^5 - t^4 = (t^4)(t) - (t^4)(1)$
$\qquad\quad = t^4(t-1)$

25) $\text{GCF} = \dfrac{1}{2}c$
$\dfrac{1}{2}c^2 + \dfrac{5}{2}c = \left(\dfrac{1}{2}c\right)(c) + \left(\dfrac{1}{2}c\right)(5)$
$\qquad\qquad = \dfrac{1}{2}c(c+5)$

27) $\text{GCF} = 5n^3$
$10n^5 - 5n^4 + 40n^3$
$= (5n^3)(2n^2) - (5n^3)(n) + (5n^3)(8)$
$= 5n^3(2n^2 - n + 8)$

29) $\text{GCF} = 2v^5$

$2v^8 - 18v^7 - 24v^6 + 2v^5$

$= \left(2v^5\right)\left(v^3\right) - \left(2v^5\right)\left(9v^2\right)$

$\qquad - \left(2v^5\right)\left(12v\right) + \left(2v^5\right)\left(1\right)$

$= 2v^5\left(v^3 - 9v^2 - 12v + 1\right)$

31) $\text{GCF} = \text{none}$

$8c^3 + 3d^2$ does not factor

33) $\text{GCF} = a^3b^2$

$a^4b^2 + 4a^3b^3$

$= \left(a^3b^2\right)\left(a\right) + \left(a^3b^2\right)\left(4b\right)$

$= a^3b^2\left(a + 4b\right)$

35) $\text{GCF} = 10x^2y$

$50x^3y^3 - 70x^3y^2 + 40x^2y$

$= \left(10x^2y\right)\left(5xy^2\right) - \left(10x^2y\right)\left(7xy\right)$

$\qquad + \left(10x^2y\right)\left(4\right)$

$= 10x^2y\left(5xy^2 - 7xy + 4\right)$

37) $\text{GCF} = n - 12$

$m\left(n-12\right) + 8\left(n-12\right)$

$= \left(n-12\right)\left(m+8\right)$

39) $\text{GCF} = 8r - 3$

$p\left(8r-3\right) - q\left(8r-3\right)$

$= \left(8r-3\right)\left(p-q\right)$

41) $\text{GCF} = z + 11$

$y\left(z+11\right) + \left(z+11\right) = y\left(z+11\right) + 1\left(z+11\right)$

$\qquad\qquad = \left(z+11\right)\left(y+1\right)$

43) $\text{GCF} = 3r + 4$

$2k^2\left(3r+4\right) - \left(3r+4\right)$

$= 2k^2\left(3r+4\right) - 1\left(3r+4\right)$

$= \left(3r+4\right)\left(2k^2 - 1\right)$

45) $-64m - 40 = \left(-8\right)\left(8m\right) + \left(-8\right)\left(5\right)$

$\qquad\qquad = -8\left(8m + 5\right)$

47) $-5t^3 + 10t^2 = \left(-5t^2\right)\left(t\right) + \left(-5t^2\right)\left(-2\right)$

$\qquad\qquad = -5t^2\left(t - 2\right)$

49) $-3a^3 + 7a^2 - a$

$= \left(-a\right)\left(3a^2\right) + \left(-a\right)\left(-7a\right) + \left(-a\right)\left(1\right)$

$= -a\left(3a^2 - 7a + 1\right)$

51) $-b + 8 = \left(-1\right)\left(b\right) + \left(-1\right)\left(-8\right)$

$\qquad\qquad = -1\left(b - 8\right)$

53) $kt + 3k + 8t + 24$

Factor out k from the first two terms.
Factor out 8 from the last two terms.

$= k\left(t+3\right) + 8\left(t+3\right)$

$= \left(t+3\right)\left(k+8\right)$

55) $fg - 7f + 4g - 28$
Factor out f from the first two terms.
Factor out 4 from the last two terms.

$= f(g-7) + 4(g-7)$

$= (g-7)(f+4)$

57) $2rs - 6r + 5s - 15$
Factor out $2r$ from the first two terms.
Factor out 5 from the last two terms.

$= 2r(s-3) + 5(s-3)$

$= (s-3)(2r+5)$

59) $3xy - 2y + 27x - 18$
Factor out y from the first two terms.
Factor out 9 from the last two terms.

$= y(3x-2) + 9(3x-2)$

$= (3x-2)(y+9)$

61) $8b^2 + 20bc + 2bc^2 + 5c^3$

$= 4b(2b+5c) + c^2(2b+5c)$

$= (2b+5c)(4b+c^2)$

63) $4a^3 - 12ab + a^2b - 3b^2$

$= 4a(a^2 - 3b) + b(a^2 - 3b)$

$= (a^2 - 3b)(4a+b)$

65) $kt + 7t - 5k - 35 = t(k+7) - 5(k+7)$

$= (k+7)(t-5)$

67) $mn - 8m - 10n + 80$

$= m(n-8) - 10(n-8)$

$= (n-8)(m-10)$

69) $dg - d + g - 1 = d(g-1) + 1(g-1)$

$= (g-1)(d+1)$

71) $5tu + 6t - 5u - 6$

$= t(5u+6) - 1(5u+6)$

$= (5u+6)(t-1)$

73) $36g^4 + 3gh - 96g^3h - 8h^2$

$= 3g(12g^3 + h) - 8h(12g^3 + h)$

$= (12g^3 + h)(3g - 8h)$

75) Answers may vary.

77)
$= 5(mn + 3m + 2n + 6)$; Group the terms and
factor out the GCF from each group;
$= 5(n+3)(m+2)$

79) $2ab + 8a + 6b + 24$

$= 2(ab + 4a + 3b + 12)$

$= 2[a(b+4) + 3(b+4)]$

$= 2(b+4)(a+3)$

81) $8s^2t - 40st + 16s^2 - 80s$

$= 8s(st - 5t + 2s - 10)$

$= 8s[t(s-5) + 2(s-5)]$

$= 8s(s-5)(t+2)$

83) $7cd + 12 + 28c + 3d$

$= 7cd + 28c + 3d + 12$

$= 7c(d+4) + 3(d+4)$

$= (d+4)(7c+3)$

85) $42k^3 + 15d^2 - 18k^2d - 35kd$

$= 42k^3 - 35kd - 18k^2d + 15d^2$

$= 7k(6k^2 - 5d) - 3d(6k^2 - 5d)$

$= (6k^2 - 5d)(7k - 3d)$

87) $9f^2j^2 + 45fj + 9fj^2 + 45f^2j$

$= 9fj(fj + 5 + j + 5f)$

$= 9fj(fj + j + 5f + 5)$

$= 9fj[j(f+1) + 5(f+1)]$

$= 9fj(f+1)(j+5)$

89) $4x^4y - 14x^3 + 28x^4 - 2x^3y$

$= 2x^3(2xy - 7 + 14x - y)$

$= 2x^3(2xy - y + 14x - 7)$

$= 2x^3[y(2x-1) + 7(2x-1)]$

$= 2x^3(2x-1)(y+7)$

91) $pq - 8p + 3q - 24$

$= p(q-8) + 3(q-8)$

$= (p+3)(q-8)$

93) $a^4b^2 + 2a^3b^3 - 8a^2b^4$

$= a^2b^2(a^2 + 2ab - 8b^2)$

$= a^2b^2(a+4b)(a-2b)$

95) $3h^3 - 8k^3 + 12h^2k^2 - 2hk$

$= 3h^3 + 12h^2k^2 - 2hk - 8k^3$

$= 3h^2(h + 4k^2) - 2k(h + 4k^2)$

$= (3h^2 - 2k)(h + 4k^2)$

97) $2c^4 + 14c^2 + 84c + 12c^3$

$= 2c^4 + 12c^3 + 14c^2 + 84c$

$= 2c(c^3 + 6c^2 + 7c + 42)$

$= 2c[c^2(c+6) + 7(c+6)]$

$= 2c(c^2 + 7)(c+6)$

99) $-16v^3 - 56v^2 + 8v$

$= -8v(2v^2 + 7v - 1)$

Section 7.2

1) They are negative.

3) Can I factor out a GCF?

5) Can I factor again?
   If so, factor again.

7) $g^2 + 8g + 12 = (g+6)(g+2)$

9) $w^2 + 13w + 42 = (w+7)(w+6)$

11) $c^2 - 13c + 36 = (c-4)(c-9)$

13) $b^2 - 2b - 8 = (b-4)(b+2)$

15) $u^2 + u - 132 = (u+12)(u-11)$

17) $q^2 - 8q + 15 = (q-5)(q-3)$

19) $y^2 + 9y + 10$ is prime

21) $p^2 - 20p + 100$
    $= (p-10)(p-10)$ or $(p-10)^2$

23) $3p^2 - 24p - 27$
    $= 3(p^2 - 8p - 9)$
    $= 3(p-9)(p+1)$

25) $2k^3 - 26k^2 + 80k$
    $= 2k(k^2 - 13k + 40)$
    $= 2k(k-8)(k-5)$

27) $a^3b + 9a^2b - 36ab$
    $= ab(a^2 + 9a - 36)$
    $= ab(a+12)(a-3)$

29) $-a^2 - 10a - 16 = -(a^2 + 10a + 16)$
    $= -(a+8)(a+2)$

31) $-h^2 + 2h + 15 = -(h^2 - 2h - 15)$
    $= -(h-5)(h+3)$

33) $-k^2 + 11k - 28 = -(k^2 - 11k + 28)$
    $= -(k-7)(k-4)$

35) $-n^2 - 14n - 49 = -(n^2 + 14n + 49)$
    $= -(n+7)(n+7)$
    or $-(n+7)^2$

37) $a^2 + 6ab + 5b^2 = (a+5b)(a+b)$

39) $m^2 + 4mn - 21n^2 = (m-3n)(m+7n)$

41) $x^2 - 15xy + 36y^2 = (x-12y)(x-3y)$

43) $f^2 - 10fg - 11g^2 = (f+g)(f-11g)$

45) $c^2 + 6cd - 55d^2 = (c-5d)(c+11d)$

47) $2r^2 + 11r + 15 = 2r^2 + 6r + 5r + 15$
$\qquad = 2r(r+3) + 5(r+3)$
$\qquad = (2r+5)(r+3)$

49) $5p^2 - 21p + 4 = 5p^2 - 20p - p + 4$
$\qquad = 5p(p-4) - 1(p-4)$
$\qquad = (5p-1)(p-4)$

51) $11m^2 - 18m - 8$
$\quad = 11m^2 - 22m + 4m - 8$
$\quad = 11m(m-2) + 4(m-2)$
$\quad = (11m+4)(m-2)$

53) $6v^2 + 11v - 7 = 6v^2 - 3v + 14v - 7$
$\qquad = 3v(2v-1) + 7(2v-1)$
$\qquad = (3v+7)(2v-1)$

55) $10c^2 + 19c + 6$
$\quad = 10c^2 + 15c + 4c + 6$
$\quad = 5c(2c+3) + 2(2c+3)$
$\quad = (5c+2)(2c+3)$

57) $9x^2 - 13xy + 4y^2$
$\quad = 9x^2 - 9xy - 4xy + 4y^2$
$\quad = 9x(x-y) - 4y(x-y)$
$\quad = (9x - 4y)(x-y)$

59) because 2 can be factored out of $2x - 4$, but cannot be factored out of $2x^2 + 13x - 24$

61) $5w^2 + 11w + 6 = (5w+6)(w+1)$

63) $3u^2 - 23u + 30 = (3u-5)(u-6)$

65) $7k^2 + 15k - 18 = (7k-6)(k+3)$

67) $8r^2 + 26r + 15 = (4r+3)(2r+5)$

69) $6v^2 - 19v + 14 = (6v-7)(v-2)$

71) $10a^2 - 13ab + 4b^2 = (5a-4b)(2a-b)$

73) $6c^2 + 31cd + 18d^2 = (3c+2d)(2c+9d)$

75) $(n+5)^2 + 6(n+5) - 27$
Let $a = (n+5)$
$\quad = a^2 + 6a - 27$
$\quad = (a+9)(a-3)$
Replace a again with $(n+5)$
$\quad = (n+5+9)(n+5-3)$
$\quad = (n+14)(n+2)$

77) $(p-6)^2+11(p-6)+28$

Let $c = (p-6)$

$= c^2+11c+28$

$= (c+7)(c+4)$

Replace c again with $(p-6)$

$= (p-6+7)(p-6+4)$

$= (p+1)(p-2)$

79) $2(w+1)^2-13(w+1)+15$

Let $f = (w+1)$

$= 2f^2-13f+15$

$= (2f-3)(f-5)$

Replace f again with $(w+1)$

$= [2(w+1)-3](w+1-5)$

$= (2w-1)(w-4)$

81) $6(2y-1)^2-5(2y-1)-4$

Let $k = (2y-1)$

$= 6k^2-5k-4$

$= (3k-4)(2k+1)$

Replace k again with $(2y-1)$

$= [3(2y-1)-4][2(2y-1)+1]$

$= (6y-3-4)(4y-2+1)$

$= (6y-7)(4y-1)$

83) $4q^3-28q^2+48q$

$= 4q(q^2-7q+12)$

$= 4q(q-3)(q-4)$

85) $6+7t+t^2$

$= t^2+7t+6$

$= (t+6)(t+1)$

87) $12c^2+15c-18$

$= 3(4c^2+5c-6)$

$= 3(4c-3)(c+2)$

89) $3(b+5)^2+4(b+5)-20$

Let $z = (b+5)$

$= 3z^2+4z-20$

$= (3z+10)(z-2)$

Replace z again with $(b+5)$

$= [3(b+5)+10](b+5-2)$

$= (3b+15+10)(b+3)$

$= (3b+25)(b+3)$

91) $7s^2-17st+6t^2$

$= (7s-3t)(s-2t)$

93) $-10z^2+19z-6$

$= -\left(10z^2-19z+6\right)$

$= -\left(5z-2\right)\left(2z-3\right)$

95) c^2+6c-5

prime

Section 7.3: Special Factoring Techniques

97) $r^2 - 11r + 18$

$= (r-9)(r-2)$

99) $12p^2(q-1)^2 - 49p(q-1)^2 + 49(q-1)^2$

$= (q-1)^2(12p^2 - 49p + 49)$

$= (q-1)^2(3p-7)(4p-7)$

Section 7.3

1) a) $6^2 = 36$　　b) $10^2 = 100$

c) $4^2 = 16$　　d) $11^2 = 121$

e) $3^2 = 9$　　f) $8^2 = 64$

g) $12^2 = 144$　　h) $\left(\dfrac{1}{2}\right)^2 = \dfrac{1}{4}$

i) $\left(\dfrac{3}{5}\right)^2 = \dfrac{9}{25}$

3) a) $\left(n^2\right)^2 = n^4$　　b) $(5t)^2 = 25t^2$

c) $(7k)^2 = 49k^2$　　d) $\left(4p^2\right)^2 = 16p^4$

e) $\left(\dfrac{1}{3}\right)^2 = \dfrac{1}{9}$　　f) $\left(\dfrac{5}{2}\right)^2 = \dfrac{25}{4}$

5) $z^2 + 18z + 81$

7) The middle term does not equal $2(3c)(-4)$. It would have to equal $-24c$ to be a perfect trinomial.

9) $t^2 + 16t + 64 = (t)^2 + (2 \cdot t \cdot 8) + (8)^2$

$= (t+8)^2$

11) $g^2 - 18g + 81 = (g)^2 - (2 \cdot g \cdot 9) + (9)^2$

$= (g-9)^2$

13) $4y^2 + 12y + 9$

$= (2y)^2 + (2 \cdot 2y \cdot 3) + (3)^2$

$= (2y+3)^2$

15) $9k^2 - 24k + 16$

$= (3k)^2 - (2 \cdot 3k \cdot 4) + (4)^2$

$= (3k-4)^2$

17) $a^2 + \dfrac{2}{3}a + \dfrac{1}{9}$

$= (a)^2 + \left(2 \cdot a \cdot \dfrac{1}{3}\right) + \left(\dfrac{1}{3}\right)^2$

$= \left(a + \dfrac{1}{3}\right)^2$

19) $v^2 - 3v + \dfrac{9}{4}$

$= (v)^2 - \left(2 \cdot v \cdot \dfrac{3}{2}\right) + \left(\dfrac{3}{2}\right)^2$

$= \left(v - \dfrac{3}{2}\right)^2$

21) $x^2 + 6xy + 9y^2$

$= (x)^2 + (2 \cdot x \cdot 3y) + (3y)^2$

$= (x + 3y)^2$

23) $36t^2 - 60tu + 25u^2$

$= (6t)^2 - (2 \cdot 6t \cdot 5u) + (5u)^2$

$= (6t - 5u)^2$

25) $4f^2 + 24f + 36$

$= 4(f^2 + 6f + 9)$

$= 4\left[(f)^2 + (2 \cdot f \cdot 3) + (3)^2\right]$

$= 4(f + 3)^2$

27) $2p^4 - 24p^3 + 72p^2$

$= 2p^2(p^2 - 12p + 36)$

$= 2p^2\left[(p)^2 - (2 \cdot p \cdot 6) + (6)^2\right]$

$= 2p^2(p - 6)^2$

29) $-18d^2 - 60d - 50$

$= -2(9d^2 + 30d + 25)$

$= -2\left[(3d)^2 + (2 \cdot 3d \cdot 5) + (5)^2\right]$

$= -2(3d + 5)^2$

31) $12c^3 + 3c^2 + 27c = 3c(4c^2 + c + 9)$

33) a) $x^2 - 16$ b) $16 - x^2$

35) $x^2 - 9 = (x)^2 - (3)^2 = (x + 3)(x - 3)$

37) $n^2 - 121 = (n)^2 - (11)^2$

$= (n + 11)(n - 11)$

39) $m^2 + 64$ is prime.

41) $y^2 - \dfrac{1}{25} = (y)^2 - \left(\dfrac{1}{5}\right)^2$

$= \left(y + \dfrac{1}{5}\right)\left(y - \dfrac{1}{5}\right)$

43) $c^2 - \dfrac{9}{16} = (c)^2 - \left(\dfrac{3}{4}\right)^2$

$= \left(c + \dfrac{3}{4}\right)\left(c - \dfrac{3}{4}\right)$

45) $36 - h^2 = (6)^2 - (h)^2 = (6 + h)(6 - h)$

Section 7.3: Special Factoring Techniques

47) $169 - a^2 = (13)^2 - (a)^2$

$= (13 + a)(13 - a)$

49) $\dfrac{49}{64} - j^2 = \left(\dfrac{7}{8}\right)^2 - (j)^2$

$= \left(\dfrac{7}{8} + j\right)\left(\dfrac{7}{8} - j\right)$

51) $100m^2 - 49 = (10m)^2 - (7)^2$

$= (10m + 7)(10m - 7)$

53) $16p^2 - 81 = (4p)^2 - (9)^2$

$= (4p + 9)(4p - 9)$

55) $4t^2 + 25$ is prime.

57) $\dfrac{1}{4}k^2 - \dfrac{4}{9} = \left(\dfrac{1}{2}k\right)^2 - \left(\dfrac{2}{3}\right)^2$

$= \left(\dfrac{1}{2}k + \dfrac{2}{3}\right)\left(\dfrac{1}{2}k - \dfrac{2}{3}\right)$

59) $b^4 - 64 = (b^2)^2 - (8)^2$

$= (b^2 + 8)(b^2 - 8)$

61) $144m^2 - n^4 = (12m)^2 - (n^2)^2$

$= (12m + n^2)(12m - n^2)$

63) $r^4 - 1 = (r^2)^2 - (1)^2$

$= (r^2 + 1)(r^2 - 1)$

$= (r^2 + 1)\left[(r)^2 - (1)^2\right]$

$= (r^2 + 1)(r + 1)(r - 1)$

65) $16h^4 - g^4$

$= (4h^2)^2 - (g^2)^2$

$= (4h^2 + g^2)(4h^2 - g^2)$

$= (4h^2 + g^2)\left[(2h)^2 - (g)^2\right]$

$= (4h^2 + g^2)(2h + g)(2h - g)$

67) $4a^2 - 100 = 4(a^2 - 25)$

$= 4\left[(a)^2 - (5)^2\right]$

$= 4(a + 5)(a - 5)$

69) $2m^2 - 128 = 2(m^2 - 64)$

$= 2\left[(m)^2 - (8)^2\right]$

$= 2(m + 8)(m - 8)$

71) $45r^4 - 5r^2 = 5r^2(9r^2 - 1)$

$= 5r^2\left[(3r)^2 - (1)^2\right]$

$= 5r^2(3r + 1)(3r - 1)$

73) a) $4^3 = 64$ b) $1^3 = 1$

c) $10^3 = 1000$ d) $3^3 = 27$

e) $5^3 = 125$ f) $2^3 = 8$

75) a) $(y)^3 = y^3$ b) $(2c)^3 = 8c^3$

c) $(5r)^3 = 125r^3$ d) $(x^2)^3 = x^6$

77) $(x^2 - 3x + 9)$

79) $d^3 + 1 = (d)^3 + (1)^3$
$= (d+1)(d^2 - d + 1)$

81) $p^3 - 27 = (p)^3 - (3)^3$
$= (p-3)(p^2 + 3p + 9)$

83) $k^3 + 64 = (k)^3 + (4)^3$
$= (k+4)(k^2 - 4k + 16)$

85) $27m^3 - 125$
$= (3m)^3 - (5)^3$
$= (3m-5)(9m^2 + 15m + 25)$

87) $125y^3 - 8 = (5y)^3 - (2)^3$
$= (5y-2)(25y^2 + 10y + 4)$

89) $1000c^3 - d^3$
$= (10c)^3 - (d)^3$
$= (10c-d)(100c^2 + 10cd + d^2)$

91) $8j^3 + 27k^3$
$= (2j)^3 + (3k)^3$
$= (2j+3k)(4j^2 - 6jk + 9k^2)$

93) $64x^3 + 125y^3$
$= (4x)^3 + (5y)^3$
$= (4x+5y)(16x^2 - 20xy + 25y^2)$

95) $6c^3 + 48 = 6(c^3 + 8)$
$= 6[(c)^3 + (2)^3]$
$= 6(c+2)(c^2 - 2c + 4)$

97) $7v^3 - 7000w^3$

$= 7(v^3 - 1000w^3)$

$= 7\left[(v)^3 - (10w)^3\right]$

$= 7(v - 10w)(v^2 + 10vw + 100w^2)$

99) $h^6 - 64 = \left(h^3\right)^2 - (8)^2 = \left(h^3 + 8\right)\left(h^3 - 8\right) = \left[(h)^3 + (2)^3\right]\left[(h)^3 - (2)^3\right]$

$= (h+2)\left(h^2 - 2h + 4\right)(h-2)\left(h^2 + 2h + 4\right)$

101) $(x+5)^2 - (x-2)^2 = \left[(x+5) + (x-2)\right]\left[(x+5) - (x-2)\right]$

$= (2x+3)(x+5-x+2)$

$= 7(2x+3)$

103) $(2p+3)^2 - (p+4)^2 = \left[(2p+3) + (p+4)\right]\left[(2p+3) - (p+4)\right]$

$= (3p+7)(2p+3-p-4) = (3p+7)(p-1)$

105) $(t+5)^3 + 8 = (t+5)^3 + (2)^3 = \left[(t+5) + 2\right]\left[(t+5)^2 - 2(t+5) + 4\right]$

$= (t+7)\left(t^2 + 10t + 25 - 2t - 10 + 4\right) = (t+7)\left(t^2 + 8t + 19\right)$

107) $(k-9)^3 - 1 = (k-9)^3 - (1)^3 = \left[(k-9) - 1\right]\left[(k-9)^2 + 1(k-9) + 1\right]$

$= (k-10)\left(k^2 - 18k + 81 + k - 9 + 1\right)$

$= (k-10)\left(k^2 - 17k + 73\right)$

Chapter 7: Putting It All Together

1) $m^2 + 16m + 60 = (m+10)(m+6)$

3) $uv + 6u + 9v + 54$
$= u(v+6) + 9(v+6)$
$= (u+9)(v+6)$

5) $3k^2 - 14k + 8 = (3k-2)(k-4)$

7) $16d^6 + 8d^5 + 72d^4$
$= 8d^4(2d^2 + d + 9)$

9) $60w^3 + 70w^2 - 50w$
$= 10w(6w^2 + 7w - 5)$
$= 10w(3w+5)(2w-1)$

11) $t^3 + 1000 = (t)^3 + (10)^3$
$= (t+10)(t^2 - 10t + 100)$

13) $49 - p^2 = (7)^2 - (p)^2$
$= (7+p)(7-p)$

15) $4x^2 + 4xy + y^2$
$= (2x)^2 + (2 \cdot 2x \cdot y) + (y)^2$
$= (2x+y)^2$

17) $3z^4 - 21z^3 - 24z^2 = 3z^2(z^2 - 7z - 8)$
$= 3z^2(z-8)(z+1)$

19) $4b^2 + 1$ is prime

21) $40x^3 - 135 = 5(8x^3 - 27)$
$= 5[(2x)^3 - (3)^3]$
$= 5(2x-3)(4x^2 + 6x + 9)$

23) $c^2 - \dfrac{1}{4} = (c)^2 - \left(\dfrac{1}{2}\right)^2$
$= \left(c+\dfrac{1}{2}\right)\left(c-\dfrac{1}{2}\right)$

25) $45s^2t + 4 - 36s^2 - 5t$
$= 45s^2t - 36s^2 - 5t + 4$
$= 9s^2(5t-4) - 1(5t-4)$
$= (9s^2 - 1)(5t-4)$
$= (3s+1)(3s-1)(5t-4)$

27) $k^2 + 9km + 18m^2 = (k+3m)(k+6m)$

29) $z^2 - 3z - 88 = (z-11)(z+8)$

31) $80y^2 - 40y + 5$

$= 5(16y^2 - 8y + 1)$

$= 5\left[(4y)^2 - (2 \cdot 4y \cdot 1) + (1)^2\right]$

$= 5(4y - 1)^2$

33) $20c^2 + 26cd + 6d^2$

$= 2(10c^2 + 13cd + 3d^2)$

$= 2(10c + 3d)(c + d)$

35) $n^4 - 16m^4 = (n^2)^2 - (4m^2)^2$

$= (n^2 + 4m^2)(n^2 - 4m^2)$

$= (n^2 + 4m^2)\left[(n)^2 - (2m)^2\right]$

$= (n^2 + 4m^2)(n + 2m)(n - 2m)$

37) $2a^2 - 10a - 72 = 2(a^2 - 5a - 36)$

$= 2(a - 9)(a + 4)$

39) $r^2 - r + \dfrac{1}{4} = (r)^2 - \left(2 \cdot r \cdot \dfrac{1}{2}\right) + \left(\dfrac{1}{2}\right)^2$

$= \left(r - \dfrac{1}{2}\right)^2$

41) $28gh + 16g - 63h - 36$

$= 4g(7h + 4) - 9(7h + 4)$

$= (4g - 9)(7h + 4)$

43) $8b^2 - 14b - 15 = (4b + 3)(2b - 5)$

45) $55a^6b^3 + 35a^5b^3 - 10a^4b - 20a^2b$

$= 5a^2b(11a^4b^2 + 7a^3b^2 - 2a^2 - 4)$

47) $2d^2 - 9d + 3$ is prime

49) $9p^2 - 24pq + 16q^2$

$= (3p)^2 - (2 \cdot 3p \cdot 4q) + (4q)^2$

$= (3p - 4q)^2$

51) $30y^2 + 37y - 7 = (6y - 1)(5y + 7)$

53) $80a^3 - 270b^3$

$= 10(8a^3 - 27b^3)$

$= 10\left[(2a)^3 - (3b)^3\right]$

$= 10(2a - 3b)(4a^2 + 6ab + 9b^2)$

55) $rt - r - t + 1 = r(t - 1) - 1(t - 1)$

$= (r - 1)(t - 1)$

57) $4g^2 - 4 = 4(g^2 - 1)$

$= 4\left[(g)^2 - (1)^2\right]$

$= 4(g + 1)(g - 1)$

59) $3c^2 - 24c + 48$

$= 3\left(c^2 - 8c + 16\right)$

$= 3\left[(c)^2 - (2 \cdot c \cdot 4) + (4)^2\right]$

$= 3(c - 4)^2$

61) $144k^2 - 121 = (12k)^2 - (11)^2$

$= (12k + 11)(12k - 11)$

63) $-48g^2 - 80g - 12$

$= -4\left(12g^2 + 20g + 3\right)$

$= -4(6g + 1)(2g + 3)$

65) $q^3 + 1 = (q)^3 + (1)^3$

$= (q + 1)\left(q^2 - q + 1\right)$

67) $81u^4 - v^4$

$= \left(9u^2\right)^2 - \left(v^2\right)^2$

$= \left(9u^2 + v^2\right)\left(9u^2 - v^2\right)$

$= \left(9u^2 + v^2\right)\left[(3u)^2 - (v)^2\right]$

$= \left(9u^2 + v^2\right)(3u + v)(3u - v)$

69) $11f^2 + 36f + 9 = (11f + 3)(f + 3)$

71) $2j^{11} - j^3 = j^3\left(2j^8 - 1\right)$

73) $w^2 - 2w - 48 = (w - 8)(w + 6)$

75) $k^2 + 100$ is prime

77) $m^2 + 4m + 4 = (m)^2 + (2 \cdot m \cdot 2) + (2)^2$

$= (m + 2)^2$

79) $9t^2 - 64 = (3t)^2 - (8)^2$

$= (3t + 8)(3t - 8)$

81) $(2z + 1)y^2 + 6(2z + 1)y - 55(2z + 1)$

$= (2z + 1)\left(y^2 + 6y - 55\right)$

$= (2z + 1)(y + 11)(y - 5)$

83) $(r - 4)^2 + 11(r - 4) + 28$

$= \left[(r - 4) + 4\right]\left[(r - 4) + 7\right]$

$= r(r + 3)$

85) $(3p - 4)^2 - 5(3p - 4) - 36$

$= \left[(3p - 4) - 9\right]\left[(3p - 4) + 4\right]$

$= 3p(3p - 13)$

87) $(4k + 1)^2 - (3k + 2)^2$

$= \left[(4k + 1) + (3k + 2)\right]\left[(4k + 1) - (3k + 2)\right]$

$= (4k + 1 + 3k + 2)(4k + 1 - 3k - 2)$

$= (7k + 3)(k - 1)$

89) $(x+y)^2 - (2x-y)^2$

$= \left[(x+y)+(2x-y)\right]\left[(x+y)-(2x-y)\right]$

$= (3x)(x+y-2x+y)$

$= 3x(-x+2y)$

$= -3x(x-2y)$

91) $n^2 + 12n + 36 - p^2$

$= \left(n^2 + 12n + 36\right) - p^2$

$= (n+6)^2 - p^2$

$= (n+6+p)(n+6-p)$

$= (n+p+6)(n-p+6)$

93) $x^2 - 2xy + y^2 - z^2$

$= \left(x^2 - 2xy + y^2\right) - z^2$

$= (x-y)^2 - z^2$

$= (x-y+z)(x-y-z)$

Section 7.4

1)

If the product of two quantities equals 0, then one or both of the quantities must be zero.

3) $(m+9)(m-8) = 0$

$m+9=0$ or $m-8=0$

$m=-9$ or $m=8$ $\{-9, 8\}$

5) $(q-4)(q-7) = 0$

$q-4=0$ or $q-7=0$

$q=4$ or $q=7$ $\{4, 7\}$

7) $(4z+3)(z-9) = 0$

$4z+3=0$ or $z-9=0$

$4z=-3$

$z=-\dfrac{3}{4}$ or $z=9$ $\left\{-\dfrac{3}{4}, 9\right\}$

9) $11s(s+15) = 0$

$11s=0$ or $s+15=0$

$s=0$ or $s=-15$ $\{-15, 0\}$

11) $(6x-5)^2 = 0$

$6x-5=0$

$6x=5$

$x=\dfrac{5}{6}$ $\left\{\dfrac{5}{6}\right\}$

13) $(4h+7)(h+3) = 0$

$4h+7=0$ or $h+3=0$

$4h=-7$

$h=-\dfrac{7}{4}$ or $h=-3$ $\left\{-3, -\dfrac{7}{4}\right\}$

15) $\left(y+\dfrac{3}{2}\right)\left(y-\dfrac{1}{4}\right)=0$

$y+\dfrac{3}{2}=0$ or $y-\dfrac{1}{4}=0$

$y=-\dfrac{3}{2}$ or $y=\dfrac{1}{4}$ $\quad\left\{-\dfrac{3}{2},\dfrac{1}{4}\right\}$

27) $14w^2+8w=0$

$2w(7w+4)=0$

$2w=0$ or $7w+4=0$

$w=0 \qquad 7w=-4$

$\qquad\qquad w=-\dfrac{4}{7} \quad \left\{-\dfrac{4}{7},0\right\}$

17) $q(q-2.5)=0$

$q=0$ or $q-2.5=0$

$q=2.5 \qquad \{0,2.5\}$

29) $\qquad d^2-15d=-54$

$d^2-15d+54=0$

$(d-6)(d-9)=0$

$d-6=0$ or $d-9=0$

$d=6$ or $\quad d=9 \qquad \{6,9\}$

19) No; the product of the factors must equal zero.

31) $\qquad t^2-49=0$

$(t+7)(t-7)=0$

$t+7=0$ or $t-7=0$

$t=-7$ or $\quad t=7 \qquad \{-7,7\}$

21) $v^2+15v+56=0$

$(v+8)(v+7)=0$

$v+8=0$ or $v+7=0$

$v=-8$ or $\quad v=-7 \quad \{-8,-7\}$

33) $36=25n^2$

$0=25n^2-36$

$0=(5n+6)(5n-6)$

$5n+6=0$ or $5n-6=0$

$5n=-6 \qquad 5n=6$

$n=-\dfrac{6}{5}$ or $\quad n=\dfrac{6}{5} \quad \left\{-\dfrac{6}{5},\dfrac{6}{5}\right\}$

23) $k^2+12k-45=0$

$(k+15)(k-3)=0$

$k+15=0$ or $\quad k-3=0$

$k=-15$ or $\quad k=3 \quad \{-15,3\}$

25) $3y^2-y-10=0$

$(3y+5)(y-2)=0$

$3y+5=0$ or $y-2=0$

$3y=-5$

$y=-\dfrac{5}{3}$ or $\quad y=2 \quad \left\{-\dfrac{5}{3},2\right\}$

35) $\qquad m^2=60-7m$

$m^2+7m-60=0$

$(m+12)(m-5)=0$

$m+12=0$ or $m-5=0$

$m=-12$ or $m=5 \quad \{-12,5\}$

Section 7.4: Solving Quadratic Equations by Factoring

37) $$55w = -20w^2 - 30$$

$20w^2 + 55w + 30 = 0 \qquad \text{divide by 5}$

$4w^2 + 11w + 6 = 0$

$(4w + 3)(w + 2) = 0$

$4w + 3 = 0 \ \text{ or } \ w + 2 = 0$

$4w = -3$

$w = -\dfrac{3}{4} \ \text{ or } \ w = -2 \qquad \left\{ -2, -\dfrac{3}{4} \right\}$

45) $-63 = 4j(j - 8)$

$-63 = 4j^2 - 32j$

$0 = 4j^2 - 32j + 63$

$0 = (2j - 7)(2j - 9)$

$2j - 7 = 0 \ \text{ or } \ 2j - 9 = 0$

$2j = 7 \qquad\qquad 2j = 9$

$j = \dfrac{7}{2} \ \text{ or } \ j = \dfrac{9}{2} \qquad \left\{ \dfrac{7}{2}, \dfrac{9}{2} \right\}$

39) $$p^2 = 11p$$

$p^2 - 11p = 0$

$p(p - 11) = 0$

$p = 0 \ \text{ or } \ p - 11 = 0$

$p = 11 \qquad\qquad \{0, 11\}$

47) $10x(x + 1) - 6x = 9(x^2 + 5)$

$10x^2 + 10x - 6x = 9x^2 + 45$

$10x^2 + 4x = 9x^2 + 45$

$x^2 + 4x - 45 = 0$

$(x + 9)(x - 5) = 0$

$x + 9 = 0 \ \text{ or } \ x - 5 = 0$

$x = -9 \ \text{ or } \ x = 5 \qquad \{-9, 5\}$

41) $$45k + 27 = 18k^2$$

$-18k^2 + 45k + 27 = 0 \qquad \text{divide by } -9$

$2k^2 - 5k - 3 = 0$

$(2k + 1)(k - 3) = 0$

$2k + 1 = 0 \ \text{ or } \ k - 3 = 0$

$2k = -1$

$k = -\dfrac{1}{2} \ \text{ or } \ k = 3 \qquad \left\{ -\dfrac{1}{2}, 3 \right\}$

49) $$3(h^2 - 4) = 5h(h - 1) - 9h$$

$3h^2 - 12 = 5h^2 - 5h - 9h$

$3h^2 - 12 = 5h^2 - 14h$

$-2h^2 + 14h - 12 = 0 \qquad \text{divide by } -2$

$h^2 - 7h + 6 = 0$

$(h - 1)(h - 6) = 0$

$h - 1 = 0 \ \text{ or } \ h - 6 = 0$

$h = 1 \ \text{ or } \quad h = 6 \qquad \{1, 6\}$

43) $$b(b - 4) = 96$$

$b^2 - 4b = 96$

$b^2 - 4b - 96 = 0$

$(b + 8)(b - 12) = 0$

$b + 8 = 0 \ \text{ or } \ b - 12 = 0$

$b = -8 \ \text{ or } \quad b = 12 \qquad \{-8, 12\}$

51) $\dfrac{1}{2}(m+1)^2 = -\dfrac{3}{4}m(m+5) - \dfrac{5}{2}$

$4\left[\dfrac{1}{2}(m+1)^2\right] = 4\left[-\dfrac{3}{4}m(m+5) - \dfrac{5}{2}\right]$

$2(m+1)^2 = -3m(m+5) - 10$

$2(m^2 + 2m + 1) = -3m^2 - 15m - 10$

$2m^2 + 4m + 2 = -3m^2 - 15m - 10$

$5m^2 + 19m + 12 = 0$

$(5m+4)(m+3) = 0$

$5m + 4 = 0$ or $m + 3 = 0$

$5m = -4$

$m = -\dfrac{4}{5}$ or $m = -3$ $\left\{-3, -\dfrac{4}{5}\right\}$

53) $3t(t-5) + 14 = 5 - t(t+3)$

$3t^2 - 15t + 14 = 5 - t^2 - 3t$

$4t^2 - 12t + 9 = 0$

$(2t-3)^2 = 0$

$2t - 3 = 0$

$2t = 3$

$t = \dfrac{3}{2}$ $\left\{\dfrac{3}{2}\right\}$

55) $33 = -m(14+m)$

$33 = -14m - m^2$

$m^2 + 14m + 33 = 0$

$(m+11)(m+3) = 0$

$m + 11 = 0$ or $m + 3 = 0$

$m = -11$ or $m = -3$ $\{-11, -3\}$

57) $(3w+2)^2 - (w-5)^2 = 0$

$(3w+2+w-5)(3w+2-(w-5)) = 0$

$(4w-3)(2w+7) = 0$

$4w - 3 = 0$ or $2w + 7 = 0$

$4w = 3$ $2w = -7$

$w = \dfrac{3}{4}$ o $w = -\dfrac{7}{2}$ $\left\{-\dfrac{7}{2}, \dfrac{3}{4}\right\}$

59) $(q+3)^2 - (2q-5)^2 = 0$

$(q+3+2q-5)(q+3-(2q-5)) = 0$

$(3q-2)(-q+8) = 0$

$3q - 2 = 0$ or $-q + 8 = 0$

$3q = 2$ $-q = -8$

$q = \dfrac{2}{3}$ or $q = 8$ $\left\{\dfrac{2}{3}, 8\right\}$

61) $8y(y+4)(2y-1) = 0$

$8y = 0$ or $y + 4 = 0$ or $2y - 1 = 0$

$y = 0$ $y = -4$ $2y = 1$

$y = \dfrac{1}{2}$

$\left\{-4, 0, \dfrac{1}{2}\right\}$

63) $(9p-2)(p^2 - 10p - 11) = 0$

$(9p-2)(p+1)(p-11) = 0$

$9p - 2 = 0$ or $p + 1 = 0$ or $p - 11 = 0$

$9p = 2$

$p = \dfrac{2}{9}$ or $p = -1$ or $p = 11$

$\left\{-1, \dfrac{2}{9}, 11\right\}$

65) $(2r-5)(r^2-6r+9)=0$

$\qquad (2r-5)(r-3)^2=0$

$\qquad 2r-5=0 \ \text{ or } \ r-3=0$

$\qquad 2r=5$

$\qquad r=\dfrac{5}{2} \ \text{ or } \ r=3 \qquad \left\{\dfrac{5}{2},3\right\}$

67) $\qquad\qquad m^3=64m$

$\qquad m^3-64m=0$

$\qquad m(m^2-64)=0$

$\qquad m(m+8)(m-8)=0$

$\qquad m=0 \ \text{ or } \ m+8=0 \ \text{ or } \ m-8=0$

$\qquad\qquad\qquad m=-8 \qquad m=8$

$\{-8,0,8\}$

69) $\qquad 5w^2+36w=w^3$

$\qquad -w^3+5w^2+36w=0$

$\qquad -w(w^2-5w-36)=0$

$\qquad -w(w+4)(w-9)=0$

$\qquad -w=0 \ \text{ or } \ w+4=0 \ \text{ or } \ w-9=0$

$\qquad w=0 \quad \text{ or } \quad w=-4 \ \text{ or } \ w=9$

$\{-4,0,9\}$

71) $\qquad\qquad 2g^3=120g-14g^2$

$\qquad 2g^3+14g^2-120g=0$

$\qquad 2g(g^2+7g-60)=0$

$\qquad 2g(g+12)(g-5)=0$

$\qquad 2g=0 \ \text{ or } \ g+12=0 \ \text{ or } \ g-5=0$

$\qquad g=0 \ \text{ or } \quad g=-12 \ \text{ or } \ g=5$

$\{-12,0,5\}$

73) $45h=20h^3$

$\qquad 0=20h^3-45h$

$\qquad 0=5h(4h^2-9)$

$\qquad 0=5h(2h+3)(2h+3)$

$\qquad 5h=0 \ \text{ or } \ 2h+3=0 \ \text{ or } \ 2h-3=0$

$\qquad h=0 \qquad\qquad 2h=-3 \qquad\qquad 2h=3$

$\qquad\qquad\qquad\qquad h=-\dfrac{3}{2} \qquad\qquad h=\dfrac{3}{2}$

$\left\{-\dfrac{3}{2},0,\dfrac{3}{2}\right\}$

75) $2s^2(3s+2)+3s(3s+2)-35(3s+2)=0$

$\qquad (2s^2+3s-35)(3s+2)=0$

$\qquad (2s-7)(s+5)(3s+2)=0$

$\qquad 2s-7=0 \ \text{ or } \ s+5=0 \ \text{ or } \ 3s+2=0$

$\qquad 2s=7 \qquad\qquad\qquad\qquad 3s=-2$

$\qquad s=\dfrac{7}{2} \ \text{ or } \ s=-5 \quad \text{ or } \quad s=-\dfrac{2}{3}$

$\left\{-5,-\dfrac{2}{3},\dfrac{7}{2}\right\}$

77)

$\qquad 10a^2(4a+3)+2(4a+3)=9a(4a+3)$

$\qquad 10a^2(4a+3)-9a(4a+3)+2(4a+3)=0$

$\qquad (10a^2-9a+2)(4a+3)=0$

$\qquad (2a-1)(5a-2)(4a+3)=0$

$\qquad 2a-1=0 \ \text{ or } \ 5a-2=0 \ \text{ or } \ 4a+3=0$

$\qquad 2a=1 \qquad\qquad 5a=2 \qquad\qquad 4a=-3$

$\qquad a=\dfrac{1}{2} \ \text{ or } \ a=\dfrac{2}{5} \ \text{ or } \ a=-\dfrac{3}{4}$

$\left\{-\dfrac{3}{4},\dfrac{2}{5},\dfrac{1}{2}\right\}$

79) $t^3 + 6t^2 - 4t - 24 = 0$

$t^2(t+6) - 4(t+6) = 0$

$(t^2 - 4)(t+6) = 0$

$(t+2)(t-2)(t+6) = 0$

$t+2 = 0$ or $t-2 = 0$ or $t+6 = 0$

$t = -2$ $t = 2$ $t = -6$

$\{-6, -2, 2\}$

81) $f(x) = 0$

$f(x) = x^2 + 10x + 21$

$0 = x^2 + 10x + 21$

$0 = (x+7)(x+3)$

$x+7 = 0$ or $x+3 = 0$

$x = -7$ or $x = -3$

83) $g(a) = 4$

$g(a) = 2a^2 - 13a + 24$

$4 = 2a^2 - 13a + 24$

$0 = 2a^2 - 13a + 20$

$0 = (2a-5)(a-4)$

$2a - 5 = 0$ or $a - 4 = 0$

$2a = 5$

$a = \dfrac{5}{2}$ or $a = 4$

85) $H(b) = 19$

$H(b) = b^2 + 3$

$19 = b^2 + 3$

$0 = b^2 - 16$

$0 = (b+4)(b-4)$

$b+4 = 0$ or $b-4 = 0$

$b = -4$ or $b = 4$

87) $h(k) = 0$

$h(k) = 5k^3 - 25k^2 + 20k$

$0 = 5k^3 - 25k^2 + 20k$

$0 = 5k(k^2 - 5k + 4)$

$0 = 5k(k-1)(k-4)$

$5k = 0$ or $k-1 = 0$ or $k-4 = 0$

$k = 0$ or $k = 1$ or $k = 4$

Section 7.5

1) $x = $ length of rectangle

$x - 9 = $ width of rectangle

Area $= ($length$)($width$)$

$36 = x(x-9)$

$36 = x^2 - 9x$

$0 = x^2 - 9x - 36$

$0 = (x-12)(x+3)$

$x - 12 = 0$ or $x + 3 = 0$

$x = 12$ $x = -3$

length $= 12$ in; width $= 12 - 9 = 3$ in

Section 7.5: Applications of Quadratic Equations

3) $2x - 1 =$ base of triangle

$x + 6 =$ height of triangle

$\text{Area} = \dfrac{1}{2}(\text{base})(\text{height})$

$12 = \dfrac{1}{2}(2x - 1)(x + 6)$

$24 = 2x^2 + 11x - 6$

$0 = 2x^2 + 11x - 30$

$0 = (2x + 15)(x - 2)$

$2x + 15 = 0 \ \text{ or } \ x - 2 = 0$

$\quad 2x = -15 \qquad x = 2$

$\quad x = -\dfrac{15}{2}$

base $= 2(2) - 1 = 3$ cm;

height $= 2 + 6 = 8$ cm

5) $x + 1 =$ base of parallelogram

$x - 2 =$ height of parallelogram

$\text{Area} = (\text{base})(\text{height})$

$18 = (x + 1)(x - 2)$

$18 = x^2 - x - 2$

$0 = x^2 - x - 20$

$0 = (x - 5)(x + 4)$

$x - 5 = 0 \ \text{ or } \ x + 4 = 0$

$\quad x = 5 \qquad\qquad x = -4$

base $= 5 + 1 = 6$ in;

width $= 5 - 2 = 3$ in

7) $3x + 1 =$ length of box

$2x =$ width of box

$\text{Volume} = (\text{length})(\text{width})(\text{height})$

$240 = (3x + 1)(2x)(4)$

$240 = (6x^2 + 2x)4$

$60 = 6x^2 + 2x$

$0 = 6x^2 + 2x - 60 \qquad$ divide by 2

$0 = 3x^2 + x - 30$

$0 = (3x + 10)(x - 3)$

$3x + 10 = 0 \ \text{ or } \ x - 3 = 0$

$\quad 3x = -10 \qquad x = 3$

$\quad x = -\dfrac{10}{3}$

length $= 3(3) + 1 = 10$ in;

width $= 2(3) = 6$ in

9) $w =$ the width of the rug

$w + 4 =$ the length of the rug

$\text{Area} = (\text{Length})(\text{Width})$

$45 = (w + 4) \cdot w$

$45 = w^2 + 4w$

$0 = w^2 + 4w - 45$

$0 = (w + 9)(w - 5)$

$w + 9 = 0 \quad \text{ or } \quad w - 5 = 0$

$\quad w = -9 \qquad\qquad w = 5$

width $= 5$ ft; length $= 5 + 4 = 9$ ft

11) l = the length of the glass

$l - 3$ = the width of the glass

Area = (Length)(Width)

$54 = l \cdot (l - 3)$

$54 = l^2 - 3l$

$0 = l^2 - 3l - 54$

$0 = (l - 9)(l + 6)$

$l - 9 = 0$ or $l + 6 = 0$

$l = 9$ $l = -6$

length = 9 in; width = 9 - 3 = 6 in

13) w = width of box

$\dfrac{w}{2}$ = height of box

Volume = (length)(width)(height)

$1440 = 20 \cdot w \cdot \dfrac{w}{2}$

$1440 = 10w^2$

$144 = w^2$

$0 = w^2 - 144$

$0 = (w + 12)(w - 12)$

$w + 12 = 0$ or $w - 12 = 0$

$w = -12$ $w = 12$

width = 12 in; height = $\dfrac{12}{2}$ = 6 in

15) b = base of triangle

$b + 3$ = height of triangle

Area = $\dfrac{1}{2}$(base)(height)

$35 = \dfrac{1}{2}(b)(b + 3)$

$70 = b^2 + 3b$

$0 = b^2 + 3b - 70$

$0 = (b + 10)(b - 7)$

$b + 10 = 0$ or $b - 7 = 0$

$b = -10$ $b = 7$

base = 7 cm; height = 7 + 3 = 10 cm

17) x = the first odd integer

$x + 2$ = the second odd integer

$x(x + 2) = 3(x + x + 2) - 1$

$x^2 + 2x = 3(2x + 2) - 1$

$x^2 + 2x = 6x + 6 - 1$

$x^2 - 4x - 5 = 0$

$(x - 5)(x + 1) = 0$

$x - 5 = 0$ or $x + 1 = 0$

$x = 5$ $x = -1$

$x = 5$, then $5 + 2 = 7$

$x = -1$, then $-1 + 2 = 1$

5 and 7   or -1 and 1

19) x = the first even integer

$x + 2$ = the second even integer

$x + 4$ = the third even integer

$x + (x + 2) = \dfrac{1}{4}(x + 2)(x + 4)$

$2x + 2 = \dfrac{1}{4}(x^2 + 6x + 8)$

$8x + 8 = x^2 + 6x + 8$

$0 = x^2 - 2x$

$0 = x(x - 2)$

$x = 0$ or $x - 2 = 0$

$x = 2$

$x = 0$, then $0 + 2 = 2$ and $0 + 4 = 4$

$x = 2$, then $2 + 2 = 4$ and $2 + 4 = 6$

0, 2, 4   or   2, 4, 6

21) $x =$ the first integer

$x+1 =$ the second integer

$x+2 =$ the third integer

$$(x+2)^2 = x(x+1)+22$$

$$x^2+4x+4 = x^2+x+22$$

$$3x-18 = 0$$

$$3x = 18$$

$$x = 6$$

$x = 6$, then $6+1 = 7$ and $6+2 = 8$

$6, 7, 8$

23) Answers may vary.

25) $a^2+b^2 = c^2$

$$a^2+(12)^2 = (15)^2$$

$$a^2+144 = 225$$

$$a^2-81 = 0$$

$$(a+9)(a-9) = 0$$

$a+9 = 0$ or $a-9 = 0$

$a = -9$ $a = 9$

The length of the missing side is 9.

27) $a^2+b^2 = c^2$

$$a^2+(8)^2 = (17)^2$$

$$a^2+64 = 289$$

$$a^2-225 = 0$$

$$(a+15)(a-15) = 0$$

$a+15 = 0$ or $a-15 = 0$

$a = -15$ $a = 15$

The length of the missing side is 15.

29) $a^2+b^2 = c^2$

$$(8)^2+(6)^2 = c^2$$

$$64+36 = c^2$$

$$100 = c^2$$

$$0 = c^2-100$$

$$0 = (c+10)(c-10)$$

$c+10 = 0$ or $c-10 = 0$

$c = -10$ $c = 10$

The length of the missing side is 10.

31) $3x-1 =$ length of the longer leg

$2x =$ length of the shorter leg

$3x+1 =$ length of the hypotenuse

$$(3x-1)^2+(2x)^2 = (3x+1)^2$$

$$9x^2-6x+1+4x^2 = 9x^2+6x+1$$

$$13x^2-6x+1 = 9x^2+6x+1$$

$$4x^2-12x = 0$$

$$4x(x-3) = 0$$

$4x = 0$ or $x-3 = 0$

$x = 0$ $x = 3$

$2x = 2(3) = 6$

$3x-1 = 3(3)-1 = 8$

$3x+1 = 3(3)+1 = 10$

The length of the longer leg is 8.

The length of the shorter leg is 6.

The length of the hypotenuse is 10.

33) $x+2 = $ length of the longer leg

$\dfrac{1}{2}x = $ length of the shorter leg

$x+3 = $ length of the hypotenuse

$$(x+2)^2 + \left(\dfrac{1}{2}x\right)^2 = (x+3)^2$$

$$x^2 + 4x + 4 + \dfrac{1}{4}x^2 = x^2 + 6x + 9$$

$$\dfrac{5}{4}x^2 + 4x + 4 = x^2 + 6x + 9$$

$$\dfrac{1}{4}x^2 - 2x - 5 = 0$$

$$4\left(\dfrac{1}{4}x^2 - 2x - 5\right) = 4(0)$$

$$x^2 - 8x - 20 = 0$$

$$(x-10)(x+2) = 0$$

$$x - 10 = 0 \quad \text{or} \quad x + 2 = 0$$

$$x = 10 \quad \text{or} \quad x = -2$$

$$\dfrac{1}{2}x = \dfrac{1}{2}(10) = 5$$

$$x + 2 = 10 + 2 = 12$$

$$x + 3 = 10 + 3 = 13$$

The length of the longer leg is 12.

The length of the shorter leg is 5.

The length of the hypotenuse is 13.

35) $x = $ length of the longer leg

$x - 2 = $ length of the shorter leg

$x + 2 = $ length of the hypotenuse

$$x^2 + (x-2)^2 = (x+2)^2$$

$$x^2 + x^2 - 4x + 4 = x^2 + 4x + 4$$

$$2x^2 - 4x + 4 = x^2 + 4x + 4$$

$$x^2 - 8x = 0$$

$$x(x-8) = 0$$

$$x = 0 \quad \text{or} \quad x - 8 = 0$$

$$x = 8$$

longer leg $= 8$ in.

37) $x = $ distance from bottom of the ladder to the wall.

$x + 7 = $ distance from the floor to the top of the ladder.

The length of the ladder is 13, so $c = 13$. Let $a = x$ and $b = x + 7$.

$$a^2 + b^2 = c^2$$

$$(x)^2 + (x+7)^2 = (13)^2$$

$$x^2 + x^2 + 14x + 49 = 169$$

$$2x^2 + 14x + 49 = 169$$

$$2x^2 + 14x - 120 = 0$$

$$x^2 + 7x - 60 = 0$$

$$(x+12)(x-5) = 0$$

$$x + 12 = 0 \quad \text{or} \quad x - 5 = 0$$

$$x = -12 \qquad\qquad x = 5$$

The distance from the bottom of the ladder to the wall is 5 ft.

Section 7.5: Applications of Quadratic Equations

39)

$a = x =$ Rana's distance from the bike shop

$b = 4 =$ Yasmeen's distance from the bike shop

$c = x + 2 =$ distance between Rana and Yasmeen

$a^2 + b^2 = c^2$

$4^2 + x^2 = (x+2)^2$

$16 + x^2 = x^2 + 4x + 4$

$\quad 12 = 4x$

$\quad\; 3 = x$

$c = 3 + 2 = 5$ miles

They are 5 miles apart.

41) a) Let $t = 0$ and solve for h.

$h = -16(0)^2 + 144$

$h = -16(0) + 144$

$h = 144$

The initial height of the rock

is 144 ft.

b) Let $h = 80$ and solve for t.

$80 = -16t^2 + 144$

$0 = -16t^2 + 64$

$0 = t^2 - 4$

$0 = (t+2)(t-2)$

$t + 2 = 0 \quad$ or $\quad t - 2 = 0$

$\quad t = -2 \qquad\qquad t = 2$

The rock will be 80 ft above the

water after 2 seconds.

c) Let $h = 0$ and solve for t.

$0 = -16t^2 + 144 \quad$ divide by -16

$0 = t^2 - 9$

$0 = (t+3)(t-3)$

$t + 3 = 0 \quad$ or $\quad t - 3 = 0$

$\quad t = -3 \qquad\qquad t = 3$

The rock will hit the water

after 3 seconds.

43) a) Let $t = 3$.

$y = -16(3)^2 + 144(3)$

$y = -16(9) + 432$

$y = 288$ ft

b) Let $t = 3$.

$x = 39(3)$

$x = 117$ ft

c) Let $t = 4.5$

$y = -16(4.5)^2 + 144(4.5)$

$y = -16(20.25) + 648$

$y = 324$ ft

d) Let $t = 4.5$

$x = 39(4.5)$

$x = 175.5 \approx 176$ ft

45) a) Let $p = 10$

$R(10) = -25(10)^2 + 600(10)$

$R(10) = -25(100) + 6000$

$R(10) = -2500 + 6000 = \$3500$

b) Let $p = 15$

$R(15) = -25(15)^2 + 600(15)$

$R(15) = -25(225) + 9000$

$R(15) = -5625 + 9000 = \$3375$

c) Let $R(p) = 3600$

$3600 = -25p^2 + 600p$

$0 = -25p^2 + 600p - 3600$

$0 = p^2 - 24p - 144$

$0 = (p - 12)^2$

$0 = p - 12$

$12 = p \qquad\qquad \$12$

47) a) Let $t = 1$

$h(1) = -16(1)^2 + 200(1)$

$h(1) = -16 + 200$

$h(1) = 184 \ \text{ft}$

b) Let $t = 4$

$h(4) = -16(4)^2 + 200(4)$

$h(4) = -16(16) + 800$

$h(4) = -256 + 800 = 544 \ \text{ft}$

c) Let $h = 400$ and solve for t.

$400 = -16t^2 + 200t$

$0 = -16t^2 + 200t - 400$

$0 = 2t^2 - 25t + 50$

$0 = (2t - 5)(t - 10)$

$2t - 5 = 0 \quad$ or $\quad t - 10 = 0$

$\qquad 2t = 5 \qquad\qquad\qquad t = 10$

$\qquad\quad t = \dfrac{5}{2}$

when $t = 2\dfrac{1}{2}$ seconds

and when $t = 10$ seconds

d) Let $h = 0$ and solve for t.

$0 = -16t^2 + 200t$

$0 = -16t^2 + 200t$

$0 = 2t^2 - 25t$

$0 = t(2t - 25)$

$t = 0 \quad$ or $\quad 2t - 25 = 0$

$\qquad\qquad\qquad\qquad 2t = 25$

$\qquad\qquad\qquad\qquad\ t = \dfrac{25}{2}$

when $t = 12\dfrac{1}{2}$ seconds

Chapter 7 Review

1) GCF of 18 and 27 is 9.

3) GCF of $33p^5q^3$, $121p^4q^3$

and $44p^7q^4$ is $11 \cdot p^4 \cdot q^3$ or $11p^4q^3$

5) $\text{GCF} = 12$

$48y + 84 = (12)(4y) + (12)(7)$
$= 12(4y + 7)$

7) $\text{GCF} = 7n^3$

$7n^5 - 21n^4 + 7n^3$
$= (7n^3)(n^2) - (7n^3)(3n) + (7n^3)(1)$
$= 7n^3(n^2 - 3n + 1)$

9) $\text{GCF} = (b + 6)$

$a(b + 6) - 2(b + 6) = (b + 6)(a - 2)$

11) $mn + 2m + 5n + 10$
$= m(n + 2) + 5(n + 2)$
$= (n + 2)(m + 5)$

13) $5qr - 10q - 6r + 12$
$= 5q(r - 2) - 6(r - 2)$
$= (r - 2)(5q - 6)$

15) $-8x^3 - 12x^2 + 4x = -4x(2x^2 + 3x - 1)$

17) $p^2 + 13p + 40 = (p + 8)(p + 5)$

19) $x^2 + xy - 20y^2 = (x + 5y)(x - 4y)$

21) $3c^2 - 24c + 36 = 3(c^2 - 8c + 12)$
$= 3(c - 6)(c - 2)$

23) $5y^2 + 11y + 6 = (5y + 6)(y + 1)$

25) $4m^2 - 16m + 15 = (2m - 5)(2m - 3)$

27) $56a^3 + 4a^2 - 16a = 4a(14a^2 + a - 4)$
$= 4a(7a + 4)(2a - 1)$

29) $3s^2 + 11st - 4t^2 = (3s - t)(s + 4t)$

31) $(3c - 5)^2 + 10(3c - 5) + 24$

Let $z = (3c - 5)$

$= z^2 + 10z + 24$
$= (z + 6)(z + 4)$

Replace z again with $(3c - 5)$

$= (3c - 5 + 6)(3c - 5 + 4)$
$= (3c + 1)(3c - 1)$

33) $n^2 - 25 = (n + 5)(n - 5)$

35) $9t^2 + 16u^2$ is prime

37) $10q^2 - 810 = 10(q^2 - 81)$
$= 10(q + 9)(q - 9)$

39) $a^2 + 16a + 64 = (a + 8)^2$

41) $h^3 + 8 = (h+2)(h^2 - 2h + 4)$

43) $27p^3 - 64q^3$
$= (3p - 4q)(9p^2 + 12pq + 16q^2)$

45) $7r^2 + 8r - 12 = (7r - 6)(r + 2)$

47) $\dfrac{9}{25} - x^2 = \left(\dfrac{3}{5} + x\right)\left(\dfrac{3}{5} - x\right)$

49) $st - 5s - 8t + 40 = s(t-5) - 8(t-5)$
$= (t-5)(s-8)$

51) $w^5 - w^2 = w^2\left(w^3 - 1\right)$
$= w^2(w-1)(w^2 + w + 1)$

53) $a^2 + 3a - 14$ is prime

55) $(a-b)^2 - (a+b)^2$
$= \left[(a-b) + (a+b)\right]\left[(a-b) - (a+b)\right]$
$= (2a)(-2b) = -4ab$

57) $6(y-2)^2 - 13(y-2) - 8$
Let $f = (y-2)$
$= 6f^2 - 13f - 8$
$= (3f - 8)(2f + 1)$
Replace f again with $(y-2)$
$= [3(y-2) - 8][2(y-2) + 1]$
$= (3y - 14)(2y - 3)$

59) $c(2c - 1) = 0$
$c = 0$ or $2c - 1 = 0$
$2c = 1$
$c = \dfrac{1}{2}$ $\qquad \left\{0, \dfrac{1}{2}\right\}$

61) $\qquad 3x^2 + x = 2$
$3x^2 + x - 2 = 0$
$(x+1)(3x - 2) = 0$
$x + 1 = 0$ or $3x - 2 = 0$
$3x = 2$
$x = -1$ or $x = \dfrac{2}{3}$ $\qquad \left\{-1, \dfrac{2}{3}\right\}$

63) $\qquad n^2 = 12n + 45$
$n^2 - 12n - 45 = 0$
$(n+3)(n-15) = 0$
$n + 3 = 0$ or $n - 15 = 0$
$n = -3$ or $n = 15$ $\qquad \{-3, 15\}$

65) $36 = 49d^2$
$0 = 49d^2 - 36$
$0 = (7d + 6)(7d - 6)$
$7d + 6 = 0$ or $7d - 6 = 0$
$7d = -6 \qquad\qquad 7d = 6$
$d = -\dfrac{6}{7}$ or $d = \dfrac{6}{7}$ $\qquad \left\{-\dfrac{6}{7}, \dfrac{6}{7}\right\}$

67)
$$8b + 64 = 2b^2$$
$$-2b^2 + 8b + 64 = 0$$
$$b^2 - 4b - 32 = 0$$
$$(b+4)(b-8) = 0$$
$$b + 4 = 0 \text{ or } b - 8 = 0$$
$$b = -4 \text{ or } b = 8 \qquad \{-4, 8\}$$

69)
$$y(5y - 9) = -4$$
$$5y^2 - 9y = -4$$
$$5y^2 - 9y + 4 = 0$$
$$(5y - 4)(y - 1) = 0$$
$$5y - 4 = 0 \text{ or } y - 1 = 0$$
$$5y = 4$$
$$y = \frac{4}{5} \text{ or } y = 1 \qquad \left\{\frac{4}{5}, 1\right\}$$

71)
$$6a^3 - 3a^2 - 18a = 0$$
$$3a(2a^2 - a - 6) = 0$$
$$3a(2a + 3)(a - 2) = 0$$
$$3a = 0 \text{ or } 2a + 3 = 0 \text{ or } a - 2 = 0$$
$$a = 0 \qquad 2a = -3 \qquad a = 2$$
$$a = -\frac{3}{2}$$
$$\left\{-\frac{3}{2}, 0, 2\right\}$$

73)
$$c(5c - 1) + 8 = 4(20 + c^2)$$
$$5c^2 - c + 8 = 80 + 4c^2$$
$$c^2 - c - 72 = 0$$
$$(c + 8)(c - 9) = 0$$
$$c + 8 = 0 \text{ or } c - 9 = 0$$
$$c = -8 \text{ or } c = 9 \qquad \{-8, 9\}$$

75)
$$p^2(6p - 1) - 10p(6p - 1) + 21(6p - 1) = 0$$
$$(6p - 1)(p^2 - 10p + 21) = 0$$
$$(6p - 1)(p - 3)(p - 7) = 0$$
$$6p - 1 = 0 \text{ or } p - 3 = 0 \text{ or } p - 7 = 0$$
$$6p = 1$$
$$p = \frac{1}{6} \text{ or } p = 3 \text{ or } p = 7$$
$$\left\{\frac{1}{6}, 3, 7\right\}$$

77) $2x - 3 = $ base of triangle
$$x + 2 = \text{height of triangle}$$
$$\text{Area} = \frac{1}{2}(\text{base})(\text{height})$$
$$15 = \frac{1}{2}(2x - 3)(x + 2)$$
$$30 = 2x^2 + x - 6$$
$$0 = 2x^2 + x - 36$$
$$0 = (2x + 9)(x - 4)$$

$$2x + 9 = 0 \text{ or } x - 4 = 0$$
$$2x = -9 \qquad x = 4$$
$$x = -\frac{9}{2} \quad \text{base} = 2(4) - 3 = 5 \text{ in;}$$
$$\text{height} = 4 + 2 = 6 \text{ in}$$

79) $3x - 1 =$ length of box
   $x =$ height of box
   Volume $= ($length$)($width$)($height$)$
   $96 = (3x - 1)(4)(x)$
   $96 = (3x^2 - x)4$
   $24 = 3x^2 - x$
   $0 = 3x^2 - x - 24$
   $0 = (3x + 8)(x - 3)$

   $3x + 8 = 0$ or $x - 3 = 0$
      $3x = -8$ $x = 3$
      $x = -\dfrac{8}{3}$
   height $= 3$ in; length $= 3(3) - 1 = 8$ in

81) $a^2 + b^2 = c^2$
      $a^2 + (5)^2 = (13)^2$
      $a^2 + 25 = 169$
      $a^2 - 144 = 0$
   $(a + 12)(a - 12) = 0$
   $a + 12 = 0$ or $a - 12 = 0$
      $a = -12$ $a = 12$
   The length of the missing side is 12.

83) $l =$ length of the countertop
      $l - 3.5 =$ width of the countertop
      Area $= ($Length$)($Width$)$
      $15 = l \cdot (l - 3.5)$
      $15 = l^2 - 3.5l$
      $0 = l^2 - 3.5l - 15$
      $0 = (l - 6)(l + 2.5)$
      $l - 6 = 0$ or $l + 2.5 = 0$
         $l = 6$ $l = -2.5$
      width $= 6 - 3.5 = 2.5$ ft; length $= 6$ ft

85) $x =$ the first integer
      $x + 1 =$ the second integer
      $x + 2 =$ the third integer
      $x + x + 1 + x + 2 = \dfrac{1}{3}(x + 1)^2$
         $3x + 3 = \dfrac{1}{3}(x^2 + 2x + 1)$
         $9x + 9 = x^2 + 2x + 1$
            $0 = x^2 - 7x - 8$
            $0 = (x - 8)(x + 1)$
      $x - 8 = 0$ or $x + 1 = 0$
         $x = 8$ $x = -1$
      $x = 8,$ then $8 + 1 = 9$ and $8 + 2 = 10$
      $x = -1,$ then $-1 + 1 = 0$ and $-1 + 2 = 1$
      $-1, 0, 1$ or $8, 9, 10$

87) $c = x =$ length of the ramp

$a = x - 1 =$ base of the ramp

$b = x - 8 =$ height of the ramp

$$(x-1)^2 + (x-8)^2 = x^2$$

$$x^2 - 2x + 1 + x^2 - 16x + 64 = x^2$$

$$2x^2 - 18x + 65 = x^2$$

$$x^2 - 18x + 65 = 0$$

$$(x-13)(x-5) = 0$$

$$x - 13 = 0 \quad \text{or} \quad x - 5 = 0$$

$$\boxed{x = 13} \qquad x = 5$$

89) a) Let $t = 0$

$$h = -16(0)^2 + 96(0) = 0 \text{ ft}$$

b) Let $h = 128$

$$128 = -16t^2 + 96t$$

$$0 = -16t^2 + 96t - 128$$

$$0 = t^2 - 6t + 8$$

$$0 = (t-2)(t-4)$$

$$t - 2 = 0 \quad \text{or} \quad t - 4 = 0$$

$$t = 2 \qquad\qquad t = 4$$

after $t = 2$ seconds

and after $t = 4$ seconds

c) Let $t = 3$

$$h = -16(3)^2 + 96(3)$$

$$h = -16(9) + 288$$

$$h = -144 + 288 = 144 \text{ ft}$$

height $= 13 - 8 = 5$ in

d) Let $h = 0$ and solve for t.

$$0 = -16t^2 + 96t$$

$$0 = t^2 - 6t$$

$$0 = t(t-6)$$

$$t = 0 \quad \text{or} \quad t - 6 = 0$$

$$t = 6$$

after $t = 6$ seconds

Chapter 7 Test

1) See if you can factor out a GCF.

3) $16 - b^2 = (4+b)(4-b)$

5) $56p^6q^6 - 77p^4q^4 + 7p^2q^3$
$= 7p^2q^3\left(8p^4q^3 - 11p^2q + 1\right)$

7) $2d^3 + 14d^2 - 36d = 2d\left(d^2 + 7d - 18\right)$
$= 2d\left(d+9\right)\left(d-2\right)$

9) $9h^2 + 24h + 16 = \left(3h+4\right)^2$

11) $s^2 - 3st - 28t^2 = \left(s-7t\right)\left(s+4t\right)$

13) $4(3p+2)^2 + 17(3p+2) - 15$
Let $f = (3p+2)$
$= 4f^2 + 17f - 15$
$= (f+5)(4f-3)$
Replace f again with $(3p+2)$
$= [3p+2+5][4(3p+2)-3]$
$= (3p+7)(12p+5)$

15) $m^{12} + m^9 = m^9\left(m^3 + 1\right)$
$= m^9\left(m+1\right)\left(m^2 - m + 1\right)$

19) $\left(c-5\right)\left(c+2\right) = 18$
$c^2 - 3c - 10 = 18$
$c^2 - 3c - 28 = 0$
$\left(c+4\right)\left(c-7\right) = 0$
$c+4 = 0$ or $c-7 = 0$
$c = -4$ o $c = 7$ $\quad \{-4, 7\}$

21) $\qquad 24y^2 + 80 = 88y$
$24y^2 - 88y + 80 = 0$
$3y^2 - 11y + 10 = 0$
$\left(3y-5\right)\left(y-2\right) = 0$
$3y - 5 = 0$ or $y - 2 = 0$
$3y = 5$
$y = \dfrac{5}{3}$ or $y = 2$ $\quad \left\{\dfrac{5}{3}, 2\right\}$

23) $\quad x = $ the first odd integer
$x + 2 = $ the second odd integer
$x + 4 = $ the third odd integer
$x + x + 2 + x + 4 = \left(x+4\right)^2 - 60$
$3x + 6 = \left(x^2 + 8x + 16\right) - 60$
$3x + 6 = x^2 + 8x - 44$
$0 = x^2 + 5x - 50$
$0 = \left(x+10\right)\left(x-5\right)$
$x + 10 = 0$ or $x - 5 = 0$
$x = -10$ $\qquad \boxed{x = 5}$
$x = 5$, then $5 + 2 = 7$ and $5 + 4 = 9$
$5, 7, 9$

25) $\qquad w = $ the width of the run
$2w + 4 = $ the length of the run
Area $= \left(\text{Length}\right)\left(\text{Width}\right)$
$96 = \left(2w+4\right)w$
$96 = 2w^2 + 4w$
$0 = 2w^2 + 4w - 96$
$0 = w^2 + 2w - 48$
$0 = \left(w+8\right)\left(w-6\right)$
$w + 8 = 0$ or $w - 6 = 0$
$w = -8$ $\qquad \boxed{w = 6}$
length $= 2\left(6\right) + 4 = 16$ ft; width $= 6$ ft

Cumulative Review: Chapters 1–7

$(x_1, y_1) = (-6, -1); \ m_{\text{perp}} = \dfrac{1}{3}$

$y - y_1 = m(x - x_1)$

$y - (-1) = \dfrac{1}{3}(x - (-6))$

$y + 1 = \dfrac{1}{3}x + 2$

$y = \dfrac{1}{3}x + 1$

1) $\dfrac{3}{8} - \dfrac{5}{6} + \dfrac{7}{12} = \dfrac{9}{24} - \dfrac{20}{24} + \dfrac{14}{24}$

$\qquad = \dfrac{9 - 20 + 14}{24}$

$\qquad = \dfrac{3}{24} = \dfrac{1}{8}$

3) $\dfrac{54t^5 u^2}{36tu^8} = \dfrac{54}{36}t^{5-1}u^{2-8} = \dfrac{3}{2}t^4 u^{-6} = \dfrac{3t^4}{2u^6}$

13) $(6y + 5)(2y - 3)$

$\qquad = 12y^2 - 18y + 10y - 15$

$\qquad = 12y^2 - 8y - 15$

5) $4.813 \times 10^5: \ 4\underset{\smile}{8}\underset{\smile}{1}\underset{\smile}{3}\underset{\smile}{0}\underset{\smile}{0} = 481,300$

7) $\qquad A = P + P\boxed{R}T$

$A - P = P\boxed{R}T$

$\dfrac{A - P}{PT} = \boxed{R}$

15) $(c + 8)^2 = c^2 + 16c + 64$

17) $2x - 5 \overline{\smash{\big)}\, 12x^4 - 30x^3 - 14x^2 + 27x + 20}$

$\ \ 6x^3 -7x - 4$

$\ \underline{-(12x^4 - 30x^3)}$

$0 \quad -14x^2 + 27x$

$\underline{-(-14x^2 + 35x)}$

$-8x + 20$

$\underline{-(-8x + 20)}$

0

9) $2 + |9 - 5n| \geq 31$

$|9 - 5n| \geq 29$

$9 - 5n \geq 29 \quad \text{or} \quad 9 - 5n \leq -29$

$-5n \geq 20 \quad \text{or} \quad -5n \leq -38$

$n \leq -4 \quad \text{or} \qquad n \geq \dfrac{38}{5}$

$(-\infty, -4] \cup \left[\dfrac{38}{5}, \infty\right)$

19) $bc + 8b - 7c - 56 = b(c + 8) - 7(c + 8)$

$\qquad = (c + 8)(b - 7)$

11) Determine the slope.

$3x + y = 4$

$y = -3x + 4$

$m = -3$

21) $y^2 + 1$ is prime

23) $x^3 - 125 = (x-5)(x^2 + 5x + 25)$

25) $-12j(1-2j) = 16(5+j)$

$\qquad -12j + 24j^2 = 80 + 16j$

$\qquad 24j^2 - 28j - 80 = 0$

$\qquad\ 6j^2 - 7j - 20 = 0$

$\qquad (3j+4)(2j-5) = 0$

$\qquad 3j+4 = 0 \ \ \text{or} \ \ 2j-5 = 0$

$\qquad\quad 3j = -4 \qquad\ \ 2j = 5$

$\qquad\qquad j = -\dfrac{4}{3} \ \ \text{or} \ \ j = \dfrac{5}{2}$

$\qquad \left\{ -\dfrac{4}{3}, \dfrac{5}{2} \right\}$

Chapter 8: Rational Expressions, Equations, and Functions

Section 8.1

1) When its numerator is equal to zero.

3) Set the denominator equal to zero and solve for the variable. Any variable that makes the denominator equal to zero is not in the domain of the function.

5) $f(x) = \dfrac{x+8}{x+6}$

   a) $f(-2) = \dfrac{x+8}{x+6} = \dfrac{-2+8}{-2+6} = \dfrac{6}{4} = \dfrac{3}{2}$

   b) Set the numerator equal to zero.
$$x+8=0$$
$$x=-8$$

   c) Set the denominator equal to zero.
$$x+6=0$$
$$x=-6$$
$$\text{Domain}: (-\infty,-6) \cup (-6,\infty)$$

7) $f(x) = \dfrac{5x-3}{x+2}$

   a)

$$f(-2) = \dfrac{5 \cdot -2 - 3}{-2+2} = \dfrac{-10-3}{0} = \text{undefined}$$

   b) Set the numerator equal to zero.
$$5x-3=0$$
$$5x=3$$
$$x=\dfrac{3}{5}$$

   c) Set the denominator equal to zero.
$$x+2=0$$
$$x=-2$$
$$\text{Domain}: (-\infty,-2) \cup (-2,\infty)$$

9) $f(x) = \dfrac{6}{x^2+6x+5}$

   a) $f(-2) = \dfrac{6}{(-2)^2 + 6 \cdot -2 + 5}$

$$= \dfrac{6}{4-12+5} = \dfrac{6}{-3} = -2$$

   b) Never equals zero.

   c) Set denominator equal to zero.
$$x^2+6x+5=0$$
$$(x+5)(x+1)=0$$
$$x+5=0 \quad \text{or} \quad x+1=0$$
$$x=-5 \qquad\qquad x=-1$$
$$(-\infty,-5) \cup (-5,-1) \cup (-1,\infty)$$

11) $p-7=0$
$$p=7$$
$$(-\infty,7) \cup (7,\infty)$$

13) $5r+2=0$
$$5r=-2$$
$$r=-\dfrac{2}{5}$$
$$\left(-\infty,-\dfrac{2}{5}\right) \cup \left(-\dfrac{2}{5},\infty\right)$$

15) $t^2 - 9t + 8 = 0$
$$(t-8)(t-1)=0$$
$$t-8=0 \quad \text{or} \quad t-1=0$$
$$t=8 \qquad\qquad t=1$$
$$(-\infty,1) \cup (1,8) \cup (8,\infty)$$

17) $w^2 - 81 = 0$
$$(w-9)(w+9)=0$$
$$w-9=0 \quad \text{or} \quad w+9=0$$
$$w=9 \qquad\qquad w=-9$$
$$(-\infty,-9) \cup (-9,9) \cup (9,\infty)$$

19) $c^2 + 6 = 0$

$(-\infty, \infty)$

21) Answers may vary.

23) $\dfrac{12d^5}{30d^8} = \dfrac{2d^{5-8}}{5} = \dfrac{2d^{-3}}{5} = \dfrac{2}{5d^3}$

25) $\dfrac{3c-12}{5c-20} = \dfrac{3(c-4)}{5(c-4)} = \dfrac{3}{5}$

27) $\dfrac{b^2+b-56}{b+8} = \dfrac{(b+8)(b-7)}{(b+8)} = b-7$

29) $\dfrac{r-4}{r^2-16} = \dfrac{(r-4)}{(r-4)(r+4)} = \dfrac{1}{r+4}$

31) $\dfrac{3k^2+28k+32}{k^2+10k+16} = \dfrac{(3k+4)(k+8)}{(k+8)(k+2)} = \dfrac{3k+4}{k+2}$

33) $\dfrac{w^3+125}{5w^2-25w+125} = \dfrac{(w+5)(w^2-5w+25)}{5(w^2-5w+25)}$

$= \dfrac{w+5}{5}$

35) $\dfrac{(4m^2-20m)+(4mn-20n)}{11m+11n}$

$= \dfrac{4m(m-5)+4n(m-5)}{11(m+n)}$

$= \dfrac{(4m+4n)(m-5)}{11(m+n)}$

$= \dfrac{4(m+n)(m-5)}{11(m+n)}$

$= \dfrac{4(m-5)}{11}$

37) $\dfrac{x^2-y^2}{x^3-y^3} = \dfrac{(x-y)(x+y)}{(x-y)(x^2+xy+y^2)}$

$= \dfrac{(x-y)(x+y)}{(x-y)(x^2+xy+y^2)}$

$= \dfrac{x+y}{x^2+xy+y^2}$

39) It reduces to -1.

41) $\dfrac{12-v}{v-12} = \dfrac{-1(12-v)}{v-12} = \dfrac{-1(v-12)}{v-12} = -1$

43) $\dfrac{k^2-49}{7-k} = \dfrac{(k+7)(k-7)}{(7-k)} = \dfrac{(k+7)(k-7)}{-1(k-7)}$

$= -(k+7)$ or $-k-7$

45) $\dfrac{30-35x}{7x^2+8x-12} = \dfrac{5(6-7x)}{(7x-6)(x+2)}$

$= \dfrac{5(7x-6)}{(7x-6)(x+2)}$

$= -\dfrac{5}{(x+2)}$

47) $\dfrac{16-4b^2}{b-2} = \dfrac{4\left(4-b^2\right)}{b-2} = \dfrac{4\,\overset{-1}{\cancel{\left(b-2\right)}}\left(b+2\right)}{\cancel{b-2}}$

$= -4\left(b+2\right)$

49) $\dfrac{8t^3-27}{9-4t^2} = \dfrac{\left(2t-3\right)\left(4t^2+6t+9\right)}{-1\left(4t^2-9\right)}$

$= \dfrac{\overset{-1}{\cancel{\left(2t-3\right)}}\left(4t^2+6t+9\right)}{\left(2t+3\right)\cancel{\left(2t-3\right)}}$

$= -\dfrac{4t^2+6t+9}{2t+3}$

51) $\dfrac{5x^2+13x+6}{x+2} = \dfrac{\left(5x+3\right)\cancel{\left(x+2\right)}}{\cancel{x+2}} = 5x+3$

53) $\dfrac{\left(c^3-2c^2\right)+\left(4c-8\right)}{c^2+4} = \dfrac{c^2\left(c-2\right)+4\left(c-2\right)}{c^2+4}$

$= \dfrac{\cancel{\left(c^2+4\right)}\left(c-2\right)}{\cancel{c^2+4}}$

$= c-2$

55) 1) $\dfrac{-\left(b+7\right)}{b-2} = \dfrac{-b-7}{b-2}$

2) $\dfrac{-\left(b+7\right)}{b-2}$

3) $\dfrac{b+7}{-\left(b-2\right)} = \dfrac{b+7}{-b+2}$

57) 1) $\dfrac{-\left(9-5t\right)}{2t-3} = \dfrac{-9+5t}{2t-3}$

2) $\dfrac{-\left(9-5t\right)}{2t-3}$

3) $\dfrac{9-5t}{-\left(2t-3\right)} = \dfrac{9-5t}{-2t+3}$

59) $\dfrac{9}{14}\cdot\dfrac{7}{6} = \dfrac{3\cdot\cancel{3}}{\cancel{7}\cdot 2}\cdot\dfrac{\cancel{7}}{\cancel{3}\cdot 2} = \dfrac{3}{4}$

61) $\dfrac{\overset{2}{\cancel{14}}\,u^5}{\underset{3}{\cancel{15}}\,v^2}\cdot\dfrac{\overset{4}{\cancel{20}}\,v^6}{\underset{1}{\cancel{7}}\,u^8} = \dfrac{8v^4}{3u^3}$

63) $\dfrac{\overset{1}{\cancel{8}}\,t^2}{\left(3t-2\right)^2}\cdot\dfrac{3t-2}{\underset{2}{\cancel{16}}\,t^3} = \dfrac{\cancel{\left(3t-2\right)}}{\cancel{\left(3t-2\right)}\left(3t-2\right)2t}$

$= \dfrac{1}{2t\left(3t-2\right)}$

65) $\dfrac{\overset{2}{\cancel{8}}}{6p+3}\cdot\dfrac{4p^2-1}{\underset{3}{\cancel{12}}} = \dfrac{2\cancel{\left(2p+1\right)}\left(2p-1\right)}{3\cdot 3\cancel{\left(2p+1\right)}}$

$= \dfrac{2\left(2p-1\right)}{9}$

67) $\dfrac{2v^2+15v+18}{3v+18}\cdot\dfrac{12v-3}{8v+12}$

$= \dfrac{\cancel{\left(2v+3\right)}\cancel{\left(v+6\right)}}{\cancel{3}\cancel{\left(v+6\right)}}\cdot\dfrac{\cancel{3}\left(4v-1\right)}{4\cancel{\left(2v+3\right)}}$

$= \dfrac{4v-1}{4}$

69) $\left(x-8\right)\cdot\dfrac{4}{x^2-8x} = \cancel{\left(x-8\right)}\cdot\dfrac{4}{x\cancel{\left(x-8\right)}} = \dfrac{4}{x}$

71) $\dfrac{r^3+27}{4t+20} \cdot \dfrac{(rt+5r)(-2t-10)}{r^2-9}$

$= \dfrac{(r+3)(r^2-3r+9)}{4(t+5)} \cdot \dfrac{rt+5r-2t-10}{(r+3)(r-3)}$

$= \dfrac{(r^2-3r+9)}{4(t+5)} \cdot \dfrac{(r-2)(t+5)}{(r-3)}$

$= \dfrac{(r-2)(r^2-3r+9)}{4(r-3)}$

73) $\dfrac{4}{5} \div \dfrac{8}{3} = \dfrac{\overset{1}{\cancel{4}}}{5} \cdot \dfrac{3}{\underset{2}{\cancel{8}}} = \dfrac{3}{10}$

75) $\dfrac{c^2}{6b} \div \dfrac{c^8}{b} = \dfrac{\cancel{c^2}}{6\cancel{b}} \cdot \dfrac{\cancel{b}}{\underset{c^6}{\cancel{c^8}}} = \dfrac{1}{6c^6}$

77)

$\dfrac{2a-1}{8a^3} \div \dfrac{(2a-1)^2}{24a^5} = \dfrac{\cancel{2a-1}}{\cancel{8a^3}} \cdot \dfrac{\overset{3a^2}{\cancel{24a^5}}}{\underset{(2a-1)}{\cancel{(2a-1)^2}}} = \dfrac{3a^2}{2a-1}$

79) $\dfrac{18y-45}{18} \div \dfrac{4y^2-25}{10}$

$= \dfrac{\cancel{9}(2y-5)}{\underset{\cancel{9}}{\cancel{18}}} \cdot \dfrac{\overset{5}{\cancel{10}}}{(2y-5)(2y+5)}$

$= \dfrac{5}{2y+5}$

81) $\dfrac{j^2-25}{5j+25} \div \dfrac{7j-35}{5}$

$= \dfrac{(j+5)(j-5)}{\cancel{5}(j+5)} \cdot \dfrac{\cancel{5}}{7(j-5)}$

$= \dfrac{1}{7}$

83) $\dfrac{z^2+18z+80}{2z+1} \div (z+8)^2$

$= \dfrac{(z+8)(z+10)}{2z+1} \cdot \dfrac{1}{\underset{(z+8)}{\cancel{(z+8)^2}}}$

$= \dfrac{z+10}{(2z+1)(z+8)}$

85) $\dfrac{36a-12}{16} \div (9a^2-1)$

$= \dfrac{\overset{3}{\cancel{12}}(3a-1)}{\underset{4}{\cancel{16}}} \cdot \dfrac{1}{(3a-1)(3a+1)}$

$= \dfrac{3}{4(3a+1)}$

87) $\dfrac{8d^2-8d+8}{25-4d^2} \div \dfrac{d^3+1}{2d^2-3d-5}$

$= \dfrac{8(d^2-d+1)}{-(4d^2-25)} \cdot \dfrac{(2d-5)(d+1)}{(d+1)(d^2-d+1)}$

$= \dfrac{8(2d-5)}{-(2d-5)(2d+5)}$

$= -\dfrac{8}{2d+5}$

Section 8.1: Simplifying, Multiplying, and Dividing Rational Expressions and Functions

89) No, it would make the problem undefined.

91) $\dfrac{a^2+4a}{6a+54}\cdot\dfrac{a^2+5a-36}{16-a^2}$

$=\dfrac{a(a+4)}{6(a+9)}\cdot\dfrac{(a+9)(a-4)}{-(a^2-16)}$

$=\dfrac{a(a+4)}{6}\cdot\dfrac{(a-4)}{-(a-4)(a+4)}$

$=-\dfrac{a}{6}$

93) $\dfrac{r^3+8}{r+2}\cdot\dfrac{7}{3r^2-6r+12}$

$=\dfrac{(r+2)(r^2-2r+4)}{r+2}\cdot\dfrac{7}{3(r^2-2r+4)}$

$=\dfrac{7}{3}$

95) $\dfrac{54x^8}{22x^3y^2}\div\dfrac{36xy^5}{11x^2y}=\dfrac{\overset{3}{54}\,x^8}{\underset{2}{22}\,x^3y^2}\cdot\dfrac{\overset{1}{11}\,x^2y}{\underset{2}{36}\,xy^5}$

$=\dfrac{3x^{10}y}{4x^4y^7}=\dfrac{3x^6}{4y^6}$

97) $\dfrac{2a^2}{a^2+a-20}\cdot\dfrac{a^3+5a^2+4a+20}{2a^2+8}$

$=\dfrac{2a^2}{(a+5)(a-4)}\cdot\dfrac{a^2(a+5)+4(a+5)}{2(a^2+4)}$

$=\dfrac{\overset{1}{2}\,a^2}{(a+5)(a-4)}\cdot\dfrac{(a^2+4)(a+5)}{\underset{1}{2}\,(a^2+4)}$

$=\dfrac{a^2}{a-4}$

99) $\dfrac{30}{4y^2-4x^2}\div\dfrac{10x^2+10xy+10y^2}{x^3-y^3}$

$=\dfrac{30}{4(y^2-x^2)}\cdot\dfrac{x^3-y^3}{10(x^2+xy+y^2)}$

$=\dfrac{\overset{3}{30}}{-4(x+y)(x-y)}\cdot\dfrac{(x-y)(x^2+xy+y^2)}{\underset{1}{10}(x^2+xy+y^2)}$

$=-\dfrac{3}{4(x+y)}$

101) $\dfrac{4j^2-21j+5}{j^3}\div\left(\dfrac{3j+2}{j^3-j^2}\cdot\dfrac{j^2-6j+5}{j}\right)$

$=\dfrac{(4j-1)(j-5)}{j^3}\div\left(\dfrac{3j+2}{j^2(j-1)}\cdot\dfrac{(j-5)(j-1)}{j}\right)$

$=\dfrac{(4j-1)(j-5)}{j^3}\div\left(\dfrac{(3j+2)(j-5)(j-1)}{j^3(j-1)}\right)$

$=\dfrac{(4j-1)(j-5)}{j^3}\cdot\dfrac{j^3}{(3j+2)(j-5)}$

$=\dfrac{4j-1}{3j+2}$

103) $l=\dfrac{A}{w}$

$\dfrac{3}{2xy^6}\div\dfrac{y^2}{12x^5}=\dfrac{3}{\underset{1}{2}\,xy^6}\cdot\dfrac{\overset{6}{12}\,x^5}{y^2}=\dfrac{18x^4}{y^8}$

216

Section 8.2

1) LCD = 40

3) LCD = c^4

5) LCD = $36p^8$

7) LCD = $24a^3b^4$

9) LCD = $2(n+4)$

11) LCD = $w(2w+1)$

13) $12a^2 - 4a$ $6a^4 - 2a^3$

    $4a(3a-1)$ $2a^3(3a-1)$

    LCD = $4a^3(3a-1)$

15) LCD = $(r+7)(r-2)$

17) $w^2 - 3w - 10 = (w-5)(w+2)$

    $w^2 - 2w - 15 = (w-5)(w+3)$

    $w^2 + 5w + 6 = (w+2)(w+3)$

    LCD = $(w-5)(w+2)(w+3)$

19) LCD = $b-4$ or $4-b$

21) Answers may vary.

23) $\dfrac{3}{t};\dfrac{8}{t^3}$ LCD = t^3

    $\dfrac{3}{t} = \dfrac{3t^2}{t^3}$ $\dfrac{8}{t^3} = \dfrac{8}{t^3}$

25) $\dfrac{9}{8n^6};\dfrac{2}{3n^2}$ LCD = $24n^6$

    $\dfrac{9}{8n^6} \cdot \dfrac{3}{3} = \dfrac{27}{24n^6}$ $\dfrac{2}{3n^2} \cdot \dfrac{8n^4}{8n^4} = \dfrac{16n^4}{24n^6}$

27) $\dfrac{1}{x^3y};\dfrac{6}{5xy^5}$ LCD = $5x^3y^5$

    $\dfrac{1}{x^3y} \cdot \dfrac{5y^4}{5y^4} = \dfrac{5y^4}{5x^3y^5}$ $\dfrac{6}{5xy^5} \cdot \dfrac{x^2}{x^2} = \dfrac{6x^2}{5x^3y^5}$

29) $\dfrac{t}{5t-6};\dfrac{10}{7}$ LCD = $7(5t-6)$

    $\dfrac{t}{5t-6} \cdot \dfrac{7}{7} = \dfrac{7t}{7(5t-6)}$

    $\dfrac{10}{7} \cdot \dfrac{(5t-6)}{(5t-6)} = \dfrac{50t-60}{7(5t-6)}$

31) $\dfrac{a}{24a+36} = \dfrac{a}{12(2a+3)}$

    $\dfrac{1}{18a+27} = \dfrac{1}{9(2a+3)}$

    LCD = $36(2a+3)$

    $\dfrac{a}{12(2a+3)} \cdot \dfrac{3}{3} = \dfrac{3a}{36(2a+3)}$

    $\dfrac{1}{9(2a+3)} \cdot \dfrac{4}{4} = \dfrac{4}{36(2a+3)}$

33) $\dfrac{4}{h+5};\dfrac{7h}{h-3}$ LCD = $(h+5)(h-3)$

    $\dfrac{4}{h+5} \cdot \dfrac{(h-3)}{(h-3)} = \dfrac{4h-12}{(h+5)(h-3)}$

    $\dfrac{7h}{h-3} \cdot \dfrac{(h+5)}{(h+5)} = \dfrac{7h^2+35h}{(h+5)(h-3)}$

35) $\dfrac{9y}{y^2-y-42}=\dfrac{9y}{(y-7)(y+6)}$

$\dfrac{3}{2y^2+12y}=\dfrac{3}{2y(y+6)}$

$\text{LCD}=2y(y-7)(y+6)$

$\dfrac{9y}{(y-7)(y+6)}\cdot\dfrac{2y}{2y}=\dfrac{18y^2}{2y(y-7)(y+6)}$

$\dfrac{3}{2y(y+6)}\cdot\dfrac{(y-7)}{(y-7)}=\dfrac{3y-21}{2y(y-7)(y+6)}$

37) $\dfrac{z}{z^2-10z+25}=\dfrac{z}{(z-5)(z-5)}$

$\dfrac{15z}{z^2-2z-15}=\dfrac{15z}{(z-5)(z+3)}$

$\text{LCD}=(z-5)(z-5)(z+3)$

$\dfrac{z}{(z-5)(z-5)}\cdot\dfrac{(z+3)}{(z+3)}=\dfrac{z^2+3z}{(z-5)(z-5)(z+3)}$

$\dfrac{15z}{(z-5)(z+3)}\cdot\dfrac{(z-5)}{(z-5)}=\dfrac{15z^2-75z}{(z-5)(z-5)(z+3)}$

39)

$\dfrac{11}{g-3};\dfrac{4}{9-g^2}=\dfrac{4}{-(g^2-9)}=\dfrac{4}{-(g-3)(g+3)}$

$\text{LCD}=-(g-3)(g+3)$

$\dfrac{11}{g-3}\cdot\dfrac{-(g+3)}{-(g+3)}=\dfrac{-11g-33}{-(g-3)(g+3)}$

$\dfrac{4}{-(g-3)(g+3)}\cdot1=\dfrac{4}{-(g-3)(g+3)}$

41) $\dfrac{4}{w^2-4w}=\dfrac{4}{w(w-4)}$

$\dfrac{6}{7w^2-28w}=\dfrac{6}{7w(w-4)}$

$\dfrac{11}{w^2-8w+16}=\dfrac{11}{(w-4)(w-4)}$

$\text{LCD}=7w(w-4)(w-4)$

$\dfrac{4}{w(w-4)}\cdot\dfrac{7(w-4)}{7(w-4)}=\dfrac{28w-112}{7w(w-4)(w-4)}$

$\dfrac{6}{7w(w-4)}\cdot\dfrac{(w-4)}{(w-4)}=\dfrac{6w-24}{7w(w-4)(w-4)}$

$\dfrac{11}{(w-4)(w-4)}\cdot\dfrac{7w}{7w}=\dfrac{77w}{7w(w-4)(w-4)}$

43) $\dfrac{7}{20}+\dfrac{9}{20}=\dfrac{16}{20}=\dfrac{4}{5}$

45) $\dfrac{8}{a}+\dfrac{2}{a}=\dfrac{10}{a}$

47) $\dfrac{8}{x+4}+\dfrac{2x}{x+4}=\dfrac{2x+8}{x+4}=\dfrac{2\cancel{(x+4)}}{\cancel{x+4}}=2$

49)

$\dfrac{7w-4}{w(3w-4)}-\dfrac{20-11w}{w(3w-4)}=\dfrac{7w-4-[20-11w]}{w(3w-4)}$

$=\dfrac{7w-4-20+11w}{w(3w-4)}$

$=\dfrac{18w-24}{w(3w-4)}$

$=\dfrac{6\cancel{(3w-4)}}{w\cancel{(3w-4)}}$

$=\dfrac{6}{w}$

51) $\dfrac{2r+15}{(r-5)(r+2)}+\dfrac{r^2-10r}{(r-5)(r+2)}$

$=\dfrac{2r+15+r^2-10r}{(r-5)(r+2)}$

$=\dfrac{r^2-8r+15}{(r-5)(r+2)}$

$=\dfrac{\cancel{(r-5)}(r-3)}{\cancel{(r-5)}(r+2)}$

$=\dfrac{r-3}{r+2}$

53) $\dfrac{8}{x-3};\dfrac{2}{x}$

a) LCD$=x(x-3)$

b) The LCD will contain each unique factor the *greatest* number of times it appears in factorization.

c) $\dfrac{8}{x-3}\cdot\dfrac{x}{x}=\dfrac{8x}{x(x-3)}$

$\dfrac{2}{x}\cdot\dfrac{(x-3)}{(x-3)}=\dfrac{2x-6}{x(x-3)}$

55) Find the product of two denominators.

57) $\dfrac{5}{8}+\dfrac{1}{6}=\dfrac{5}{8}\cdot\dfrac{6}{6}+\dfrac{1}{6}\cdot\dfrac{8}{8}=\dfrac{30}{48}+\dfrac{8}{48}=\dfrac{38}{48}=\dfrac{19}{24}$

59) $\dfrac{5x}{12}-\dfrac{4x}{15}=\dfrac{5x}{12}\cdot\dfrac{5}{5}-\dfrac{4x}{15}\cdot\dfrac{4}{4}=\dfrac{25x}{60}-\dfrac{16x}{60}$

$=\dfrac{9x}{60}=\dfrac{3x}{20}$

61) $\dfrac{3}{2a}+\dfrac{6}{7a^2}=\dfrac{3}{2a}\cdot\dfrac{7a}{7a}+\dfrac{6}{7a^2}\cdot\dfrac{2}{2}$

$=\dfrac{21a}{14a^2}+\dfrac{12}{14a^2}=\dfrac{21a+12}{14a^2}$

$=\dfrac{3(7a+4)}{14a^2}$

63) $\dfrac{15}{d-8}-\dfrac{4}{d}=\dfrac{15}{d-8}\cdot\dfrac{d}{d}-\dfrac{4}{d}\cdot\dfrac{(d-8)}{(d-8)}$

$=\dfrac{15d}{d(d-8)}-\dfrac{4d-32}{d(d-8)}$

$=\dfrac{15d-[4d-32]}{d(d-8)}$

$=\dfrac{15d-4d+32}{d(d-8)}$

$=\dfrac{11d+32}{d(d-8)}$

65) $\dfrac{1}{z+6}+\dfrac{4}{z+2}$

$=\dfrac{1}{(z+6)}\cdot\dfrac{(z+2)}{(z+2)}+\dfrac{4}{(z+2)}\cdot\dfrac{(z+6)}{(z+6)}$

$=\dfrac{z+2}{(z+2)(z+6)}+\dfrac{4z+24}{(z+2)(z+6)}$

$=\dfrac{z+2+4z+24}{(z+2)(z+6)}=\dfrac{5z+26}{(z+2)(z+6)}$

67) $\dfrac{x}{2x+1}-\dfrac{3}{x+5}$

$=\dfrac{x}{2x+1}\cdot\dfrac{(x+5)}{(x+5)}-\dfrac{3}{x+5}\cdot\dfrac{(2x+1)}{(2x+1)}$

$=\dfrac{x^2+5x}{(x+5)(2x+1)}-\dfrac{6x+3}{(x+5)(2x+1)}$

$=\dfrac{x^2+5x-[6x+3]}{(x+5)(2x+1)}=\dfrac{x^2+5x-6x-3}{(x+5)(2x+1)}$

$=\dfrac{x^2-x-3}{(x+5)(2x+1)}$

69) $\dfrac{t}{t+7}+\dfrac{11t-21}{t^2-49}=\dfrac{t}{t+7}+\dfrac{11t-21}{(t+7)(t-7)}$

$=\dfrac{t}{t+7}\cdot\dfrac{(t-7)}{(t-7)}+\dfrac{11t-21}{(t+7)(t-7)}$

$=\dfrac{t^2-7t+11t-21}{(t+7)(t-7)}=\dfrac{t^2+4t-21}{(t+7)(t-7)}$

$=\dfrac{\cancel{(t+7)}(t-3)}{\cancel{(t+7)}(t-7)}=\dfrac{(t-3)}{(t-7)}$

71) $\dfrac{b}{b^2-16}+\dfrac{10}{b^2-5b-36}=\dfrac{b}{(b-4)(b+4)}+\dfrac{10}{(b-9)(b+4)}$

$=\dfrac{b}{(b-4)(b+4)}\cdot\dfrac{(b-9)}{(b-9)}+\dfrac{10}{(b-9)(b+4)}\cdot\dfrac{(b-4)}{(b-4)}=\dfrac{b^2-9b+10b-40}{(b-9)(b+4)(b-4)}$

$=\dfrac{b^2+b-40}{(b-9)(b+4)(b-4)}$

73) $\dfrac{3c}{c^2+4c-12}-\dfrac{2c-5}{c^2+2c-24}=\dfrac{3c}{(c+6)(c-2)}\cdot\dfrac{(c-4)}{(c-4)}-\dfrac{2c-5}{(c-4)(c+6)}\cdot\dfrac{(c-2)}{(c-2)}$

$=\dfrac{3c^2-12c-\left[2c^2-9c+10\right]}{(c+6)(c-2)(c-4)}=\dfrac{3c^2-12c-2c^2+9c-10}{(c+6)(c-2)(c-4)}$

$=\dfrac{c^2-3c-10}{(c+6)(c-2)(c-4)}=\dfrac{(c-5)(c+2)}{(c+6)(c-2)(c-4)}$

75) $\dfrac{4b+1}{3b-12}+\dfrac{5b}{b^2-b-12}=\dfrac{4b+1}{3(b-4)}\cdot\dfrac{(b+3)}{(b+3)}+\dfrac{5b}{(b-4)(b+3)}\cdot\dfrac{3}{3}=\dfrac{4b^2+13b+3+15b}{3(b-4)(b+3)}$

$=\dfrac{4b^2+28b+3}{3(b-4)(b+3)}$

77) No, $7-x$ can be rewritten as $-(x-7)$; therefore
the LCD $=x-7$

79) $\dfrac{9}{z-6}+\dfrac{2}{6-z}=\dfrac{9}{z-6}\cdot\dfrac{-1}{-1}+\dfrac{2}{-(z-6)}=\dfrac{-9+2}{-(z-6)}=\dfrac{-7}{-(z-6)}=\dfrac{7}{z-6}$

81) $\dfrac{2c}{12b-7c}-\dfrac{13}{7c-12b}=\dfrac{2c}{12b-7c}\cdot\dfrac{-1}{-1}-\dfrac{13}{-(12b-7c)}=-\dfrac{2c+13}{12b-7c}$

83) $\dfrac{5}{8-t}+\dfrac{10}{t^2-64}=\dfrac{5}{-(t-8)}\cdot\dfrac{(t+8)}{(t+8)}+\dfrac{10}{(t-8)(t+8)}\cdot\dfrac{-1}{-1}=\dfrac{5t+40-10}{-(t-8)(t+8)}=\dfrac{5t+30}{-(t-8)(t+8)}$

$$=-\dfrac{5(t+6)}{(t-8)(t+8)}$$

85) $\dfrac{a}{4a^2-9}-\dfrac{4}{3-2a}=\dfrac{a}{(2a-3)(2a+3)}\cdot\dfrac{-1}{-1}-\dfrac{4}{-(2a-3)}\cdot\dfrac{(2a+3)}{(2a+3)}=\dfrac{-a-[8a+12]}{-(2a-3)(2a+3)}$

$$=\dfrac{-a-8a-12}{-(2a-3)(2a+3)}=\dfrac{-9a-12}{-(2a-3)(2a+3)}=\dfrac{-3(3a-4)}{-(2a-3)(2a+3)}$$

$$=\dfrac{3(3a+4)}{(2a-3)(2a+3)}$$

87) $\dfrac{2}{j^2+8j}+\dfrac{2j}{j+8}-\dfrac{1}{3j}=\dfrac{2}{j(j+8)}\cdot\dfrac{3}{3}+\dfrac{2j}{j+8}\cdot\dfrac{3j}{3j}-\dfrac{1}{3j}\cdot\dfrac{(j+8)}{(j+8)}=\dfrac{6+6j^2-[j+8]}{3j(j+8)}=\dfrac{6+6j^2-j-8}{3j(j+8)}$

$$=\dfrac{6j^2-j-2}{3j(j+8)}$$

89) $\dfrac{c}{c^2-8c+16}-\dfrac{5}{c^2-c-12}=\dfrac{c}{(c-4)(c-4)}\cdot\dfrac{(c+3)}{(c+3)}-\dfrac{5}{(c-4)(c+3)}\cdot\dfrac{(c-4)}{(c-4)}$

$$=\dfrac{c^2+3c-[5c-20]}{(c-4)(c-4)(c+3)}=\dfrac{c^2+3c-5c+20}{(c-4)(c-4)(c+3)}=\dfrac{c^2-2c+20}{(c-4)^2(c+3)}$$

91) $\dfrac{1}{x+y}+\dfrac{x}{x^2-y^2}-\dfrac{4}{2x-2y}=\dfrac{1}{x+y}\cdot\dfrac{2(x-y)}{2(x-y)}+\dfrac{x}{(x+y)(x-y)}\cdot\dfrac{2}{2}-\dfrac{4}{2(x-y)}\cdot\dfrac{(x+y)}{(x+y)}$

$$=\dfrac{2x-2y}{2(x+y)(x-y)}+\dfrac{2x}{2(x+y)(x-y)}-\dfrac{4x+4y}{2(x+y)(x-y)}$$

$$=\dfrac{2x-2y+2x-[4x+4y]}{2(x+y)(x-y)}=\dfrac{2x-2y+2x-4x-4y}{2(x+y)(x-y)}$$

$$=\dfrac{-6y}{2(x+y)(x-y)}=\dfrac{-3y}{2(x+y)(x-y)}$$

93) $\dfrac{n+5}{4n^2+7n-2}-\dfrac{n-4}{3n^2+7n+2}=\dfrac{n+5}{(4n-1)(n+2)}\cdot\dfrac{(3n+1)}{(3n+1)}-\dfrac{n-4}{(3n+1)(n+2)}\cdot\dfrac{(4n-1)}{(4n-1)}=$

$$=\dfrac{3n^2+16n+5-\left[4n^2-17n+4\right]}{(4n-1)(n+2)(3n+1)}=\dfrac{3n^2+16n+5-4n^2+17n-4}{(4n-1)(n+2)(3n+1)}$$

$$=\dfrac{-n^2+33n+1}{(4n-1)(n+2)(3n+1)}$$

95) $\dfrac{y+6}{y^2-4y}+\dfrac{y}{2y^2-13y+20}-\dfrac{1}{2y^2-5y}$

$$=\dfrac{y+6}{y(y-4)}\cdot\dfrac{(2y-5)}{(2y-5)}+\dfrac{y}{(2y-5)(y-4)}\cdot\dfrac{y}{y}-\dfrac{1}{y(2y-5)}\cdot\dfrac{(y-4)}{(y-4)}$$

$$=\dfrac{2y^2+7y-30+y^2-[y-4]}{y(2y-5)(y-4)}=\dfrac{2y^2+7y-30+y^2-y+4}{y(2y-5)(y-4)}=\dfrac{3y^2+6y-26}{y(2y-5)(y-4)}$$

97) a) $A=lw=\left(\dfrac{x+1}{\cancel{2}}\right)\left(\dfrac{\cancel{4}^{\,2}}{x-3}\right)=\dfrac{2(x+1)}{x-3}$ sq units

b) $p=2l+2w=\cancel{2}\left(\dfrac{x+1}{\cancel{2}}\right)+2\left(\dfrac{4}{x-3}\right)=(x+1)\cdot\dfrac{(x-3)}{(x-3)}+\dfrac{8}{x-3}=\dfrac{x^2-2x-3+8}{x-3}=\dfrac{x^2-2x+5}{x-3}$

99) a) $A=lw=\left(\dfrac{w}{w+2}\right)\left(\dfrac{1}{w^2-4}\right)=\left(\dfrac{w}{w+2}\right)\left(\dfrac{1}{(w-2)(w+2)}\right)=\dfrac{w}{(w+2)^2(w-2)}$ sq units

b) $p=2l+2w=2\left(\dfrac{w}{w+2}\right)+2\left(\dfrac{1}{w^2-4}\right)=\dfrac{2w}{w+2}\cdot\dfrac{(w-2)}{(w-2)}+\dfrac{2}{(w+2)(w-2)}=\dfrac{2w^2-4w+2}{(w+2)(w-2)}$

$$=\dfrac{2\left(w^2-2w+1\right)}{(w+2)(w-2)}=\dfrac{(2w-1)^2}{(w+2)(w-2)}\text{ units}$$

101) $f(x) = \dfrac{6}{x}$ $g(x) = \dfrac{3x+12}{x+5}$

$$(f+g)(x) = \frac{6}{x} + \frac{3x+12}{x+5} = \frac{6}{x} \cdot \frac{x+5}{x+5} + \frac{3x+12}{x+5} \cdot \frac{x}{x} = \frac{6x+30 - \left[3x^2 + 12x\right]}{x(x+5)}$$

$$= \frac{6x+30 - 3x^2 - 12x}{x(x+5)} = \frac{-3x^2 - 6x + 30}{x(x+5)} = \frac{3\left(x^2 - 2x + 10\right)}{x(x+5)}$$

Domain: $(-\infty, -5) \cup (-5, 0) \cup (0, \infty)$

103) $f(x) = \dfrac{6}{x}$ $g(x) = \dfrac{3x+12}{x+5}$

$$(f \cdot g)(x) = \frac{6}{x} \cdot \frac{3x+12}{x+5} = \frac{6}{x} \cdot \frac{3(x+4)}{x+5}$$

$$= \frac{18(x+4)}{x(x+5)}$$

Domain: $(-\infty, -5) \cup (-5, 0) \cup (0, \infty)$

Section 8.3

1) Rewrite the problem as division

$$\frac{\dfrac{2}{9}}{\dfrac{5}{18}} = \frac{2}{9} \div \frac{5}{18} = \frac{2}{\cancel{9}} \cdot \frac{\cancel{18}^{\,2}}{5} = \frac{4}{5}$$

Multiply both numerators by the LCD.

$$\frac{\dfrac{2}{9}}{\dfrac{5}{18}} = \frac{\cancel{18}^{\,2} \cdot \dfrac{2}{\cancel{9}}}{\cancel{18} \cdot \dfrac{5}{\cancel{18}}} = \frac{4}{5}$$

3) $\dfrac{\dfrac{7}{10}}{\dfrac{5}{4}} = \dfrac{7}{10} \div \dfrac{5}{4} = \dfrac{7}{\cancel{10}_{5}} \cdot \dfrac{\cancel{4}^{\,2}}{5} = \dfrac{14}{25}$

5) $\dfrac{\dfrac{a^2}{b}}{\dfrac{a}{b^3}} = \dfrac{\cancel{b^3}^{\,b^2} \cdot \dfrac{a^2}{\cancel{b}}}{\cancel{b^3} \cdot \dfrac{a}{\cancel{b^3}}} = \dfrac{a^2 b^2}{a} = ab^2$

7) $\dfrac{\dfrac{s^3}{t^3}}{\dfrac{s^4}{t}} = \dfrac{\cancel{t} \cdot \dfrac{s^3}{\cancel{t^3}^{\,t^2}}}{\cancel{t} \cdot \dfrac{s^4}{\cancel{t}}} = \dfrac{s^3}{s^4 t^2} = \dfrac{1}{st^2}$

9) $\dfrac{\dfrac{14m^5 n^4}{9}}{\dfrac{35mn^6}{3}} = \dfrac{14m^5 n^4}{9} \div \dfrac{35mn^6}{3}$

$$= \frac{\cancel{14m^5 n^4}^{\,2m^4}}{\cancel{9}_{3}} \cdot \frac{\cancel{3}}{\cancel{35mn^6}_{5n^2}} = \frac{2m^4}{15n^2}$$

11) $\dfrac{\dfrac{t-6}{5}}{\dfrac{t-6}{t}} = \dfrac{t-6}{5} \div \dfrac{t-6}{t} = \dfrac{\cancel{t-6}}{5} \cdot \dfrac{t}{\cancel{t-6}} = \dfrac{t}{5}$

13) $\dfrac{\dfrac{8}{y^2 - 64}}{\dfrac{6}{y+8}} = \dfrac{\dfrac{8}{(y-8)(y+8)}}{\dfrac{6}{y+8}}$

$$= \frac{8}{(y-8)(y+8)} \div \frac{6}{y+8}$$

$$= \frac{\cancel{8}^{\,4}}{(y-8)\cancel{(y+8)}} \cdot \frac{\cancel{y+8}}{\cancel{6}_{3}}$$

$$= \frac{4}{3(y-8)}$$

Section 8.3: Simplifying Complex Fractions

15) $\dfrac{\dfrac{25w-35}{w^5}}{\dfrac{30w-42}{w}} = \dfrac{w^5 \cdot \dfrac{25w-35}{w^5}}{w^5 \cdot \dfrac{30w-42}{w}} = \dfrac{25w-35}{w^4(30w-42)}$

$$= \dfrac{5(5w-7)}{6w^4(5w-7)} = \dfrac{5}{6w^4}$$

17) $\dfrac{\dfrac{2x}{x+7}}{\dfrac{2}{x^2+4x-21}} = \dfrac{2x}{x+7} \div \dfrac{2}{\underset{(x+7)(x-3)}{x^2+4x-21}}$

$$= \dfrac{2x}{x+7} \cdot \dfrac{(x+7)(x-3)}{2}$$

$$= x(x-3)$$

19) $\dfrac{\dfrac{1}{4}+\dfrac{3}{2}}{\dfrac{2}{3}+\dfrac{1}{2}} = \dfrac{12\cdot\left(\dfrac{1}{4}+\dfrac{3}{2}\right)}{12\cdot\left(\dfrac{2}{3}+\dfrac{1}{2}\right)} = \dfrac{\dfrac{12}{4}+\dfrac{36}{2}}{\dfrac{24}{3}+\dfrac{12}{2}} = \dfrac{3+18}{8+6}$

$$= \dfrac{21}{14} = \dfrac{3}{2}$$

$\dfrac{\dfrac{1}{4}+\dfrac{3}{2}}{\dfrac{2}{3}+\dfrac{1}{2}} = \left(\dfrac{1}{4}+\dfrac{3}{2}\right) \div \left(\dfrac{2}{3}+\dfrac{1}{2}\right) = \dfrac{1+3\cdot2}{4} \div \dfrac{2\cdot2+3}{6}$

$$= \dfrac{7}{4} \div \dfrac{7}{6} = \dfrac{7}{4} \cdot \dfrac{6}{7} = \dfrac{6}{4} = \dfrac{3}{2}$$

21) $\dfrac{\dfrac{7}{c}+\dfrac{2}{d}}{1-\dfrac{5}{c}} = \dfrac{cd\cdot\left(\dfrac{7}{c}+\dfrac{2}{d}\right)}{cd\cdot\left(1-\dfrac{5}{c}\right)} = \dfrac{\dfrac{7cd}{c}+\dfrac{2cd}{d}}{cd-\dfrac{5cd}{c}} = \dfrac{7d+2c}{cd-5d} = \dfrac{7d+2c}{d(c-5)}$

$\dfrac{\dfrac{7}{c}+\dfrac{2}{d}}{1-\dfrac{5}{c}} = \left(\dfrac{7}{c}+\dfrac{2}{d}\right) \div \left(1-\dfrac{5}{c}\right) = \dfrac{7d+2c}{cd} \div \dfrac{c-5}{c} = \dfrac{7d+2c}{cd} \cdot \dfrac{c}{c-5} = \dfrac{7d+2c}{d(c-5)}$

23) $\dfrac{\dfrac{5}{z-2}-\dfrac{1}{z+1}}{\dfrac{1}{z-2}+\dfrac{4}{z+1}} = \dfrac{(z+1)(z-2)\left(\dfrac{5}{z-2}-\dfrac{1}{z+1}\right)}{(z+1)(z-2)\left(\dfrac{1}{z-2}+\dfrac{4}{z+1}\right)} = \dfrac{5(z+1)-(z-2)}{(z+1)+4(z-2)} = \dfrac{5z+5-z+2}{z+1+4z-8} = \dfrac{4z+7}{5z-7}$

$\dfrac{\dfrac{5}{z-2}-\dfrac{1}{z+1}}{\dfrac{1}{z-2}+\dfrac{4}{z+1}} = \left(\dfrac{5}{z-2}-\dfrac{1}{z+1}\right) \div \left(\dfrac{1}{z-2}+\dfrac{4}{z+1}\right) = \dfrac{5(z+1)-(z-2)}{(z-2)(z+1)} \div \dfrac{z+1+4(z-2)}{(z-2)(z+1)}$

$= \dfrac{5z+5-z+2}{(z-2)(z+1)} \div \dfrac{z+1+4z-8}{(z-2)(z+1)} = \dfrac{4z+7}{\cancel{(z-2)}\cancel{(z+1)}} \cdot \dfrac{\cancel{(z-2)}\cancel{(z+1)}}{5z-7} = \dfrac{4z+7}{5z-7}$

25) $\dfrac{9+\dfrac{5}{y}}{\dfrac{9y+5}{8}} = \dfrac{8y\cdot\left(9+\dfrac{5}{y}\right)}{8y\cdot\left(\dfrac{9y+5}{8}\right)} = \dfrac{72y+40}{9y^2+5y} = \dfrac{8\cancel{(9y+5)}}{y\cancel{(9y+5)}} = \dfrac{8}{y}$

27) $\dfrac{x-\dfrac{7}{x}}{x-\dfrac{11}{x}} = \dfrac{x\left(x-\dfrac{7}{x}\right)}{x\left(x-\dfrac{11}{x}\right)} = \dfrac{x^2-7}{x^2-11}$

29) $\dfrac{\dfrac{4}{3}+\dfrac{2}{5}}{\dfrac{1}{6}-\dfrac{2}{3}} = \dfrac{30\left(\dfrac{4}{3}+\dfrac{2}{5}\right)}{30\left(\dfrac{1}{6}-\dfrac{2}{3}\right)} = \dfrac{\dfrac{120}{3}+\dfrac{60}{5}}{\dfrac{30}{6}-\dfrac{60}{3}}$

$= \dfrac{40+12}{5-20} = -\dfrac{52}{15}$

31) $\dfrac{\dfrac{2}{a}-\dfrac{2}{b}}{\dfrac{1}{a^2}-\dfrac{1}{b^2}} = \dfrac{a^2b^2\cdot\left(\dfrac{2}{a}-\dfrac{2}{b}\right)}{a^2b^2\cdot\left(\dfrac{1}{a^2}-\dfrac{1}{b^2}\right)}$

$= \dfrac{2\cancel{a}^{\,a}b^2 - 2a^2\cancel{b}^{\,b}}{\cancel{a}\quad\cancel{b}}{b^2-a^2}$

$= \dfrac{2ab^2-2a^2b}{(b-a)(b+a)} = \dfrac{2ab\cancel{(b-a)}}{\cancel{(b-a)}(b+a)}$

$= \dfrac{2ab}{b+a}$

33) $\dfrac{\dfrac{r}{s^2}+\dfrac{1}{rs}}{\dfrac{s}{r}+\dfrac{1}{r^2}} = \dfrac{s^2r^2\cdot\left(\dfrac{r}{s^2}+\dfrac{1}{rs}\right)}{s^2r^2\cdot\left(\dfrac{s}{r}+\dfrac{1}{r^2}\right)} = \dfrac{r^3+sr}{s^3r+s^2}$

$= \dfrac{r\left(r^2+s\right)}{s^2\left(sr+1\right)}$

Section 8.3: Simplifying Complex Fractions

35) $\dfrac{1-\dfrac{4}{t+5}}{\dfrac{4}{t^2-25}+\dfrac{t}{t-5}}$

$= \dfrac{(t-5)(t+5)\left(1-\dfrac{4}{t+5}\right)}{(t-5)(t+5)\left(\dfrac{4}{(t-5)(t+5)}+\dfrac{t}{t-5}\right)}$

$= \dfrac{(t-5)(t+5)-4(t-5)}{4+t(t+5)} = \dfrac{t^2-25-4t+20}{4+t^2+5t}$

$= \dfrac{t^2-4t-5}{t^2+5t+4} = \dfrac{(t-5)\,\cancel{(t+1)}}{(t+4)\,\cancel{(t+1)}} = \dfrac{t-5}{t+4}$

37) $\dfrac{b+\dfrac{1}{b}}{b-\dfrac{3}{b}} = \dfrac{b\cdot\left(b+\dfrac{1}{b}\right)}{b\cdot\left(b-\dfrac{3}{b}\right)} = \dfrac{b^2+1}{b^2-3}$

39) $\dfrac{\dfrac{m}{n^2}}{\dfrac{m^4}{n}} = \dfrac{m}{n^2}\div\dfrac{m^4}{n} = \dfrac{\cancel{m}}{\underset{n}{\cancel{n^2}}}\cdot\dfrac{\cancel{n}}{\underset{m^3}{\cancel{m^4}}} = \dfrac{1}{m^3 n}$

41) $\dfrac{\dfrac{h^2-1}{4h-12}}{\dfrac{7h+7}{h^2-9}} = \dfrac{h^2-1}{4h-12}\div\dfrac{7h+7}{h^2-9}$

$\qquad = \dfrac{(h-1)(h+1)}{4(h-3)}\div\dfrac{7(h+1)}{(h-3)(h+3)}$

$\qquad = \dfrac{(h-1)\,\cancel{(h+1)}}{4\,\cancel{(h-3)}}\cdot\dfrac{\cancel{(h-3)}\,(h+3)}{7\,\cancel{(h+1)}}$

$\qquad = \dfrac{(h-1)(h+3)}{28}$

43) $\dfrac{\dfrac{6}{x+3}-\dfrac{4}{x-1}}{\dfrac{2}{x-1}+\dfrac{1}{x+2}}=\left(\dfrac{6}{x+3}-\dfrac{4}{x-1}\right)\div\left(\dfrac{2}{x-1}+\dfrac{1}{x+2}\right)=\dfrac{6(x-1)-4(x+3)}{(x+3)(x-1)}\div\dfrac{2(x+2)+x-1}{(x-1)(x+2)}$

$=\dfrac{6x-6-4x-12}{(x+3)(x-1)}\cdot\dfrac{(x-1)(x+2)}{2x+4+x-1}=\dfrac{2x-18}{(x+3)(x-1)}\cdot\dfrac{(x-1)(x+2)}{3x+3}$

$=\dfrac{2(x-9)}{(x+3)(x-1)}\cdot\dfrac{(x-1)(x+2)}{3(x+1)}=\dfrac{2(x-9)(x+2)}{3(x+1)(x+3)}$

45) $\dfrac{\dfrac{r^2-6}{20}}{r-\dfrac{6}{r}}=\dfrac{20r\cdot\left(\dfrac{r^2-6}{20}\right)}{20r\cdot\left(r-\dfrac{6}{r}\right)}=\dfrac{r^3-6r}{20r^2-120}=\dfrac{r(r^2-6)}{20(r^2-6)}=\dfrac{r}{20}$

47) $\dfrac{\dfrac{a-4}{12}}{\dfrac{a-4}{a}}=\dfrac{a-4}{12}\div\dfrac{a-4}{a}=\dfrac{a-4}{12}\cdot\dfrac{a}{a-4}=\dfrac{a}{12}$

49) $\dfrac{\dfrac{5}{6}}{\dfrac{9}{15}}=\dfrac{5}{6}\div\dfrac{9}{15}=\dfrac{5}{6}\cdot\dfrac{\overset{5}{15}}{9}=\dfrac{25}{18}$

51) $\dfrac{\dfrac{5}{2n+1}+1}{\dfrac{1}{n+3}+\dfrac{2}{2n+1}}=\left(\dfrac{5}{2n+1}+1\right)\div\left(\dfrac{1}{n+3}+\dfrac{2}{2n+1}\right)=\dfrac{5+2n+1}{2n+1}\div\dfrac{2n+1+2(n+3)}{(n+3)(2n+1)}$

$=\dfrac{2n+6}{2n+1}\cdot\dfrac{(n+3)(2n+1)}{2n+1+2n+6}=\dfrac{2(n+3)}{2n+1}\cdot\dfrac{(n+3)(2n+1)}{4n+7}=\dfrac{2(n+3)^2}{4n+7}$

53) $\dfrac{w^{-1}-v^{-1}}{2w^{-2}+v^{-1}}=\dfrac{w^2v\cdot\left(\dfrac{1}{w}-\dfrac{1}{v}\right)}{w^2v\cdot\left(\dfrac{2}{w^2}+\dfrac{1}{v}\right)}=\dfrac{wv-w^2}{2v+w^2}=\dfrac{w(v-w)}{2v+w^2}$

Section 8.3: Simplifying Complex Fractions

55) $\dfrac{8x^{-2}}{x^{-1}-y^{-2}} = \dfrac{x^2y^2\cdot\left(\dfrac{8}{x^2}\right)}{x^2y^2\cdot\left(\dfrac{1}{x}-\dfrac{1}{y^2}\right)} = \dfrac{8y^2}{xy^2-x^2} = \dfrac{8y^2}{x\left(y^2-x\right)}$

57) $\dfrac{a^{-3}+b^{-2}}{2b^{-2}-7} = \dfrac{a^3b^2\cdot\left(\dfrac{1}{a^3}+\dfrac{1}{b^2}\right)}{a^3b^2\cdot\left(\dfrac{2}{b^2}-7\right)} = \dfrac{b^2+a^3}{2a^3-7a^3b^2} = \dfrac{a^3+b^2}{a^3\left(2-7b^2\right)}$

59) $\dfrac{4m^{-1}-n^{-1}}{n^{-1}+m} = \dfrac{mn\cdot\left(\dfrac{4}{m}-\dfrac{1}{n}\right)}{mn\cdot\left(\dfrac{1}{n}+m\right)} = \dfrac{4n-m}{m+m^2n} = \dfrac{4n-m}{m\left(1+mn\right)}$

61) 0

x	$f(x)$
1	1
2	$\dfrac{1}{2}$
3	$\dfrac{1}{3}$
10	$\dfrac{1}{10}$
100	$\dfrac{1}{100}$
1000	$\dfrac{1}{1000}$

Section 8.4

1) Eliminate the denominator.

3) Sum: $\dfrac{3m-14}{8}$

$$\frac{m}{8}+\frac{m-7}{4}=\frac{m}{8}+\frac{m-7}{4}\cdot\frac{2}{2}=\frac{m}{8}+\frac{2(m-7)}{8}$$

$$=\frac{m+2m-14}{8}$$

$$=\frac{3m-14}{8}$$

5) Equation: {-9}

$$\frac{2f-19}{20}=\frac{f}{4}+\frac{2}{5}$$

$$20\cdot\left(\frac{2f-19}{20}\right)=\overset{5}{20}\cdot\left(\frac{f}{\cancel{4}}\right)+\overset{4}{20}\cdot\left(+\frac{2}{\cancel{5}}\right)$$

$$2f-19=5f+8$$

$$2f-5f=8+19$$

$$-3f=27$$

$$f=-9$$

7) Difference: $\dfrac{z^2-4z+24}{z(z-6)}$

$$\frac{z}{z-6}-\frac{4}{z}=\frac{z\cdot z-4(z-6)}{z(z-6)}=\frac{z^2-4z+24}{z(z-6)}$$

9) Equation: {3}

$$1+\frac{4}{c+2}=\frac{9}{c+2}$$

$$(c+2)\left(1+\frac{4}{c+2}\right)=(c+2)\left(\frac{9}{c+2}\right)$$

$$c+2+4=9$$

$$c+6=9$$

$$c=9-6$$

$$c=3$$

11) $t\neq-10,0$

13) $d^2-81=(d-9)(d+9)$

$$d\neq-9,0,9$$

15) $v^2-13v+36=(v-4)(v-9)$

and

$$3v-12=3(v-4)$$

$$v\neq4,9$$

17) $\quad\dfrac{y}{3}-\dfrac{1}{2}=\dfrac{1}{6}$

$$6\cdot\frac{y}{3}-6\cdot\frac{1}{2}=6\cdot\frac{1}{6}$$

$$2y-3=1$$

$$2y=1+3$$

$$2y=4$$

$$y=2\qquad\{2\}$$

19) $\quad\dfrac{1}{2}h+h=-3$

$$2\cdot\frac{1}{2}h+2\cdot h=2\cdot-3$$

$$h+2h=-6$$

$$3h=-6$$

$$h=-2\qquad\{-2\}$$

21) $\quad\dfrac{7u+12}{15}=\dfrac{2u}{5}-\dfrac{3}{5}$

$$15\left(\frac{7u+12}{15}\right)=15\left(\frac{2u}{5}-\frac{3}{5}\right)$$

$$7u+12=6u-9$$

$$7u-6u=-9-12$$

$$u=-21\qquad\{-21\}$$

Section 8.4: Solving Rational Equations

23) $\dfrac{4}{3t+2} = \dfrac{2}{2t-1}$

$4(2t-1) = 2(3t+2)$

$8t-4 = 6t+4$

$8t-6t = 4+4$

$2t = 8$

$t = 4 \qquad \{4\}$

25) $\dfrac{w}{3} = \dfrac{2w-5}{12}$

$12w = 3(2w-5)$

$12w = 6w-15$

$12w-6w = -15$

$6w = -15$

$w = -\dfrac{15}{6} = -\dfrac{5}{2} \qquad \left\{-\dfrac{5}{2}\right\}$

27) $\dfrac{12}{a} - 2 = \dfrac{6}{a}$

$a\left(\dfrac{12}{a}\right) - a\cdot 2 = a\left(\dfrac{6}{a}\right)$

$12-2a = 6$

$-2a = 6-12$

$-2a = -6$

$a = 3 \qquad \{3\}$

29) $\dfrac{n}{n+2} + 3 = \dfrac{8}{n+2}$

$(n+2)\left(\dfrac{n}{n+2}+3\right) = (n+2)\left(\dfrac{8}{n+2}\right)$

$n+3(n+2) = 8$

$n+3n+6 = 8$

$4n = 8-6$

$4n = 2$

$n = \dfrac{2}{4} = \dfrac{1}{2} \qquad \left\{\dfrac{1}{2}\right\}$

31) $\dfrac{2}{s+6} + 4 = \dfrac{2}{s+6}$

$(s+6)\left(\dfrac{2}{s+6}+4\right) = (s+6)\left(\dfrac{2}{s+6}\right)$

$2+4(s+6) = 2$

$2+4s+24 = 2$

$4s = 2-2-24$

$4s = -24$

$s = -6 \qquad \{-6\}$

s cannot be -6, therefore $\varnothing$

33) $\dfrac{c}{c-7} - 4 = \dfrac{10}{c-7}$

$(c-7)\left(\dfrac{c}{c-7}-4\right) = (c-7)\left(\dfrac{10}{c-7}\right)$

$c-4(c-7) = 10$

$c-4c+28 = 10$

$-3c = 10-28$

$-3c = -18$

$c = 6 \qquad \{6\}$

35) $\dfrac{32}{g} + 10 = -\dfrac{8}{g}$

$g\left(\dfrac{32}{g}+10\right) = g\left(-\dfrac{8}{g}\right)$

$32+10g = -8$

$10g = -8-32$

$10g = -40$

$g = -4 \qquad \{-4\}$

37) $\dfrac{1}{m-1}+\dfrac{2}{m+3}=\dfrac{4}{m+3}$

$(m-1)(m+3)\left(\dfrac{1}{m-1}+\dfrac{2}{m+3}\right)$

$=(m-1)(m+3)\left(\dfrac{4}{m+3}\right)$

$m+3+2(m-1)=4(m-1)$

$m+3+2m-2=4m-4$

$m+2m-4m=-4-3+2$

$-m=-5$

$m=5 \qquad \{5\}$

39) $\dfrac{4}{w-8}-\dfrac{10}{w+8}=\dfrac{40}{w^2-64} \Rightarrow (w+8)(w-8)\left(\dfrac{4}{w-8}-\dfrac{10}{w+8}\right)=(w+8)(w-8)\left(\dfrac{40}{(w+8)(w-8)}\right)$

$4(w+8)-10(w-8)=40$

$4w+32-10w+80=40$

$-6w=40-80-32$

$-6w=-72$

$w=12 \qquad \{12\}$

41) $\dfrac{3}{a+3}+\dfrac{14}{a^2-4a-21}=\dfrac{5}{a-7} \Rightarrow (a+3)(a-7)\left(\dfrac{3}{a+3}+\dfrac{14}{(a+3)(a-7)}\right)=(a+3)(a-7)\left(\dfrac{5}{a-7}\right)$

$3(a-7)+14=5(a+3)$

$3a-21+14=5a+15$

$3a-5a=15-14+21$

$-2a=22$

$a=-11 \qquad \{-11\}$

43) $\dfrac{9}{t+4}+\dfrac{8}{t^2-16}=\dfrac{1}{t-4} \Rightarrow (t+4)(t-4)\left(\dfrac{9}{t+4}+\dfrac{8}{(t+4)(t-4)}\right)=(t+4)(t-4)\left(\dfrac{1}{t-4}\right)$

$$9(t-4)+8=t+4$$
$$9t-36+8=t+4$$
$$9t-t=4+36-8$$
$$8t=32$$
$$t=4$$

t cannot be 4, therefore $\varnothing$

g cannot equal 3, therefore $\varnothing$

45) $\dfrac{4}{x^2+2x-15}=\dfrac{8}{x-3}+\dfrac{2}{x+5} \Rightarrow (x-3)(x+5)\left(\dfrac{4}{(x-3)(x+5)}\right)=(x-3)(x+5)\left(\dfrac{8}{x-3}+\dfrac{2}{x+5}\right)$

$$4=8(x+5)+2(x-3)$$
$$4=8x+40+2x-6$$
$$4-40+6=10x$$
$$-30=10x$$
$$-3=x \qquad \{-3\}$$

47) $\dfrac{k^2}{3}=\dfrac{k^2+2k}{4}$

$$4k^2=3\left(k^2+2k\right)$$
$$4k^2=3k^2+6k$$
$$4k^2-3k^2-6k=0$$
$$k^2-6k=0$$
$$k(k-6)=0$$
$$k=0,6 \qquad \{0,6\}$$

49) $\dfrac{5}{m^2-25} = \dfrac{4}{m^2+5m}$

$$5\left(m^2+5m\right) = 4\left(m^2-25\right)$$

$$5m\left(m+5\right) = 4\left(m+5\right)\left(m-5\right)$$

$$5m^2+25m = 4\left(m^2-25\right)$$

$$5m^2+25m = 4m^2-100$$

$$5m^2-4m^2+25m+100 = 0$$

$$m^2+25m+100 = 0$$

$$\left(m+20\right)\left(m+5\right) = 0$$

$$m = -20, -5 \qquad \{-20, -5\}$$

m cannot be -5, therefore $m = -20$

51) $\dfrac{10v}{\cancel{3v-12}} - \dfrac{v+6}{v-4} = \dfrac{v}{3}$
$\quad\; {\scriptstyle 3(v-4)}$

$$3(v-4)\left(\dfrac{10v}{3(v-4)} - \dfrac{v+6}{v-4}\right) = 3(v-4)\left(\dfrac{v}{3}\right)$$

$$10v - 3(v+6) = v(v-4)$$

$$10v - 3v - 18 = v^2 - 4v$$

$$0 = v^2 - 4v - 10v + 3v + 18$$

$$0 = v^2 - 11v + 18$$

$$0 = (v-2)(v-9)$$

$$2, 9 = v \qquad \{2, 9\}$$

53) $\dfrac{w}{5} = \dfrac{w-3}{w+1} + \dfrac{12}{\cancel{5w+5}}$
$\qquad\qquad\qquad {\scriptstyle 5(w+1)}$

$$5(w+1)\left(\dfrac{w}{5}\right) = 5(w+1)\left(\dfrac{w-3}{w+1} + \dfrac{12}{5(w+1)}\right)$$

$$w(w+1) = 5(w-3) + 12$$

$$w^2 + w = 5w - 15 + 12$$

$$w^2 + w - 5w + 15 - 12 = 0$$

$$w^2 - 4w + 3 = 0$$

$$(w-3)(w-1) = 0$$

$$x = 3, 1 \qquad\qquad \{3, 1\}$$

55)

$$\dfrac{8}{p+2} + \dfrac{p}{p+1} = \dfrac{5p+2}{\cancel{p^2+3p+2}}$$
$\qquad\qquad\qquad\qquad {\scriptstyle (p+2)(p+1)}$

$$(p+1)(p+2)\left(\dfrac{8}{p+2} + \dfrac{p}{p+1}\right)$$

$$= (p+1)(p+2)\left(\dfrac{5p+2}{(p+2)(p+1)}\right)$$

$$8(p+1) + p(p+2) = 5p+2$$

$$8p + 8 + p^2 + 2p - 5p - 2 = 0$$

$$p^2 + 5p + 6 = 0$$

$$(p+3)(p+2) = 0$$

$$p = -3, -2$$

p cannot be -2, therefore $p = -3$

Section 8.4: Solving Rational Equations

57) $\dfrac{11}{c+9} = \dfrac{c}{c-4} - \dfrac{36-8c}{\underset{(c+9)(c-4)}{\cancel{c^2+5c-36}}}$

$(c+9)(c-4)\left(\dfrac{11}{c+9}\right)$

$\quad = (c+9)(c-4)\left(\dfrac{c}{c-4} - \dfrac{36-8c}{(c+9)(c-4)}\right)$

$11(c-4) = c(c+9) - (36-8c)$

$11c - 44 = c^2 + 9c - 36 + 8c$

$0 = c^2 + 9c - 36 + 8c - 11c + 44$

$0 = c^2 + 6c + 8$

$0 = (c+4)(c+2)$

$-4, -2 = c$

59) $\dfrac{8}{\underset{(3g+2)(g-3)}{\cancel{3g^2-7g-6}}} + \dfrac{4}{g-3} = \dfrac{8}{3g+2}$

$(g-3)(3g+2)\left(\dfrac{8}{(g-3)(3g+2)} + \dfrac{4}{g-3}\right)$

$\quad = (g-3)(3g+2)\left(\dfrac{8}{3g+2}\right)$

$8 + 4(3g+2) = 8(g-3)$

$8 + 12g + 8 = 8g - 24$

$12g - 8g = -24 - 8 - 8$

$4g = -40$

$g = -10$

61) $\dfrac{h}{h^2+2h-8} + \dfrac{4}{h^2+8h-20} = \dfrac{4}{h^2+14h+40}$

$\dfrac{h}{(h+4)(h-2)} + \dfrac{4}{(h+10)(h-2)} = \dfrac{4}{(h+10)(h+4)}$

$(h+4)(h-2)(h+10)\left[\dfrac{h}{(h+4)(h-2)} + \dfrac{4}{(h+10)(h-2)} = \dfrac{4}{(h+10)(h+4)}\right]$

$h(h+10) + 4(h+4) = 4(h-2)$

$h^2 + 10h + 4h + 16 = 4h - 8$

$h^2 + 14h + 16 - 4h + 8 = 0$

$h^2 + 10h + 24 = 0$

$(h+6)(h+4) = 0$

$h = -4, -6$

h cannot be -4, therefore $h = -6$

63) $\dfrac{u}{8} = \dfrac{2}{10-u}$

$u(10-u) = 2\cdot 8$

$10u - u^2 = 16$

$0 = u^2 - 10u + 16$

$0 = (u-8)(u-2)$

$2, 8 = u \qquad\qquad \{2, 8\}$

65) $\dfrac{5}{r+4} - \dfrac{2}{r} = -1$

$r(r+4)\dfrac{5}{r+4} - \dfrac{2}{r} = r(r+4)(-1)$

$5r - 2(r+4) = -r^2 - 4r$

$5r - 2r - 8 = -r^2 - 4r$

$r^2 + 7r - 8 = 0$

$(r+8)(r-1) = u \qquad\qquad \{1, -8\}$

67) $\dfrac{q}{q^2 + 4q - 32} + \dfrac{2}{q^2 - 14q + 40} = \dfrac{6}{q^2 - 2q - 80}$

$(q+8)(q-4)(q-10)\left[\dfrac{q}{(q+8)(q-4)} + \dfrac{2}{(q-10)(q-4)} = \dfrac{6}{(q-10)(q+8)}\right]$

$q(q-10) + 2(q+8) = 6(q-4)$

$q^2 - 10q + 2q + 16 = 6q - 24$

$q^2 - 10q + 2q + 16 - 6q + 24 = 0$

$q^2 - 14q + 40 = 0$

$(q-10)(q-4) = 0$

$q = 4, 10; \quad q \text{ cannot be 4 or 10; therefore } q = \varnothing$

69) $A(x) = 6 - \dfrac{1500}{x}$

$x(4.00) = x\left(6 - \dfrac{1500}{x}\right)$

$4x = 6x - 1500$

$4x - 6x = -1500$

$-2x = -1500$

$x = 750 \text{ lbs}$

71) $P = \dfrac{1}{f}$

$2.5 = \dfrac{1}{f}$

$2.5f = 1$

$f = \dfrac{1}{2.5} = 0.4 \text{ m}$

73) $V = \dfrac{nRT}{P}$

$VP = nRT$

$P = \dfrac{nRT}{V}$

75) $y = \dfrac{kx}{z}$

$yz = kx$

$z = \dfrac{kx}{y}$

77) $B = \dfrac{t+u}{3x}$

$3Bx = t+u$

$x = \dfrac{t+u}{3B}$

79) $z = \dfrac{a}{b+c}$

$z(b+c) = a$

$zb + zc = a$

$zb = a - zc$

$b = \dfrac{a - zc}{z}$

81) $A = \dfrac{4r}{q-t}$

$A(q-t) = 4r$

$Aq - At = 4r$

$-At = 4r - Aq$

$At = -4r + Aq$

$t = \dfrac{Aq - 4r}{A}$

83) $w = \dfrac{na}{kc+b}$

$w(kc+b) = na$

$wkc + wb = na$

$wkc = na - wb$

$c = \dfrac{na - wb}{wk}$

85) $\dfrac{1}{t} = \dfrac{1}{r} - \dfrac{1}{s}$

$srt\left[\dfrac{1}{t} = \dfrac{1}{r} - \dfrac{1}{s}\right]$

$sr = st - rt$

$sr + rt = st$

$r(s+t) = st$

$r = \dfrac{st}{s+t}$

87) $\dfrac{2}{A} + \dfrac{1}{C} = \dfrac{3}{B}$

$ABC\left(\dfrac{2}{A} + \dfrac{1}{C}\right) = ABC\left(\dfrac{3}{B}\right)$

$\quad 2BC + AB = 3AC$

$\quad 2BC - 3AC = -AB$

$\quad C(2B - 3A) = -AB$

$\qquad C = \dfrac{-AB}{2B - 3A} \text{ or } C = \dfrac{AB}{3A - 2B}$

Chapter 8: Putting It All Together

1) $f(x) = \dfrac{4x}{x^2 - 9}$

    a) $f(2) = \dfrac{4 \cdot 2}{2^2 - 9} = \dfrac{8}{4 - 9} = -\dfrac{8}{5}$

    b) $0 = \dfrac{4x}{x^2 - 9}$

      $0 = 4x$

      $0 = x$

    c) Domain: $(-\infty, -3) \cup (-3, 3) \cup (3, \infty)$

3) $f(x) = \dfrac{12}{2x - 1}$

    a) $f(2) = \dfrac{12}{2 \cdot 2 - 1} = \dfrac{12}{4 - 1} = \dfrac{12}{3} = 4$

    b) $0 = \dfrac{12}{2x - 1}$

      $0 \neq 12$

      Never equals zero.

    c) $2x - 1 = 0$

      $2x = 1$

      $x = \dfrac{1}{2}$

      Domain: $\left(-\infty, \dfrac{1}{2}\right) \cup \left(\dfrac{1}{2}, \infty\right)$

5) $f(x) = \dfrac{x - 1}{x^2 + 2}$

    a) $f(2) = \dfrac{2 - 1}{2^2 + 2} = \dfrac{1}{4 + 2} = \dfrac{1}{6}$

    b) $0 = \dfrac{x - 1}{x^2 + 2}$

      $0 = x - 1$

      $1 = x$

    c) Domain: $(-\infty, \infty)$

7) $\dfrac{36n^9}{27n^{12}} = \dfrac{4n^{-3}}{3} = \dfrac{4}{3n^3}$

9) $\dfrac{2j + 5}{2j^2 - 3j - 20} = \dfrac{2\cancel{j + 5}}{(2\cancel{j + 5})(j - 4)} = \dfrac{1}{j - 4}$

11) $\dfrac{12 - 15n}{5n^2 + 6n - 8} = \dfrac{3(4 - 5n)}{(5n - 4)(n + 2)}$

$\qquad = \dfrac{-3(\cancel{5n - 4})}{(\cancel{5n - 4})(n + 2)} = \dfrac{-3}{n + 2}$

13) $\dfrac{5}{f + 8} - \dfrac{2}{f} = \dfrac{5f - 2(f + 8)}{f(f + 8)} = \dfrac{5f - 2f - 16}{f(f + 8)}$

$\qquad = \dfrac{3f - 16}{f(f + 8)}$

15) $\dfrac{\cancel{9}\,\overset{a^2}{\cancel{a^5}}}{\cancel{10}\,\cancel{b}} \cdot \dfrac{\overset{4}{\cancel{40}}\,\overset{b}{\cancel{b^2}}}{\underset{9}{\cancel{81}}\,\cancel{a}} = \dfrac{4a^2 b}{9}$

17) $\dfrac{3}{q^2-q-20}+\dfrac{8q}{q^2+11q+28}$

$=\dfrac{3}{(q-5)(q+4)}+\dfrac{8q}{(q+4)(q+7)}$

$=\dfrac{3(q+7)+8q(q-5)}{(q-5)(q+4)(q+7)}$

$=\dfrac{3q+21+8q^2-40q}{(q-5)(q+4)(q+7)}$

$=\dfrac{8q^2-37q+21}{(q-5)(q+4)(q+7)}$

19) $\dfrac{16-m^2}{m+4}\div\dfrac{8m-32}{m+7}$

$=\dfrac{-(m^2-16)}{m+4}\cdot\dfrac{m+7}{8(m-4)}$

$=\dfrac{-(m-4)(m+4)}{m+4}\cdot\dfrac{m+7}{8(m-4)}$

$=-\dfrac{m+7}{8}$

21) $\dfrac{13}{r-8}+\dfrac{4}{8-r}=\dfrac{13}{r-8}+\dfrac{4}{-(r-8)}$

$=\dfrac{-13+4}{-(r-8)}=\dfrac{-9}{-(r-8)}=\dfrac{9}{r-8}$

23) $\dfrac{a^2-4}{a^3+8}\cdot\dfrac{5a^2-10a+20}{3a-6}$

$=\dfrac{(a+2)(a-2)}{(a+2)(a^2-2a+4)}\cdot\dfrac{5(a^2-2a+4)}{3(a-2)}$

$=\dfrac{5}{3}$

25)

$\dfrac{10}{x-8}+\dfrac{4}{x+3}=\dfrac{10(x+3)+4(x-8)}{(x-8)(x+3)}$

$=\dfrac{10x+30+4x-32}{(x-8)(x+3)}$

$=\dfrac{14x-2}{(x-8)(x+3)}=\dfrac{2(7x-1)}{(x-8)(x+3)}$

27) $\dfrac{13}{5z}-\dfrac{1}{3z}=\dfrac{13\cdot3-5}{15z}=\dfrac{39-5}{15z}=\dfrac{34}{15z}$

29) $\dfrac{10q}{8p-10q}+\dfrac{8p}{10q-8p}$

$=\dfrac{10q}{8p-10q}+\dfrac{8p}{-(8p-10q)}$

$=\dfrac{-10q+8p}{-(8p-10q)}$

$=\dfrac{8p-10q}{-(8p-10q)}$

$=-1$

31) $\dfrac{6u+1}{3u^2-2u}-\dfrac{u}{3u^2+u-2}+\dfrac{10}{u^2+u}$

$=\dfrac{6u+1}{u(3u-2)}-\dfrac{u}{(3u-2)(u+1)}+\dfrac{10}{u(u+1)}$

$=\dfrac{(6u+1)(u+1)-[u\cdot u]+10(3u-2)}{u(3u-2)(u+1)}$

$=\dfrac{6u^2+7u+1-u^2+30u-20}{u(3u-2)(u+1)}$

$=\dfrac{5u^2+37u-19}{u(3u-2)(u+1)}$

33) $\dfrac{x}{2x^2-7x-4}-\dfrac{x+3}{4x^2+4x+1}$

$=\dfrac{x}{(2x+1)(x-4)}-\dfrac{x+3}{(2x+1)(2x+1)}$

$=\dfrac{x(2x+1)-[(x+3)(x-4)]}{(2x+1)^2(x-4)}$

$=\dfrac{2x^2+x-[x^2-x-12]}{(2x+1)^2(x-4)}$

$=\dfrac{2x^2+x-x^2+x+12}{(2x+1)^2(x-4)}$

$=\dfrac{x^2+2x+12}{(2x+1)^2(x-4)}$

35) $\left(\dfrac{2c}{c+8}+\dfrac{4}{c-2}\right)\div\dfrac{6}{5c+40}$

$=\left(\dfrac{2c(c-2)+4(c+8)}{(c+8)(c-2)}\right)\div\dfrac{6}{5c+40}$

$=\dfrac{2c^2\cancel{-4c}\cancel{+4c}+32}{(c+8)(c-2)}\cdot\dfrac{5(c+8)}{6}$

$=\dfrac{2c^2+32}{\cancel{(c+8)}(c-2)}\cdot\dfrac{5\cancel{(c+8)}}{6}$

$=\dfrac{\overset{5}{\cancel{10}}(c^2+16)}{\underset{3}{\cancel{6}}(c-2)}$

$=\dfrac{5(c^2+16)}{3(c-2)}$

37)

$\dfrac{3}{\underset{w(w-1)}{\cancel{w^2-w}}}+\dfrac{4}{5w}-\dfrac{3}{w-1}$

$=\dfrac{3\cdot5+4(w-1)-3\cdot5w}{5w(w-1)}=\dfrac{15+4w-4-15w}{5w(w-1)}$

$=\dfrac{-11w+11}{5w(w-1)}=\dfrac{-11\cancel{(w-1)}}{5w\cancel{(w-1)}}=-\dfrac{11}{5w}$

39) $l=\dfrac{x-3}{4};\quad w=\dfrac{x}{2}$

a) $A=lw=\left(\dfrac{x-3}{4}\right)\left(\dfrac{x}{2}\right)=\dfrac{x^2-3x}{8}$

$=\dfrac{x(x-3)}{8}$ sq units

b)

$P=2l+2w=\cancel{2}\left(\dfrac{x-3}{\underset{2}{\cancel{4}}}\right)+2\left(\dfrac{x}{2}\right)$

$=\dfrac{x-3+2x}{2}=\dfrac{3x-3}{2}=\dfrac{3(x-1)}{2}$ units

41) $f(x)+g(x)=\dfrac{x}{3x-1}+\dfrac{2x^2}{9x^2-1}$

$=\dfrac{x}{3x-1}+\dfrac{2x^2}{(3x-1)(3x+1)}=\dfrac{x(3x+1)+2x^2}{(3x-1)(3x+1)}$

$=\dfrac{3x^2+x+2x^2}{(3x-1)(3x+1)}=\dfrac{5x^2+x}{(3x-1)(3x+1)}$

$=\dfrac{x(5x+1)}{(3x-1)(3x+1)}$

Domain: $\left(-\infty,-\dfrac{1}{3}\right)\cup\left(-\dfrac{1}{3},\dfrac{1}{3}\right)\cup\left(\dfrac{1}{3},\infty\right)$

43) $\dfrac{\dfrac{3c^3}{8c+24}}{\dfrac{6c}{c+3}} = \dfrac{3c^3}{8c+24} \div \dfrac{6c}{c+3}$

$= \dfrac{\cancel{3}\,\cancel{c}^{\,c^2}}{8(\cancel{c+3})} \cdot \dfrac{\cancel{c+3}}{\cancel{6}\,\cancel{c}_{\,2}} = \dfrac{c^2}{16}$

45) $\dfrac{\dfrac{5}{m}+\dfrac{2}{m-3}}{1-\dfrac{4}{m}}$

$= \dfrac{m(m-3)\left(\dfrac{5}{m}+\dfrac{2}{m-3}\right)}{m(m-3)\left(1-\dfrac{4}{m}\right)} = \dfrac{5(m-3)+2m}{m(m-3)-4(m-3)}$

$= \dfrac{5m-15+2m}{m^2-3m-4m+12} = \dfrac{7m-15}{m^2-7m+12}$

$= \dfrac{7m-15}{(m-4)(m-3)}$

47) $\dfrac{\dfrac{25t^2}{6u}}{\dfrac{10t}{9u^4}} = \dfrac{18u^4\left(\dfrac{25t^2}{6u}\right)}{18u^4\left(\dfrac{10t}{9u^4}\right)} = \dfrac{75t^2u^3}{20t} = \dfrac{15tu^3}{4}$

49) $h(x) = \dfrac{f(x)}{g(x)} = \dfrac{6x^2+43x-40}{x+8}$

$= \dfrac{(6x-5)\,\cancel{(x+8)}}{\cancel{x+8}} = 6x-5$

Domain: $(-\infty,-8)\cup(-8,\infty)$

51) $\dfrac{3y}{y+7}-6 = \dfrac{3}{y+7}$

$y+7\left(\dfrac{3y}{y+7}-6\right) = y+7\left(\dfrac{3}{y+7}\right)$

$3y-6(y+7) = 3$

$3y-6y-42 = 3$

$-3y = 3+42$

$-3y = 45$

$y = -15 \qquad \{-15\}$

53) $\quad k-\dfrac{28}{k} = 3$

$k\left(k-\dfrac{28}{k}\right) = 3k$

$k^2-28 = 3k$

$k^2-3k-28 = 0$

$(k-7)(k+4) = 0$

$k = -4, 7 \qquad \{-4,7\}$

55)

$\dfrac{3}{d+2}+\dfrac{10}{\underset{(d-8)(d+2)}{d^2-6d-16}} = \dfrac{5}{d-8}$

$(d-8)(d+2)\left[\dfrac{3}{d+2}+\dfrac{10}{\underset{(d-8)(d+2)}{d^2-6d-16}} = \dfrac{5}{d-8}\right]$

$3(d-8)+10 = 5(d+2)$

$3d-24+10 = 5d+10$

$3d-5d = 10-10+24$

$-2d = 24$

$d = -12 \qquad \{-12\}$

57) $$\frac{4}{3x-1}-7=\frac{4}{3x-1}$$

$$3x-1\left(\frac{4}{3x-1}-7\right)=3x-1\left(\frac{4}{3x-1}\right)$$

$$4-7(3x-1)=4$$

$$4-21x+7=4$$

$$-21x+11=4-7$$

$$-20x=-3-11$$

$$x=\frac{14}{20}=\frac{7}{10} \qquad \left\{\frac{7}{10}\right\}$$

Section 8.5

1) Let x = cost for 5 packs of batteries

$$\frac{3}{4.26}=\frac{5}{x}$$

$$3x=5\cdot4.26$$

$$3x=21.30$$

$$x=\$7.10$$

3) Let x = caffeine in an 18-oz serving of Mountain Dew

$$\frac{12}{55}=\frac{18}{x}$$

$$12x=990$$

$$x=82.5 \text{ mg}$$

5) $$\frac{1000\,\text{ml}}{8\,\text{hr}}=\frac{x}{3\,\text{hr}}$$

$$8x=1000\cdot3$$

$$8x=3000$$

$$\frac{8}{8}x=\frac{3000}{8}$$

$$x=375\,\text{ml}$$

7) $$\frac{10 \text{ male}}{3 \text{ female}}=\frac{370 \text{ male}}{f \text{ female}}$$

$$10f=370\cdot3$$

$$10f=1110$$

$$f=111$$

9) $$\frac{2 \text{ potato flour}}{1 \text{ tapioca flour}}=\frac{x+3 \text{ potato four}}{x \text{ tapioca flour}}$$

$$2x=x+3$$

$$2x-x=3$$

$$x=3$$

3 cups of tapioca flour and 6 cups of potato-starch flour.

11) $$\frac{8 \text{ length}}{5 \text{ width}}=\frac{x \text{ length}}{x-18 \text{ width}}$$

$$8(x-18)=5x$$

$$8x-144=5x$$

$$8x-5x=144$$

$$3x=144$$

$$x=48$$

Length is 48 ft and width is 30 ft.

13) $$\frac{3 \text{ bonds}}{2 \text{ stocks}}=\frac{x+4000 \text{ bonds}}{x \text{ stocks}}$$

$$3x=2(x+4000)$$

$$3x=2x+8000$$

$$x=8000$$

$8000 in bonds and $12,000 in stocks

15) $\dfrac{5 \text{ with pets}}{4 \text{ without pets}} = \dfrac{x+271 \text{ with pets}}{x \text{ without pets}}$

$$5x = 4(x+271)$$
$$5x = 4x + 1084$$
$$x = 1084$$

1355 houses have pets.

17) Boat in still water is 10 mph.

a) $10 - 3 = 7$ mph

b) $10 + 3 = 13$ mph

19) Airplane constant rate of x mph.

a) $x + 30$ mph

b) $x - 30$ mph

21)

	D	R	T
Downstream	20 mi	$x+5$	$\dfrac{20}{x+5}$
Upstream	12 mi	$x-5$	$\dfrac{12}{x-5}$

$$\frac{20}{x+5} = \frac{12}{x-5}$$
$$20(x-5) = 12(x+5)$$
$$20x - 100 = 12x + 60$$
$$20x - 12x = 60 + 100$$
$$8x = 160$$
$$x = 20 \text{ mph}$$

23)

	D	R	T
downstream	15 mi	$16+c$ mph	$\dfrac{15}{16+c}$
upstream	9 mi	$16-c$ mph	$\dfrac{9}{16-c}$

$$\frac{15}{16+c} = \frac{9}{16-c}$$
$$15(16-c) = 9(16+c)$$
$$240 - 15c = 144 + 9c$$
$$240 - 144 = 9c + 15c$$
$$96 = 24c$$
$$4 \text{ mph} = c$$

25)

	D	R	T
with	350 mi	$x+20$	$\dfrac{350}{x+20}$
against	300 mi	$x-20$	$\dfrac{350}{x-20}$

$$\frac{350}{x+20} = \frac{300}{x-20}$$
$$350(x-20) = 300(x+20)$$
$$350x - 7000 = 300x + 6000$$
$$350x - 300x = 6000 + 7000$$
$$50x = 13{,}000$$
$$x = 260 \text{ mph}$$

27)

	D	R	T
With	6 mi	$10+c$	$\dfrac{6}{10+c}$
against	4 mi	$10-c$	$\dfrac{4}{10-c}$

$$\frac{6}{10+c}=\frac{4}{10-c}$$
$$6(10-c)=4(10+c)$$
$$60-6c=40+4c$$
$$60-40=4c+6c$$
$$20=10c$$
$$2 \text{ mph}=c$$

29) $\dfrac{1}{4}$ job/hr

31) $\dfrac{1}{t}$ job/hr

33) Arlene $=\dfrac{1}{2}$ Andre $=\dfrac{1}{3}$

$$\frac{1}{2}t+\frac{1}{3}t=1$$
$$6\left(\frac{1}{2}t+\frac{1}{3}t\right)=6\cdot1$$
$$3t+2t=6$$
$$5t=6$$
$$t=\frac{6}{5} \text{ hr or } 1\frac{1}{5} \text{ hr}$$

35) Jermain $=\dfrac{1}{5}$ Sue $=\dfrac{1}{8}$

$$\frac{1}{5}t+\frac{1}{8}t=1$$
$$40\left(\frac{1}{5}t+\frac{1}{8}t\right)=40\cdot1$$
$$8t+5t=40$$
$$13t=40$$
$$t=\frac{40}{13} \text{ hr or } 3\frac{1}{13} \text{ hr}$$

37) fill $=\dfrac{1}{12}$ empty $=\dfrac{1}{30}$

$$\frac{1}{12}t-\frac{1}{30}t=1$$
$$60\left(\frac{1}{12}t-\frac{1}{30}t\right)=60\cdot1$$
$$5t-2t=60$$
$$3t=60$$
$$t=20 \text{ min}$$

39) Fatima $=\dfrac{1}{2t}$ Antonio $=\dfrac{1}{t}$

$$\frac{1}{2t}+\frac{1}{t}=\frac{1}{2}$$
$$2t\left(\frac{1}{2t}+\frac{1}{t}\right)=2t\left(\frac{1}{2}\right)$$
$$1+2=t$$
$$3 \text{ hr }=t$$

41) new employee $=\dfrac{1}{6}$ experienced $=\dfrac{1}{t}$

$$\frac{1}{6}+\frac{1}{t}=\frac{1}{2}$$
$$6t\left(\frac{1}{6}+\frac{1}{t}\right)=6t\left(\frac{1}{2}\right)$$
$$t+6=3t$$
$$6=3t-t$$
$$6=2t$$
$$3 \text{ hr }=t$$

43)

	D	R	T
with	126 ft	$x+2$	$\dfrac{126}{x+2}$
without	66 ft	x	$\dfrac{66}{x}$

$$\dfrac{126}{x+2}=\dfrac{66}{x}$$
$$126x=66(x+2)$$
$$126x=66x+132$$
$$126x-66x=132$$
$$60x=132$$
$$x=\dfrac{132}{60}\ \text{ft/sec or }2\dfrac{1}{5}\ \text{ft/sec}$$

45) $\dfrac{7}{5}=\dfrac{14}{x}$
$$7x=5\cdot14$$
$$7x=70$$
$$x=10$$

47) $\dfrac{25}{15}=\dfrac{\frac{65}{3}}{x}$
$$25x=\dfrac{65}{3}\cdot15$$
$$25x=65\cdot5$$
$$25x=325$$
$$x=13$$

Section 8.6

1) Increases

3) direct

5) inverse

7) combined

9) $M=kn$

11) $h=\dfrac{k}{j}$

13) $T=\dfrac{k}{c^2}$

15) $s=krt$

17) $Q=\dfrac{k\sqrt{z}}{m}$

19) $z=kx$
   a) $63=k(7)$ b) $z=9x$
         $k=9$
   c) $z=9\cdot6=54$

21) $N=\dfrac{k}{y}$
   a) $4=\dfrac{k}{12}$
        $48=k$
   b) $N=\dfrac{48}{y}$
   c) $N=\dfrac{48}{3}=16$

23) $Q=\dfrac{kr^2}{w}$
   a) $25=\dfrac{k\cdot10^2}{20}$ b) $Q=\dfrac{5r^2}{w}$
        $500=100k$
         $5=k$
   c) $Q=\dfrac{5r^2}{w}=\dfrac{5\cdot6^2}{4}=\dfrac{5\cdot36}{4}=45$

25) $B = kr$ $B = 7 \cdot 8 = 56$

     $35 = k \cdot 5$

     $7 = k$

27) $L = \dfrac{k}{h^2}$ $L = \dfrac{72}{2^2} = \dfrac{72}{4} = 18$

     $8 = \dfrac{k}{3^2}$

     $72 = k$

29) $y = kxz$ $y = 5(7)(2) = 70$

     $60 = k(4)(3)$

     $60 = 12k$

     $5 = k$

31) $w = kh$ $w = kh$

     $437.5 = k(35)$ $w = 12.5(40)$

     $\$12.5 \text{ per hour} = k$ $w = \$500$

33) $t = \dfrac{d}{r}$ $t = \dfrac{840}{70} = 12 \text{ hrs}$

     $14 = \dfrac{d}{60}$

     $840 = d$

35) $P = kCR^2$ $P = kCR^2$

     $100 = k(4)(5)^2$ $P = 1(5)(6)^2$

     $1 = k$ $P = 180 \text{ watts}$

37) $KE = kmv^2$

     $112500 = k(1000)(15)^2$

     $.5 = k$

     $KE = .5(1000)(18)^2$

     $= 162{,}000 \text{ J}$

39) $f = \dfrac{k}{l}$ $f = \dfrac{500}{2.5}$

     $100 = \dfrac{k}{5}$ $f = 200 \text{ cycles/sec}$

     $500 = k$

41) $R = \dfrac{kl}{A}$

     $2 = \dfrac{k(40)}{0.05}$

     $0.0025 = k$

     $R = \dfrac{0.0025(60)}{0.05}$

     $R = 3 \text{ ohms}$

43) $f = kd$ $f = 40 \cdot 8$

     $200 = k5$ $f = 320 \text{ lb}$

     $40 = k$

Chapter 8 Review

1) Determine what values make the denominator equal zero.

3) $f(x) = \dfrac{x+9}{5x-1}$

     a) $f(5) = \dfrac{5+9}{5 \cdot 5 - 1} = \dfrac{14}{25-1} = \dfrac{14}{24} = \dfrac{7}{12}$

     b) $0 = \dfrac{x+9}{5x-1}$

         $0 = x+9$

         $-9 = x$

     c) $5x-1 = 0$

         $5x = 1$

         $x = \dfrac{1}{5}$

     Domain: $\left(-\infty, \dfrac{1}{5}\right) \cup \left(\dfrac{1}{5}, \infty\right)$

5) $a^2 - 2a - 24 = 0$

$\quad (a-6)(a+4) = 0$

$\quad\quad a = 6, -4$

Domain: $(-\infty, -4) \cup (-4, 6) \cup (6, \infty)$

7) $\dfrac{63a^2}{9a^{11}} = \dfrac{7}{a^9}$

9) $\dfrac{2z-7}{6z^2 - 19z - 7} = \dfrac{2\cancel{z-7}}{(2\cancel{z-7})(3z+1)} = \dfrac{1}{3z+1}$

11) $\dfrac{y^2 + 9y - yz - 9z}{yz - 12y - z^2 + 12z} = \dfrac{y(y+9) - z(y+9)}{y(z-12) - z(z-12)}$

$\quad\quad = \dfrac{(y+9)(\cancel{y-z})}{(\cancel{y-z})(z-12)}$

$\quad\quad = \dfrac{y+9}{z-12}$

13) $A = lw; \quad l = \dfrac{A}{w}$

$\quad l = \dfrac{2l^2 - 5l - 3}{l-3} = \dfrac{(2l+1)(\cancel{l-3})}{\cancel{l-3}} = 2l+1$

15) $\dfrac{16k^4}{3m^2} \div \dfrac{4k^2}{27m} = \dfrac{\overset{4}{\cancel{16}}k^4}{\cancel{3}m^2} \cdot \dfrac{\overset{9}{\cancel{27}}m}{\cancel{4}k^2} = \dfrac{36k^2}{m}$

17) $\dfrac{6w-1}{6w^2 + 5w - 1} \cdot \dfrac{3w+3}{12w}$

$\quad = \dfrac{\cancel{6w-1}}{(\cancel{6w-1})(w+1)} \cdot \dfrac{\cancel{3}(\cancel{w+1})}{\underset{4}{\cancel{12}}w} = \dfrac{1}{4w}$

19)

$\dfrac{25 - a^2}{4a^2 + 12a} \div \dfrac{a^3 - 125}{a^2 + 3a}$

$= \dfrac{-(a^2 - 25)}{4a(\cancel{a+3})} \cdot \dfrac{a(\cancel{a+3})}{(a-5)(a^2 + 5a + 25)}$

$= \dfrac{-(\cancel{a-5})(a+5)}{4(\cancel{a-5})(a^2 + 5a + 25)} = -\dfrac{a+5}{4(a^2 + 5a + 25)}$

21) LCD: k^2

23) LCD: $m(m+5)$

25) LCD: $(3x+7)(x-9)$

27) $c^2 + 10c + 24 = (c+6)(c+4)$

$\quad c^2 - 3c - 28 = (c-7)(c+4)$

$\quad$ LCD: $(c+6)(c+4)(c-7)$

29) $a^2 - 13a + 40 = (a-8)(a-5)$

$\quad a^2 - 7a - 8 = (a-8)(a+1)$

$\quad a^2 - 4a - 5 = (a-5)(a+1)$

$\quad$ LCD: $(a-8)(a-5)(a+1)$

31) $\dfrac{6}{5r} \cdot \dfrac{4r^2}{4r^2} = \dfrac{24r^2}{20r^3}$

33) $\dfrac{t-3}{2t+1} = \dfrac{(t-3)(t+5)}{(2t+1)(t+5)} = \dfrac{t^2 + 2t - 15}{(2t+1)(t+5)}$

35) $\dfrac{3}{8a^3b}; \dfrac{6}{5ab^2}$

$\quad$ LCD: $40a^3b^2$

$\quad \dfrac{3}{8a^3b} = \dfrac{15b}{40a^3b^2}; \dfrac{6}{5ab^2} = \dfrac{48a^2}{40a^3b^2}$

37) $\dfrac{9c}{\underset{(c+8)(c-2)}{c^2+6c-16}}$; $\dfrac{4}{\underset{(c-2)(c-2)}{c^2-4c+4}}$

  LCD: $(c+8)(c-2)(c-2)$

  $\dfrac{9c}{c^2+6c-16}=\dfrac{9c^2-18c}{(c+8)(c-2)(c-2)}$;

  $\dfrac{4}{c^2-4c+4}=\dfrac{4c+32}{(c+8)(c-2)(c-2)}$

39) $\dfrac{3}{8c}+\dfrac{7}{8c}=\dfrac{10}{8c}=\dfrac{5}{4c}$

41) $\dfrac{2}{5z^2}+\dfrac{9}{10z}=\dfrac{2\cdot 2+9z}{10z^2}=\dfrac{4+9z}{10z^2}$

43) $\dfrac{5}{y-2}-\dfrac{6}{y+3}=\dfrac{5(y+3)-[6(y-2)]}{(y-2)(y+3)}$

  $=\dfrac{5y+15-[6y-12]}{(y-2)(y+3)}$

  $=\dfrac{5y+15-6y+12}{(y-2)(y+3)}$

  $=\dfrac{27-y}{(y-2)(y+3)}$

45)

  $\dfrac{10p+3}{\underset{4(p+1)}{4p+4}}-\dfrac{8}{\underset{(p-7)(p+1)}{p^2-6p-7}}=\dfrac{(10p+3)(p-7)-8\cdot 4}{4(p-7)(p+1)}$

  $=\dfrac{10p^2-67p-21-32}{4(p-7)(p+1)}$

  $=\dfrac{10p^2-67p-53}{4(p-7)(p+1)}$

47) $\dfrac{2}{m-11}+\dfrac{19}{\underset{-(m-11)}{11-m}}=\dfrac{-2+19}{-(m-11)}=-\dfrac{17}{m-11}$

49) $\dfrac{x^2}{\underset{(x-y)(x+y)}{x^2-y^2}}+\dfrac{x}{\underset{-(x-y)}{y-x}}=\dfrac{-x^2+x(x+y)}{-(x-y)(x+y)}$

  $=\dfrac{-x^2+x^2+xy}{-(x-y)(x+y)}$

  $=-\dfrac{xy}{(x-y)(x+y)}$

51) $\dfrac{3}{\underset{g(g-7)}{g^2-7g}}+\dfrac{2g}{\underset{5(g-7)}{5g-35}}-\dfrac{6}{5g}$

  $=\dfrac{15+2g\cdot g-[6(g-7)]}{5g(g-7)}$

  $=\dfrac{3\cdot 5+2g^2-[6g-42]}{5g(g-7)}$

  $=\dfrac{15+2g^2-6g+42}{5g(g-7)}$

  $=\dfrac{2g^2-6g+57}{5g(g-7)}$

53) $A=lw$; $P=2l+2w$

  a) $A=lw=\dfrac{\overset{3}{\cancel{12}}}{x-4}\cdot\dfrac{x}{\underset{2}{\cancel{8}}}=\dfrac{3x}{2(x-4)}$ sq. units

  b) $P=2\left(\dfrac{12}{x-4}\right)+2\left(\dfrac{x}{8}\right)=\dfrac{24}{x-4}+\dfrac{2x}{8}$

   $=\dfrac{24\cdot 8+2x(x-4)}{8(x-4)}=\dfrac{192+2x^2-8x}{8(x-4)}$

   $=\dfrac{2(x^2-4x+96)}{8(x-4)}$

   $=\dfrac{x^2-4x+96}{4(x-4)}$ units

55) $(f+g)(x) = \dfrac{5x+3}{x-2} + \dfrac{4}{x}$

$= \dfrac{(5x+3)(x) + 4(x-2)}{x(x-2)}$

$= \dfrac{5x^2 + 3x + 4x - 8}{x(x-2)}$

$= \dfrac{5x^2 + 7x - 8}{x(x-2)}$

Domain: $(-\infty, 0) \cup (0, 2) \cup (2, \infty)$

57) $\dfrac{\frac{2}{5}}{\frac{7}{15}} = \dfrac{15 \cdot \frac{2}{5}}{15 \cdot \frac{7}{15}} = \dfrac{6}{7}$

59) $\dfrac{p + \frac{6}{p}}{\frac{8}{p} + p} = \dfrac{p\left(p + \frac{6}{p}\right)}{p\left(\frac{8}{p} + p\right)} = \dfrac{p^2 + 6}{p^2 + 8}$

61) $\dfrac{\frac{6n+48}{n^2}}{\frac{4n+32}{}} = \dfrac{n}{\underset{6(n+8)}{6n+48}} \div \dfrac{n^2}{\underset{4(n+8)}{4n+32}}$

$= \dfrac{n}{\underset{3}{\cancel{6}\,(\cancel{n+8})}} \cdot \dfrac{\overset{2}{\cancel{4}}\,(\cancel{n+8})}{n^2} = \dfrac{2}{3n}$

63) $\dfrac{1 - \frac{1}{y-9}}{\frac{2}{y+3} + 1} = \left(1 - \dfrac{1}{y-9}\right) \div \left(\dfrac{2}{y+3} + 1\right)$

$= \dfrac{y-9-1}{y-9} \div \dfrac{2+y+3}{y+3}$

$= \dfrac{y-10}{y-9} \cdot \dfrac{y+3}{y+5}$

$= \dfrac{(y-10)(y+3)}{(y-9)(y+5)}$

65) $\dfrac{\frac{c}{c+2} + \frac{1}{c^2-4}}{1 - \frac{3}{c+2}} = (c-2)(c+2) \left[\dfrac{\frac{c}{c+2} + \frac{1}{(c-2)(c+2)}}{1 - \frac{3}{c+2}} \right] = \dfrac{c(c-2)+1}{(c-2)(c+2) - 3(c-2)}$

$= \dfrac{c^2 - 2c + 1}{c^2 - 4 - 3c + 6} = \dfrac{(c-1)(c-1)}{c^2 - 3c + 2} = \dfrac{(c-1)\,(\cancel{c-1})}{(c-2)\,(\cancel{c-1})} = \dfrac{c-1}{c-2}$

67) $\dfrac{2x^{-2} + y^{-1}}{x^{-1} - y^{-2}} = x^2 y^2 \left[\dfrac{\frac{2}{x^2} + \frac{1}{y}}{\frac{1}{x} - \frac{1}{y^2}} \right] = \dfrac{2y^2 + x^2 y}{xy^2 - x^2} = \dfrac{y(2y + x^2)}{x(y^2 - x)}$

69) $\dfrac{5w}{6} - \dfrac{1}{2} = -\dfrac{1}{6}$

$6\left[\dfrac{5w}{6} - \dfrac{1}{2}\right] = 6\left(-\dfrac{1}{6}\right)$

$5w - 3 = -1$

$5w = -1 + 3$

$5w = 2$

$w = \dfrac{2}{5} \qquad \left\{\dfrac{2}{5}\right\}$

71) $\dfrac{4}{y-6} = \dfrac{12}{y+2}$

$4(y+2) = 12(y-6)$

$4y + 8 = 12y - 72$

$4y - 12y = -72 - 8$

$-8y = -80$

$y = 10 \qquad \{10\}$

73) $\dfrac{16}{9t-27} + \dfrac{2t-4}{t-3} = \dfrac{t}{9}$
    $\underset{9(t-3)}{}$

$9(t-3)\left(\dfrac{16}{9(t-3)} + \dfrac{2t-4}{t-3}\right) = 9(t-3)\left(\dfrac{t}{9}\right)$

$16 + 9(2t-4) = t(t-3)$

$16 + 18t - 36 = t^2 - 3t$

$0 = t^2 - 3t - 18t - 16 + 36$

$0 = t^2 - 21t + 20$

$0 = (t-20)(t-1)$

$1, 20 = t \qquad \{1, 20\}$

75)

$\dfrac{3}{b+2} = \dfrac{16}{\underset{(b-2)(b+2)}{b^2-4}} - \dfrac{4}{b-2}$

$(b-2)(b+2)\left[\dfrac{3}{b+2} = \dfrac{16}{(b-2)(b+2)} - \dfrac{4}{b-2}\right]$

$3(b-2) = 16 - 4(b+2)$

$3b - 6 = 16 - 4b - 8$

$3b + 4b = 8 + 6$

$7b = 14$

$b = 2$

b cannot be 2, therefore $\varnothing$

77) $\dfrac{3k}{k+9} = \dfrac{3}{k+1}$

$3k(k+1) = 3(k+9)$

$3k^2 + 3k = 3k + 27$

$3k^2 + 3k - 3k - 27 = 0$

$3k^2 - 27 = 0$

$3(k^2 - 9) = 0$

$3(k+3)(k-3) = 0$

$k = -3, 3$

79) $A = \dfrac{2p}{c}$

$cA = 2p$

$c = \dfrac{2p}{A}$

81) $n = \dfrac{t}{a+b}$; $n(a+b) = t$

$a + b = \dfrac{t}{n}$

$a = \dfrac{t}{n} - b$ or $a = \dfrac{t - bn}{n}$

83) $\dfrac{1}{r} = \dfrac{1}{s} + \dfrac{1}{t}$

$rst \cdot \left(\dfrac{1}{r} - \dfrac{1}{t} = \dfrac{1}{s} \right)$

$st - rs = rt$

$s(t - r) = rt$

$s = \dfrac{rt}{t - r}$

85) $\dfrac{2 \text{ sat. fat}}{3 \text{ total fat}} = \dfrac{x \text{ sat. fat}}{x + 4 \text{ total fat}}$

$2(x + 4) = 3x$

$2x + 8 = 3x$

$8 = 3x - 2x$

$8 = x$

Therefore total fat is $x + 4 = 8 + 4 = 12g$

87)

	D	R	T
with	800 mi	$x+40$	$\dfrac{800}{x+40}$
against	600 mi	$x-40$	$\dfrac{600}{x-40}$

$\dfrac{800}{x+40} = \dfrac{600}{x-40}$

$800(x - 40) = 600(x + 40)$

$800x - 32,000 = 600x + 24,000$

$800x - 600x = 24,000 + 32,000$

$200x = 56,000$

$x = 280$ mph

89) $c = km$ $\qquad\qquad$ $c = km$

$56 = k \cdot 8$ $\qquad\qquad$ $c = 7 \cdot 3$

$7 = k$ $\qquad\qquad$ $c = 21$

91) $z = \dfrac{k}{w^3}$ $\qquad\qquad$ $z = \dfrac{128}{4^3}$

$16 = \dfrac{k}{2^3}$ $\qquad\qquad$ $z = 2$

$16 = \dfrac{k}{8}$

$128 = k$

93) $w = kr^3$ $\qquad\qquad$ $w = kr^3$

$0.96 = k(2)^3$ $\qquad\qquad$ $w = 0.12(3)^3$

$0.96 = 8k$ $\qquad\qquad$ $w = 0.12(27)$

$k = 0.12$ $\qquad\qquad$ $w = 3.24$ lbs

Chapter 8 Test

1) $f(x) = \dfrac{x^2+4}{x^2-2x-48}$

  a)

$$f(-2) = \dfrac{(-2)^2+4}{(-2)^2-2(-2)-48} = \dfrac{4+4}{4+4-48}$$

$$= -\dfrac{8}{40} = -\dfrac{1}{5}$$

  b) $0 = \dfrac{x^2+4}{x^2-2x-48};\ \ 0 \neq x^2+4$

    Never equals zero

  c) $x^2 - 2x - 48 = (x-8)(x+6)$

$$x = 8, -6$$

    Domain: $(-\infty,-6)\cup(-6,8)\cup(8,\infty)$

3) $\dfrac{54w^3}{24w^8} = \dfrac{9}{4w^5}$

5) $\dfrac{9-h}{2h-3} = -\dfrac{h-9}{2h-3} = \dfrac{h-9}{-2h+3} = \dfrac{-(h-9)}{2h-3}$

7) $\dfrac{7}{12z} + \dfrac{5}{12z} = \dfrac{7+5}{12z} = \dfrac{12}{12z} = \dfrac{1}{z}$

9) $\dfrac{r}{2r+1} + \dfrac{3}{r+5} = \dfrac{r(r+5)+3(2r+1)}{(2r+1)(r+5)}$

$$= \dfrac{r^2+5r+6r+3}{(2r+1)(r+5)}$$

$$= \dfrac{r^2+11r+3}{(2r+1)(r+5)}$$

11) $\dfrac{c-3}{c-15} + \dfrac{c+8}{\underset{-(c-15)}{15-c}} = \dfrac{-(c-3)+c+8}{-(c-15)}$

$$= \dfrac{-c+3+c+8}{-(c-15)} = -\dfrac{11}{c-15}$$

13) $h(x) = f(x) - g(x) = \dfrac{2}{x} - \dfrac{x-5}{x+7}$

$$= \dfrac{2(x+7)-[x(x-5)]}{x(x+7)}$$

$$= \dfrac{2x+14-[x^2-5x]}{x(x+7)} = \dfrac{2x+14-x^2+5x}{x(x+7)}$$

$$= \dfrac{-x^2+7x+14}{x(x+7)}$$

Domain: $(-\infty,-7)\cup(-7,0)\cup(0,\infty)$

15) $\dfrac{\frac{15}{7}}{\frac{20}{21}} = \dfrac{21\left(\frac{15}{7}\right)}{21\left(\frac{20}{21}\right)} = \dfrac{45}{20} = \dfrac{9}{4}$

17) $A = \dfrac{1}{2}bh;\ \ \dfrac{2A}{h} = b$

$$\dfrac{2(12k^2+28k)}{8k} = b$$

$$\dfrac{24k^2+56k}{8k} = b$$

$$3k+7 = b$$

19) $\dfrac{7t}{12} + \dfrac{t-4}{6} = \dfrac{7}{3}$

$$12\left(\dfrac{7t}{12} + \dfrac{t-4}{6}\right) = 12\left(\dfrac{7}{3}\right)$$

$$7t + 2(t-4) = 28$$

$$7t + 2t - 8 = 28$$

$$9t = 28+8$$

$$9t = 36$$

$$t = 4 \qquad \{4\}$$

21)
$$\frac{w+2}{6} = \frac{2w-3}{4}$$
$$4(w+2) = 6(2w-3)$$
$$4w+8 = 12w-18$$
$$12w-4w = 18+8$$
$$8w = 26$$
$$w = \frac{26}{8} \text{ or } \frac{13}{4}$$

23) $y = \dfrac{kxz}{c}$

$$c = \frac{kxz}{y}$$
$$cy = kxz$$

25)   Let x be Twitter followers and $x + 540$ be Facebook friends.

The ratio of Khole's Facebook friends to Twitter followers is 19:4.
$$19:4 = x+540:x$$
$$19x = 4(x+540)$$
$$19x = 4x+2160$$
$$19x-4x = 2160$$
$$15x = 2160$$
$$x = \frac{2160}{15} = 144$$
Twitter followers: 144 and Facebook friends: $x + 540 = 684$.

27)
$$n = krs^2 \qquad n = krs^2$$
$$72 = k(2)(3)^2 \qquad n = 4(3)(5)^2$$
$$72x = 18k \qquad n = 300$$
$$4 = k$$

Cumulative Review: Chapters 1–8

1) $-5(3w^4)^2 = -5 \cdot 9w^8 = -45w^8$

3) $-\dfrac{12}{7}c - 7 = 20$
$$-\frac{12}{7}c = 20+7$$
$$-12c = 27 \cdot 7$$
$$c = -\frac{189}{12} = -\frac{45}{4} \qquad \left\{ -\frac{45}{4} \right\}$$

5)

x-int: $(2, 0)$, y-int: $(0, -5)$

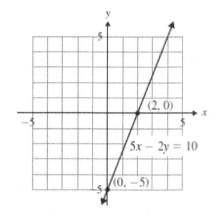

7) $m = \dfrac{y_2 - y_1}{x_2 - x_1} = \dfrac{2-8}{1-(-5)} = \dfrac{-6}{6} = -1$
$$y - y_1 = m(x - x_1)$$
$$y - 2 = -1(x-1)$$
$$y - 2 = -x+1$$
$$x + y = 3$$

9)
$$P = 2l + 2w$$
$$76 = 2(2w-7) + 2w$$
$$76 = 4w-14+2w$$
$$76+14 = 6w$$
$$90 = 6w$$
$$15 = w$$
Length : 23 cm   Width: 15 cm

11) $|6p+13|=8$

$\quad 6p+13=8$

$\qquad 6p=8-13$

$\qquad 6p=-5$

$\qquad p=-\dfrac{5}{6}$

$\quad 6p+13=-8$

$\qquad 6p=-8-13$

$\qquad 6p=-21$

$\qquad p=-\dfrac{21}{6}=-\dfrac{7}{2}$

$$\left\{-\dfrac{5}{6},-\dfrac{7}{2}\right\}$$

13)

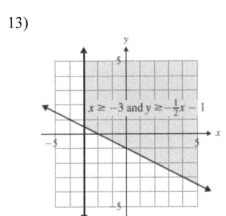

$x \geq -3$ and $y \geq -\frac{1}{2}x - 1$

15)

$\left(4q^2+11q-2\right)-3\left(6q^2-5q+4\right)+2\left(-10q-3\right)$

$=4q^2+11q-2-18q^2+15q-12-20q-6$

$=-14q^2+6q-20$

17) $\dfrac{8a^4b^4-20a^3b^2+56ab+8b}{8a^3b^3}$

$\quad \dfrac{8a^4b^4}{8a^3b^3}-\dfrac{20a^3b^2}{8a^3b^3}+\dfrac{56ab}{8a^3b^3}+\dfrac{8b}{8a^3b^3}$

$\quad ab-\dfrac{5}{2b}+\dfrac{7}{a^2b^2}+\dfrac{1}{a^3b^2}$

19) $\dfrac{7}{c-6}-\dfrac{4}{c}=\dfrac{7c-4(c-6)}{c(c-6)}=\dfrac{7c-4c+24}{c(c-6)}$

$\qquad =\dfrac{3c+24}{c(c-6)}=\dfrac{3(c+8)}{c(c-6)}$

21) $3xy^2+15xy-72x=3x\left(y^2+5y-24\right)$

$\qquad\qquad\qquad\qquad =3x(y+8)(y-3)$

23) $8-a=0$

$\qquad 8=a$

$\quad$ Domain: $(-\infty,8)\cup(8,\infty)$

25) Leticia $=\dfrac{t}{3}\quad$ Betty $=\dfrac{t}{5}$

$\qquad \dfrac{t}{3}+\dfrac{t}{5}=1$

$\qquad 15\left(\dfrac{t}{3}+\dfrac{t}{5}\right)=15\cdot1$

$\qquad\qquad 5t+3t=15$

$\qquad\qquad\quad 8t=15$

$\qquad\qquad\quad t=\dfrac{15}{8}$

Chapter 9: Radicals and Rational Exponents

Section 9.1

1) False; the $\sqrt{}$ symbol means to find only the positive square root of 121. $\sqrt{121} = 11$

3) False; the square root of a negative number is not a real number.

5) 12 and -12

7) $\dfrac{6}{5}$ and $-\dfrac{6}{5}$

9) $\sqrt{49} = 7$

11) $\sqrt{-4}$ is not real.

13) $\sqrt{\dfrac{81}{25}} = \dfrac{9}{5}$

15) $-\sqrt{36} = -6$

17) True

19) False: the only even root of zero is zero.

21) $\sqrt[3]{64}$ is the number you cube to get 64. $\sqrt[3]{64} = 4$

23) No; the even root of a negative number is not a real number.

25) $\sqrt[3]{125} = 5$

27) $.\sqrt[3]{-1} = -1$.

29) $\sqrt[4]{81} = 3$

31) $\sqrt[4]{-1}$ is not real.

33) $-\sqrt[4]{16} = -2$

35) $\sqrt[5]{-32} = -2$

37) $-\sqrt[3]{-27} = -(-3) = 3$

39) $\sqrt[6]{-64}$ is not real

41) $\sqrt[3]{\dfrac{8}{125}} = \dfrac{2}{5}$

59) $\sqrt[3]{z^3} = z$

43) $\sqrt{60-11} = \sqrt{49} = 7$

61) $\sqrt[4]{h^4} = |h|$

45) $\sqrt[3]{9-36} = \sqrt[3]{-27} = -3$

63) $\sqrt{(x+7)^2} = |x+7|$

47) $\sqrt{5^2+12^2} = \sqrt{25+144} = \sqrt{169} = 13$

65) $\sqrt[3]{(2t-1)^3} = 2t-1$

49)

If a is negative and we did not use the absolute value, the result would be negative.

67) $\sqrt[4]{(3n+2)^4} = |3n+2|$

This is incorrect because if a is negative and n is even, then $a^n > 0$. Using absolute values ensures a positive result.

69) $\sqrt[7]{(d-8)^7} = d-8$

51) $\sqrt{8^2} = \sqrt{64} = 8$

71) No, because $\sqrt{-1}$ is not a real number.

73) Set up an inequality so that the radicand is greater than or equal to 0. Solve for the variable. These are the real numbers in the domain of the function.

53) $\sqrt{(-6)^2} = \sqrt{36} = 6$

55) $\sqrt{y^2} = |y|$

75) $f(100) = \sqrt{100} = 10$

57) $\sqrt[3]{5^3} = \sqrt[3]{125} = 5$

77) $f(-49) = \sqrt{-49} =$ not a real number

79) $g(-1) = \sqrt{3(-1)+4} = \sqrt{1} = 1$

81) $g(-5) = \sqrt{3(-5)+4} = \sqrt{-11}$
   not a real number

83) $f(a) = \sqrt{a}$

85) $f(t+4) = \sqrt{t+4}$

87) $g(2n-1) = \sqrt{3(2n-1)+4}$
    $= \sqrt{6n-3+4}$
    $= \sqrt{6n+1}$

89) $f(64) = \sqrt[3]{64} = 4$

91) $f(-27) = \sqrt[3]{-27} = -3$

93) $g(-4) = \sqrt[3]{4(-4)-1} = \sqrt[3]{-17}$

95) $g(r) = \sqrt[3]{4(r)-1} = \sqrt[3]{4r-1}$

97) $f(c+8) = \sqrt[3]{c+8}$

99) $g(2a-3) = \sqrt[3]{4(2a-3)-1}$
    $= \sqrt[3]{8a-12-1}$
    $= \sqrt[3]{8a-13}$

101) $n+2 \geq 0$
     $n \geq -2$
     $[-2, \infty)$

103) $a-8 \geq 0$
     $a \geq 8$
     $[8, \infty)$

105) $(-\infty, \infty)$

107) $2x-5 \geq 0$
     $2x \geq 5$
     $x \geq \dfrac{5}{2}$
     $\left[\dfrac{5}{2}, \infty\right)$

109) $(-\infty, \infty)$

111) $-t \geq 0$
     $t \leq 0$
     $(-\infty, 0]$

113) $9-7a\geq 0$

$-7a\geq -9$

$a\leq \dfrac{9}{7}$

$\left(-\infty,\dfrac{9}{7}\right]$

115) $x-1\geq 0$

$x\geq 1$

$[1,\infty)$

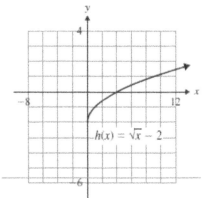

117) $x+3\geq 0$

$x\geq -3$

$[-3,\infty)$

119) $x\geq 0$

$x\geq 0$

$[0,\infty)$

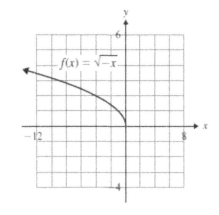

121) $-x\geq 0$

$x\leq 0$

$(-\infty,0]$

123) $(-\infty, \infty)$

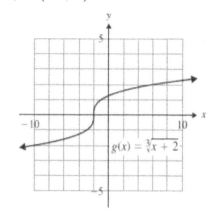

125) $(-\infty, \infty)$

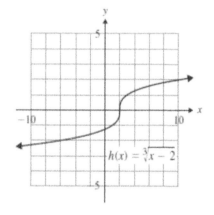

127) $(-\infty, \infty)$

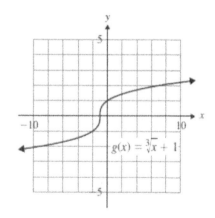

129) $V = 8 \text{ ft}^3, s = \sqrt[3]{V}$

   $\therefore s = \sqrt[3]{8 \text{ ft}^3} = 2 \text{ ft}$

131) $h = 0, t = \sqrt{\dfrac{160 - h}{16}}$

   $\therefore t = \sqrt{\dfrac{160 - 0}{16}} = \sqrt{10} \text{ sec} \approx 3.16 \text{ sec}$

133) $d = 40 \text{ ft}, S(d) = \sqrt{22.5d}$

   $\therefore S(40) = \sqrt{22.5(40)}$

   $= \sqrt{900} \text{ mph} \approx 30 \text{ mph}$

   Yes, the car was traveling at 30 mph.

135) $L = 8 \text{ ft}, T(L) = 2\pi \sqrt{\dfrac{L}{32}}$

   $\therefore T(8) = 2\pi \sqrt{\dfrac{8}{32}}$

   $= \pi \text{ sec} \approx 3.14 \text{ sec}$

137)

   $L = \dfrac{1}{2} \text{ ft}, T(L) = 2\pi \sqrt{\dfrac{L}{32}}$

   $\therefore T\left(\dfrac{1}{2}\right) = 2\pi \sqrt{\dfrac{\frac{1}{2}}{32}}$

   $= \dfrac{\pi}{4} \text{ sec}$

   A $\dfrac{1}{2}$ ft pendulum has a period of $\dfrac{\pi}{4}$ sec.

   This is approximately 0.79 sec.

139) $L = 30 \text{ in} = 2.5 \text{ ft}, T(L) = 2\pi\sqrt{\dfrac{L}{32}}$

$\therefore T(2.5) = 2\pi\sqrt{\dfrac{2.5}{32}}$

$= \dfrac{\sqrt{5}}{4}\pi \text{ sec} \approx 1.76 \text{ sec}$

Section 9.2

1) The denominator of 2 becomes the index of the radical. $25^{1/2} = \sqrt{25}$

3) $9^{1/2} = \sqrt{9} = 3$

5) $1000^{1/3} = \sqrt[3]{1000} = 10$

7) $32^{1/5} = \sqrt[5]{32} = 2$

9) $-125^{1/3} = -\sqrt[3]{125} = -5$

11) $\left(\dfrac{4}{121}\right)^{1/2} = \sqrt{\dfrac{4}{121}} = \dfrac{2}{11}$

13) $\left(\dfrac{125}{64}\right)^{1/3} = \sqrt[3]{\dfrac{125}{64}} = \dfrac{5}{4}$

15) $-\left(\dfrac{36}{169}\right)^{1/2} = -\sqrt{\dfrac{36}{169}} = -\dfrac{6}{13}$

17) $(-81)^{1/4} = \sqrt[4]{-81} = $ not a real number.

19) $(-1)^{1/7} = \sqrt[7]{-1} = -1$

21) The denominator of 4 becomes the index of the radical. The numerator of 3 is the power to which we raise the radical expression.

$16^{3/4} = \left(\sqrt[4]{16}\right)^3$

23) $8^{4/3} = \left(8^{1/3}\right)^4 = \left(\sqrt[3]{8}\right)^4 = 2^4 = 16$

25) $64^{5/6} = \left(64^{1/6}\right)^5 = \left(\sqrt[6]{64}\right)^5 = 2^5 = 32$

27) $(-125)^{2/3} = \left((-125)^{1/3}\right)^2 = \left(\sqrt[3]{-125}\right)^2$

$= (-5)^2 = 25$

29) $-36^{3/2} = -\left(\sqrt{36}\right)^3 = -(6)^3 = -216$

Section 9.2: Rational Exponents

31) $(-81)^{3/4} = \left((-81)^{1/4}\right)^3 = \left(\sqrt[4]{-81}\right)^3$

$\qquad\qquad$ = not a real number

33) $\left(\dfrac{16}{81}\right)^{3/4} = \left(\sqrt[4]{\dfrac{16}{81}}\right)^3 = \left(\dfrac{2}{3}\right)^3 = \dfrac{8}{27}$

35) $-\left(\dfrac{1000}{27}\right)^{2/3} = -\left(\sqrt[3]{\dfrac{1000}{27}}\right)^2$

$\qquad\qquad\qquad = -\left(\dfrac{10}{3}\right)^2 = -\dfrac{100}{9}$

37) False; the negative exponent does not make the result negative.

$\quad 81^{-1/2} = \dfrac{1}{9}$

39) $64^{-1/2} = \left(\dfrac{1}{64}\right)^{1/2}$ The reciprocal

$\quad$ of 64 is $\boxed{\dfrac{1}{64}}$

$\qquad = \sqrt{\dfrac{1}{64}}$

$\boxed{\text{The denominator of the fractional}}$
$\boxed{\text{exponent is the index of the radical}}$

$\qquad = \boxed{\dfrac{1}{8}}\quad$ Simplify

41) $\qquad 49^{-1/2} = \left(\dfrac{1}{49}\right)^{1/2} = \sqrt{\dfrac{1}{49}} = \dfrac{1}{7}$

43) $1000^{-1/3} = \left(\dfrac{1}{1000}\right)^{1/3} = \sqrt[3]{\dfrac{1}{1000}} = \dfrac{1}{10}$

45) $\left(\dfrac{1}{81}\right)^{-1/4} = (81)^{1/4} = \sqrt[4]{81} = 3$

47) $-\left(\dfrac{1}{64}\right)^{-1/3} = -(64)^{1/3} = -\sqrt[3]{64} = -4$

49) $64^{-5/6} = \left(\dfrac{1}{64}\right)^{5/6} = \left(\sqrt[6]{\dfrac{1}{64}}\right)^5$

$\qquad\qquad = \left(\dfrac{1}{2}\right)^5 = \dfrac{1}{32}$

51) $125^{-2/3} = \left(\dfrac{1}{125}\right)^{2/3} = \left(\sqrt[3]{\dfrac{1}{125}}\right)^2$

$\qquad\qquad = \left(\dfrac{1}{5}\right)^2 = \dfrac{1}{25}$

53) $\left(\dfrac{25}{4}\right)^{-3/2} = \left(\dfrac{4}{25}\right)^{3/2} = \left(\sqrt{\dfrac{4}{25}}\right)^3$

$\qquad\qquad = \left(\dfrac{2}{5}\right)^3 = \dfrac{8}{125}$

55) $\left(\dfrac{64}{125}\right)^{-2/3} = \left(\dfrac{125}{64}\right)^{2/3} = \left(\sqrt[3]{\dfrac{125}{64}}\right)^2$

$\qquad\qquad = \left(\dfrac{5}{4}\right)^2 = \dfrac{25}{16}$

57) $2^{2/3} \cdot 2^{7/3} = 2^{2/3+7/3} = 2^{9/3} = 2^3 = 8$

59) $\left(9^{1/4}\right)^2 = 9^{\frac{1}{4} \cdot 2} = 9^{1/2} = \sqrt{9} = 3$

61) $8^{7/5} \cdot 8^{-3/5} = 8^{7/5+(-3/5)} = 8^{4/5}$

63) $\dfrac{2^{23/4}}{2^{3/4}} = 2^{23/4-3/4} = 2^{20/4} = 2^5 = 32$

65) $\dfrac{4^{2/5}}{4^{6/5} \cdot 4^{3/5}} = \dfrac{4^{2/5}}{4^{6/5+3/5}} = \dfrac{4^{2/5}}{4^{9/5}}$

$= 4^{2/5-9/5} = 4^{-7/5} = \dfrac{1}{4^{7/5}}$

67) $z^{1/6} \cdot z^{5/6} = z^{1/6+5/6} = z^{6/6} = z^1 = z$

69) $\left(-9v^{5/8}\right)\left(8v^{3/4}\right) = -72v^{5/8+3/4}$

$= -72v^{5/8+6/8}$

$= -72v^{11/8}$

71) $\dfrac{a^{5/9}}{a^{4/9}} = a^{5/9-4/9} = a^{1/9}$

73) $\dfrac{20c^{-2/3}}{72c^{5/6}} = \dfrac{5}{18}c^{-2/3-5/6} = \dfrac{5}{18}c^{-4/6-5/6}$

$= \dfrac{5}{18}c^{-9/6} = \dfrac{5}{18}c^{-3/2} = \dfrac{5}{18c^{3/2}}$

75) $\left(x^{-2/9}\right)^3 = x^{-\frac{2}{9} \cdot 3} = x^{-2/3} = \dfrac{1}{x^{2/3}}$

77) $\left(z^{1/5}\right)^{2/3} = z^{\frac{1}{5} \cdot \frac{2}{3}} = z^{2/15}$

79) $\left(81u^{8/3}v^4\right)^{3/4}$

$= 81^{3/4} \cdot \left(u^{8/3}\right)^{3/4} \cdot \left(v^4\right)^{3/4}$

$= \left(\sqrt[4]{81}\right)^3 \cdot u^2 \cdot v^3$

$= (3)^3 \cdot u^2v^3 = 27u^2v^3$

81) $\left(32r^{1/3}s^{4/9}\right)^{3/5}$

$= 32^{3/5} \cdot \left(r^{1/3}\right)^{3/5} \cdot \left(s^{4/9}\right)^{3/5}$

$= \left(\sqrt[5]{32}\right)^3 \cdot r^{1/5} \cdot s^{4/15}$

$= (2)^3 \cdot r^{1/5}s^{4/15} = 8r^{1/5}s^{4/15}$

83) $\left(\dfrac{f^{6/7}}{27g^{-5/3}}\right)^{1/3} = \dfrac{\left(f^{6/7}\right)^{1/3}}{(27)^{1/3}\left(g^{-5/3}\right)^{1/3}}$

$= \dfrac{f^{2/7}}{3g^{-5/9}} = \dfrac{f^{2/7}g^{5/9}}{3}$

85) $\left(\dfrac{x^{-5/3}}{w^{3/2}}\right)^{-6} = \left(\dfrac{w^{3/2}}{x^{-5/3}}\right)^6 = \dfrac{\left(w^{3/2}\right)^6}{\left(x^{-5/3}\right)^6}$

$= \dfrac{w^9}{x^{-10}} = w^9x^{10}$

87) $\dfrac{y^{1/2}\cdot y^{-1/3}}{y^{5/6}}=\dfrac{y^{3/6}\cdot y^{-2/6}}{y^{5/6}}=\dfrac{y^{1/6}}{y^{5/6}}$

$$=y^{1/6-5/6}=y^{-4/6}$$

$$=y^{-2/3}=\dfrac{1}{y^{2/3}}$$

89) $\left(\dfrac{a^{4}b^{3}}{32a^{-2}b^{4}}\right)^{2/5}$

$$=\left(\dfrac{1}{32}a^{4-(-2)}b^{3-4}\right)^{2/5}$$

$$=\left(\dfrac{1}{32}a^{6}b^{-1}\right)^{2/5}$$

$$=\left(\dfrac{1}{32}\right)^{2/5}\cdot\left(a^{6}\right)^{2/5}\cdot\left(b^{-1}\right)^{2/5}$$

$$=\left(\sqrt[5]{\dfrac{1}{32}}\right)^{2}\cdot a^{12/5}\cdot b^{-2/5}$$

$$=\left(\dfrac{1}{2}\right)^{2}\cdot a^{12/5}b^{-2/5}=\dfrac{a^{12/5}}{4b^{2/5}}$$

91) $\left(\dfrac{r^{4/5}t^{-2}}{r^{2/3}t^{5}}\right)^{-3/2}$

$$=\left(\dfrac{r^{2/3}t^{5}}{r^{4/5}t^{-2}}\right)^{3/2}$$

$$=\left(r^{\frac{2}{3}-\frac{4}{5}}t^{5+2}\right)^{3/2}$$

$$=\left(r^{\frac{10-12}{15}}\right)^{3/2}\cdot\left(t^{7}\right)^{3/2}$$

$$=r^{-\frac{6}{30}}t^{\frac{21}{2}}$$

$$=\dfrac{t^{21/2}}{r^{1/5}}$$

93) $\left(\dfrac{h^{-2}k^{5/2}}{h^{-8}k^{5/6}}\right)^{-5/6}$

$$=\left(\dfrac{h^{-8}k^{5/6}}{h^{-2}k^{5/2}}\right)^{5/6}$$

$$=\left(h^{-6}k^{\frac{5}{6}-\frac{5}{2}}\right)^{5/6}$$

$$=\left(h^{-6}\right)^{5/6}\left(k^{-10/6}\right)^{5/6}$$

$$=h^{-5}k^{-25/18}$$

$$=\dfrac{1}{h^{5}k^{25/18}}$$

95) $p^{1/2}\left(p^{2/3}+p^{1/2}\right)$

$$=p^{1/2}p^{2/3}+p^{1/2}p^{1/2}$$

$$=p^{\frac{1}{2}+\frac{2}{3}}+p^{\frac{1}{2}+\frac{1}{2}}$$

$$=p^{7/6}+p$$

97) $\sqrt[12]{25^{6}}=\boxed{25^{6/12}}\quad$ Write with a rational exponent.

$$=\boxed{25^{1/2}}\text{ Reduce the exponent.}$$

$$=5\quad\boxed{\text{Evaluate}}$$

99) $\sqrt[6]{49^{3}}=\left(49^{3}\right)^{1/6}=49^{1/2}=\sqrt{49}=7$

101) $\sqrt[4]{81^{2}}=\left(81^{2}\right)^{1/4}=81^{1/2}=\sqrt{81}=9$

103) $\left(\sqrt{5}\right)^2 = 5$

105) $\left(\sqrt[3]{12}\right)^3 = 12$

107) $\left(\sqrt[3]{x^{12}}\right) = x^{12/3} = x^4$

109) $\left(\sqrt[6]{k^2}\right) = k^{2/6} = k^{1/3} = \sqrt[3]{k}$

111) $\left(\sqrt[4]{z^2}\right) = z^{2/4} = z^{1/2} = \sqrt{z}$

113) $\sqrt{d^4} = d^{4/2} = d^2$

115) a) $WC = 35.74 + 0.6215(20)$
$- 35.7(5)^{4/25}$
$+ 0.4275(20)(5)^{4/25}$

$= 13 \text{ degrees F}$

b) $WC = 35.74 + 0.6215(20)$
$- 35.7(15)^{4/25}$
$+ 0.4275(20)(15)^{4/25}$

$= 6 \text{ degrees F}$

Section 9.3

1) $\sqrt{3} \cdot \sqrt{7} = \sqrt{3 \cdot 7} = \sqrt{21}$

3) $\sqrt{10} \cdot \sqrt{3} = \sqrt{10 \cdot 3} = \sqrt{30}$

5) $\sqrt{6} \cdot \sqrt{y} = \sqrt{6 \cdot y} = \sqrt{6y}$

7) False; 20 contains the factor 4, which is a perfect square.

9) True; 42 does not have any factors (other than 1) that are perfect squares.

11) $\sqrt{60} = \sqrt{4 \cdot 15}$ $\boxed{\text{Factor.}}$

$= \boxed{\sqrt{4} \cdot \sqrt{15}}$ Product Rule.

$= \boxed{2\sqrt{15}}$ Simplify.

13) $\sqrt{20} = \sqrt{4 \cdot 5} = \sqrt{4} \cdot \sqrt{5} = 2\sqrt{5}$

15) $\sqrt{54} = \sqrt{9 \cdot 6} = \sqrt{9} \cdot \sqrt{6} = 3\sqrt{6}$

17) $\sqrt{33}$; simplified

19) $\sqrt{80} = \sqrt{16 \cdot 5} = \sqrt{16} \cdot \sqrt{5} = 4\sqrt{5}$

21) $\sqrt{98} = \sqrt{49 \cdot 2} = \sqrt{49} \cdot \sqrt{2} = 7\sqrt{2}$

23) $\sqrt{38}$; simplified

25) $\sqrt{400} = 20$

27) $\sqrt{750} = \sqrt{25}\sqrt{30} = 5\sqrt{30}$

Section 9.3: Simplifying Expressions Containing Square Roots

29) $\sqrt{\dfrac{144}{25}} = \dfrac{\sqrt{144}}{\sqrt{25}} = \dfrac{12}{5}$

31) $\dfrac{\sqrt{4}}{\sqrt{49}} = \dfrac{2}{7}$

33) $\dfrac{\sqrt{54}}{\sqrt{6}} = \sqrt{\dfrac{54}{6}} = \sqrt{9} = 3$

35) $\sqrt{\dfrac{60}{5}} = \sqrt{12} = \sqrt{4 \cdot 3} = \sqrt{4} \cdot \sqrt{3} = 2\sqrt{3}$

37) $\dfrac{\sqrt{120}}{\sqrt{6}} = \sqrt{\dfrac{120}{6}} = \sqrt{20} = \sqrt{4 \cdot 5}$
$\qquad = \sqrt{4} \cdot \sqrt{5} = 2\sqrt{5}$

39) $\dfrac{\sqrt{30}}{\sqrt{2}} = \sqrt{\dfrac{30}{2}} = \sqrt{15}$

41) $\sqrt{\dfrac{6}{49}} = \dfrac{\sqrt{6}}{\sqrt{49}} = \dfrac{\sqrt{6}}{7}$

43) $\sqrt{\dfrac{45}{16}} = \dfrac{\sqrt{45}}{\sqrt{16}} = \dfrac{\sqrt{9 \cdot 5}}{4} = \dfrac{\sqrt{9} \cdot \sqrt{5}}{4}$
$\qquad = \dfrac{3\sqrt{5}}{4}$

45) $\sqrt{x^8} = x^{8/2} = x^4$

47) $\sqrt{w^{14}} = w^{14/2} = w^7$

49) $\sqrt{100c^2} = \sqrt{100} \cdot \sqrt{c^2} = 10c^{2/2} = 10c$

51) $\sqrt{64k^6 m^{10}} = \sqrt{64} \cdot \sqrt{k^6} \sqrt{m^{10}}$
$\qquad = 8k^{6/2} m^{10/2} = 8k^3 m^5$

53) $\sqrt{28r^4} = \sqrt{28} \cdot \sqrt{r^4} = \sqrt{4} \cdot \sqrt{7} \cdot r^{4/2}$
$\qquad = 2\sqrt{7} \cdot r^2 = 2r^2 \sqrt{7}$

55) $\sqrt{300q^{22} t^{16}} = \sqrt{300} \cdot \sqrt{q^{22}} \sqrt{t^{16}}$
$\qquad = \sqrt{100} \cdot \sqrt{3} \cdot q^{22/2} \cdot t^{16/2}$
$\qquad = 10\sqrt{3} \cdot q^{11} t^8 = 10q^{11} t^8 \sqrt{3}$

57) $\sqrt{\dfrac{81}{c^6}} = \dfrac{\sqrt{81}}{\sqrt{c^6}} = \dfrac{9}{c^{6/2}} = \dfrac{9}{c^3}$

59) $\dfrac{\sqrt{40}}{\sqrt{t^8}} = \dfrac{\sqrt{4} \cdot \sqrt{10}}{t^{8/2}} = \dfrac{2\sqrt{10}}{t^4}$

61) $\sqrt{\dfrac{75x^2}{y^{12}}} = \dfrac{\sqrt{75}\sqrt{x^2}}{\sqrt{y^{12}}} = \dfrac{\sqrt{25} \cdot x \cdot \sqrt{3}}{y^{12/2}} = \dfrac{5x\sqrt{3}}{y^6}$

63) $\sqrt{w^9} = \sqrt{w^8 \cdot w^1}$ $\boxed{\text{Factor.}}$

$\qquad = \boxed{\sqrt{w^8} \cdot \sqrt{w^1}}$ Product Rule.

$\qquad = w^4 \sqrt{w}$ $\boxed{\text{Simplify.}}$

65) $\sqrt{a^5} = \sqrt{a^4} \cdot \sqrt{a} = a^{4/2} \cdot \sqrt{a} = a^2\sqrt{a}$

67) $\sqrt{g^{13}} = \sqrt{g^{12}} \cdot \sqrt{g} = g^6\sqrt{g}$

69) $\sqrt{b^{25}} = \sqrt{b^{24}} \cdot \sqrt{b} = b^{12}\sqrt{b}$

71) $\sqrt{72x^3} = \sqrt{72} \cdot \sqrt{x^3}$
$= \sqrt{36} \cdot \sqrt{2} \cdot \sqrt{x^2} \cdot \sqrt{x}$
$= 6\sqrt{2} \cdot x\sqrt{x} = 6x\sqrt{2x}$

73) $\sqrt{13q^7} = \sqrt{13} \cdot \sqrt{q^7}$
$= \sqrt{13} \cdot \sqrt{q^6} \cdot \sqrt{q}$
$= \sqrt{13} \cdot q^3\sqrt{q} = q^3\sqrt{13q}$

75) $\sqrt{75t^{11}} = \sqrt{75} \cdot \sqrt{t^{11}}$
$= \sqrt{25} \cdot \sqrt{3} \cdot \sqrt{t^{10}} \cdot \sqrt{t}$
$= 5\sqrt{3} \cdot t^5\sqrt{t} = 5t^5\sqrt{3t}$

77) $\sqrt{c^8 d^2} = \sqrt{c^8} \cdot \sqrt{d^2} = c^4 d$

79) $\sqrt{a^4 b^3} = \sqrt{a^4} \cdot \sqrt{b^3} = a^2 \cdot \sqrt{b^2} \cdot \sqrt{b}$
$= a^2 b\sqrt{b}$

81) $\sqrt{u^5 v^7} = \sqrt{u^5} \cdot \sqrt{v^7}$
$= \sqrt{u^4} \cdot \sqrt{u} \cdot \sqrt{v^6} \cdot \sqrt{v}$
$= u^2\sqrt{u} \cdot v^3\sqrt{v} = u^2 v^3\sqrt{uv}$

83) $\sqrt{36m^9 n^4} = \sqrt{36} \cdot \sqrt{m^9} \cdot \sqrt{n^4}$
$= 6 \cdot \sqrt{m^8} \cdot \sqrt{m} \cdot n^2$
$= 6m^4 n^2\sqrt{m}$

85) $\sqrt{44x^{12} y^5} = \sqrt{44} \cdot \sqrt{x^{12}} \cdot \sqrt{y^5}$
$= \sqrt{4} \cdot \sqrt{11} \cdot x^6 \cdot \sqrt{y^4} \cdot \sqrt{y}$
$= 2\sqrt{11} \cdot x^6 \cdot y^2\sqrt{y}$
$= 2x^6 y^2\sqrt{11y}$

87) $\sqrt{32t^5 u^7}$
$= \sqrt{32} \cdot \sqrt{t^5} \cdot \sqrt{u^7}$
$= \sqrt{16} \cdot \sqrt{2} \cdot \sqrt{t^4} \cdot \sqrt{t} \cdot \sqrt{u^6} \cdot \sqrt{u}$
$= 4\sqrt{2} \cdot t^2\sqrt{t} \cdot u^3\sqrt{u} = 4t^2 u^3\sqrt{2tu}$

89) $\sqrt{\dfrac{a^7}{81b^6}} = \dfrac{\sqrt{a^6} \cdot \sqrt{a}}{\sqrt{81b^6}} = \dfrac{a^3\sqrt{a}}{9b^3}$

91) $\sqrt{\dfrac{3r^9}{s^2}} = \dfrac{\sqrt{3r^9}}{\sqrt{s^2}} = \dfrac{\sqrt{3} \cdot \sqrt{r^9}}{s}$
$= \dfrac{\sqrt{3} \cdot \sqrt{r^8} \cdot \sqrt{r}}{s} = \dfrac{r^4\sqrt{3r}}{s}$

93) $\sqrt{5} \cdot \sqrt{10} = \sqrt{50} = \sqrt{25} \cdot \sqrt{2} = 5\sqrt{2}$

95) $\sqrt{21} \cdot \sqrt{3} = \sqrt{63} = \sqrt{9} \cdot \sqrt{7} = 3\sqrt{7}$

97) $\sqrt{w} \cdot \sqrt{w^5} = \sqrt{w^6} = w^3$

99) $\sqrt{n^3} \cdot \sqrt{n^4} = \sqrt{n^7}$

$= \sqrt{n^6} \cdot \sqrt{n} = n^3 \sqrt{n}$

101) $\sqrt{2k} \cdot \sqrt{8k^5} = \sqrt{16k^6} = \sqrt{16} \cdot \sqrt{k^6}$

$= 4k^3$

103) $\sqrt{6x^4 y^3} \cdot \sqrt{2x^5 y^2}$

$= \sqrt{12x^9 y^5} = \sqrt{12} \cdot \sqrt{x^9} \cdot \sqrt{y^5}$

$= \sqrt{4} \cdot \sqrt{3} \cdot \sqrt{x^8} \cdot \sqrt{x} \cdot \sqrt{y^4} \cdot \sqrt{y}$

$= 2\sqrt{3} \cdot x^4 \sqrt{x} \cdot y^2 \sqrt{y} = 2x^4 y^2 \sqrt{3xy}$

105) $\sqrt{8c^9 d^2} \cdot \sqrt{5cd^7}$

$= \sqrt{40c^{10} d^9} = \sqrt{40} \cdot \sqrt{c^{10}} \cdot \sqrt{d^9}$

$= \sqrt{4} \cdot \sqrt{10} \cdot c^5 \cdot \sqrt{d^8} \cdot \sqrt{d}$

$= 2\sqrt{10} \cdot c^5 \cdot d^4 \sqrt{d} = 2c^5 d^4 \sqrt{10d}$

107) $\dfrac{\sqrt{18k^{11}}}{\sqrt{2k^3}} = \sqrt{\dfrac{18k^{11}}{2k^3}} = \sqrt{9k^8}$

$= \sqrt{9} \cdot \sqrt{k^8} = 3k^4$

109) $\dfrac{\sqrt{120h^8}}{\sqrt{3h^2}} = \sqrt{\dfrac{120h^8}{3h^2}} = \sqrt{40h^6}$

$= \sqrt{4} \cdot \sqrt{10} \cdot \sqrt{h^6}$

$= 2\sqrt{10} \cdot h^3 = 2h^3 \sqrt{10}$

111) $\dfrac{\sqrt{50a^{16} b^9}}{\sqrt{5a^7 b^4}} = \sqrt{\dfrac{50a^{16} b^9}{5a^7 b^4}} = \sqrt{10a^9 b^5}$

$= \sqrt{10} \cdot \sqrt{a^8} \cdot \sqrt{a} \cdot \sqrt{b^4} \cdot \sqrt{b}$

$= \sqrt{10} \cdot a^4 \sqrt{a} \cdot b^2 \sqrt{b}$

$= a^4 b^2 \sqrt{10ab}$

113) $v = \sqrt{\dfrac{2KE}{m}} = \sqrt{\dfrac{2 \cdot 120,000}{600}}$

$= \sqrt{4000}$

$= 20$ m/s

Section 9.4

1) To multiply radicals with the same indices, multiply the radicands and put the product under a radical with the same index.

3) i)   Its radicand will not contain any factors that are perfect cubes.

ii) There will be no radical in the denominator of a fraction.

iii) The radicand will not contain fractions.

5) $\sqrt[3]{5} \cdot \sqrt[3]{4} = \sqrt[3]{20}$

7) $\sqrt[5]{9} \cdot \sqrt[5]{m^2} = \sqrt[5]{9m^2}$

9) $\sqrt[3]{a^2} \cdot \sqrt[3]{b} = \sqrt[3]{a^2 b}$

11) $\sqrt[3]{56} = \sqrt[3]{8 \cdot 7}$ $\boxed{\text{Factor.}}$

$= \boxed{\sqrt[3]{8} \cdot \sqrt[3]{7}}$ Product Rule.

$= \boxed{2\sqrt[3]{7}}$ Simplify.

13) $\sqrt[3]{24} = \sqrt[3]{8} \cdot \sqrt[3]{3} = 2\sqrt[3]{3}$

15) $\sqrt[4]{64} = \sqrt[4]{16} \cdot \sqrt[4]{4} = 2\sqrt[4]{4}$

17) $\sqrt[3]{54} = \sqrt[3]{27} \cdot \sqrt[3]{2} = 3\sqrt[3]{2}$

19) $\sqrt[3]{2000} = \sqrt[3]{1000} \cdot \sqrt[3]{2} = 10\sqrt[3]{2}$

21) $\sqrt[5]{64} = \sqrt[5]{32} \cdot \sqrt[5]{2} = 2\sqrt[5]{2}$

23) $\sqrt[4]{\dfrac{1}{16}} = \dfrac{\sqrt[4]{1}}{\sqrt[4]{16}} = \dfrac{1}{2}$

25) $\sqrt[3]{-\dfrac{54}{2}} = \sqrt[3]{-27} = -3$

27) $\dfrac{\sqrt[3]{48}}{\sqrt[3]{2}} = \sqrt[3]{\dfrac{48}{2}} = \sqrt[3]{24} = \sqrt[3]{8} \cdot \sqrt[3]{3} = 2\sqrt[3]{3}$

29) $\dfrac{\sqrt[4]{240}}{\sqrt[4]{3}} = \sqrt[4]{\dfrac{240}{3}} = \sqrt[4]{80}$

$= \sqrt[4]{16} \cdot \sqrt[4]{5} = 2\sqrt[4]{5}$

31) $\sqrt[3]{d^6} = d^{6/3} = d^2$

33) $\sqrt[4]{n^{20}} = n^{20/4} = n^5$

35) $\sqrt[5]{x^5 y^{15}} = x^{5/5} y^{15/5} = xy^3$

37) $\sqrt[3]{w^{14}} = \sqrt[3]{w^{12}} \cdot \sqrt[3]{w^2} = w^{12/3} \cdot \sqrt[3]{w^2}$

$= w^4 \sqrt[3]{w^2}$

39) $\sqrt[4]{y^9} = \sqrt[4]{y^8} \cdot \sqrt[4]{y} = y^{8/4} \cdot \sqrt[4]{y} = y^2 \sqrt[4]{y}$

41) $\sqrt[3]{d^5} = \sqrt[3]{d^3} \cdot \sqrt[3]{d^2} = d\sqrt[3]{d^2}$

43) $\sqrt[3]{u^{10} v^{15}} = \sqrt[3]{u^{10}} \cdot \sqrt[3]{v^{15}}$

$= \sqrt[3]{u^9} \cdot \sqrt[3]{u} \cdot v^{15/3}$

$= u^{9/3} \sqrt[3]{u} \cdot v^5 = u^3 v^5 \sqrt[3]{u}$

45) $\sqrt[3]{b^{16} c^5} = \sqrt[3]{b^{16}} \cdot \sqrt[3]{c^5}$

$= \sqrt[3]{b^{15}} \cdot \sqrt[3]{b} \cdot \sqrt[3]{c^3} \cdot \sqrt[3]{c^2}$

$= b^{15/3} \cdot \sqrt[3]{b} \cdot c^{3/3} \cdot \sqrt[3]{c^2}$

$= b^5 c \sqrt[3]{bc^2}$

Section 9.4: Simplifying Expressions Containing Higher Roots

47) $\sqrt[4]{m^3 n^{18}} = \sqrt[4]{m^3} \cdot \sqrt[4]{n^{18}}$

$\qquad = \sqrt[4]{m^3} \cdot \sqrt[4]{n^{16}} \cdot \sqrt[4]{n^2}$

$\qquad = \sqrt[4]{m^3} \cdot n^{16/4} \cdot \sqrt[4]{n^2}$

$\qquad = n^4 \sqrt[4]{m^3 n^2}$

49) $\sqrt[3]{24 x^{10} y^{12}} = \sqrt[3]{24} \cdot \sqrt[3]{x^{10}} \cdot \sqrt[3]{y^{12}}$

$\qquad = \sqrt[3]{8} \cdot \sqrt[3]{3} \cdot \sqrt[3]{x^9} \cdot \sqrt[3]{x} \cdot y^{12/3}$

$\qquad = 2\sqrt[3]{3} \cdot x^{9/3} \sqrt[3]{x} \cdot y^4$

$\qquad = 2 x^3 y^4 \sqrt[3]{3x}$

51) $\sqrt[3]{250 w^4 x^{16}}$

$\qquad = \sqrt[3]{250} \cdot \sqrt[3]{w^4} \cdot \sqrt[3]{x^{16}}$

$\qquad = \sqrt[3]{125} \cdot \sqrt[3]{2} \cdot \sqrt[3]{w^3} \cdot \sqrt[3]{w} \cdot \sqrt[3]{x^{15}} \cdot \sqrt[3]{x}$

$\qquad = 5\sqrt[3]{2} \cdot w^{3/3} \sqrt[3]{w} \cdot x^{15/3} \sqrt[3]{x}$

$\qquad = 5 w x^5 \sqrt[3]{2wx}$

53) $\sqrt[4]{\dfrac{m^8}{81}} = \dfrac{\sqrt[4]{m^8}}{\sqrt[4]{81}} = \dfrac{m^{8/4}}{3} = \dfrac{m^2}{3}$

55) $\sqrt[5]{\dfrac{32 a^{23}}{b^{15}}} = \dfrac{\sqrt[5]{32 a^{23}}}{\sqrt[5]{b^{15}}}$

$\qquad = \dfrac{\sqrt[5]{32} \cdot \sqrt[5]{a^{20}} \cdot \sqrt[5]{a^3}}{b^3}$

$\qquad = \dfrac{2 a^4 \sqrt[5]{a^3}}{b^3}$

57) $\sqrt[4]{\dfrac{t^9}{81 s^{24}}} = \dfrac{\sqrt[4]{t^9}}{\sqrt[4]{81 s^{24}}} = \dfrac{\sqrt[4]{t^8} \cdot \sqrt[4]{t}}{\sqrt[4]{81} \cdot \sqrt[4]{s^{24}}}$

$\qquad = \dfrac{t^2 \sqrt[4]{t}}{3 s^6}$

59) $\sqrt[3]{\dfrac{u^{28}}{v^3}} = \dfrac{\sqrt[3]{u^{28}}}{\sqrt[3]{v^3}} = \dfrac{\sqrt[3]{u^{27}} \cdot \sqrt[3]{u}}{v} = \dfrac{u^9 \sqrt[3]{u}}{v}$

61) $\sqrt[3]{6} \cdot \sqrt[3]{4} = \sqrt[3]{24} = \sqrt[3]{8} \cdot \sqrt[3]{3} = 2\sqrt[3]{3}$

63) $\sqrt[3]{9} \cdot \sqrt[3]{12} = \sqrt[3]{108} = \sqrt[3]{27} \cdot \sqrt[3]{4}$

$\qquad = 3\sqrt[3]{4}$

65) $\sqrt[3]{20} \cdot \sqrt[3]{4} = \sqrt[3]{80} = \sqrt[3]{8} \cdot \sqrt[3]{10} = 2\sqrt[3]{10}$

67) $\sqrt[3]{m^4} \cdot \sqrt[3]{m^5} = \sqrt[3]{m^9} = m^3$

69) $\sqrt[4]{k^7} \cdot \sqrt[4]{k^9} = \sqrt[4]{k^{16}} = k^4$

71) $\sqrt[3]{r^7} \cdot \sqrt[3]{r^4} = \sqrt[3]{r^{11}} = \sqrt[3]{r^9} \cdot \sqrt[3]{r^2}$

$\qquad = r^3 \sqrt[3]{r^2}$

73) $\sqrt[5]{p^{14}} \cdot \sqrt[5]{p^9} = \sqrt[5]{p^{23}} = \sqrt[5]{p^{20}} \cdot \sqrt[5]{p^3}$

$\qquad = p^4 \sqrt[5]{p^3}$

75) $\sqrt[3]{9 z^{11}} \cdot \sqrt[3]{3 z^8} = \sqrt[3]{27 z^{19}} = \sqrt[3]{27} \cdot \sqrt[3]{z^{19}}$

$\qquad = 3 \cdot \sqrt[3]{z^{18}} \cdot \sqrt[3]{z} = 3 z^6 \sqrt[3]{z}$

77) $\sqrt[3]{\dfrac{h^{14}}{h^2}} = \sqrt[3]{h^{12}} = h^4$

79) $\sqrt[3]{\dfrac{c^{11}}{c^4}} = \sqrt[3]{c^7} = \sqrt[3]{c^6} \cdot \sqrt[3]{c} = c^2 \sqrt[3]{c}$

81) $\sqrt[4]{\dfrac{162d^{21}}{2d^2}} = \sqrt[4]{81d^{19}} = \sqrt[4]{81} \cdot \sqrt[4]{d^{19}}$

$\qquad = 3 \cdot \sqrt[4]{d^{16}} \cdot \sqrt[4]{d^3} = 3d^4 \sqrt[4]{d^3}$

83) $\sqrt{a}\sqrt[4]{a^3} = a^{1/2} \cdot a^{3/4}$

Change radicals to fractional exponents.

$= a^{2/4} \cdot a^{3/4}$

Rewrite exponents with a common denominator.

$= \boxed{a^{5/4}}$

Add exponents.

$= \sqrt[4]{a^5}$

Rewrite in radical form.

$= \boxed{a\sqrt[4]{a}}$

Simplify.

85) $\sqrt{p} \cdot \sqrt[3]{p} = p^{1/2} \cdot p^{1/3} = p^{3/6} \cdot p^{2/6}$

$\qquad = p^{5/6} = \sqrt[6]{p^5}$

87) $\sqrt[4]{n^3} \cdot \sqrt{n} = n^{3/4} \cdot n^{1/2} = n^{3/4} \cdot n^{2/4}$

$\qquad = n^{5/4} = n^{4/4} \cdot n^{1/4} = n\sqrt[4]{n}$

89) $\sqrt[5]{c^3} \cdot \sqrt[3]{c^2} = c^{3/5} \cdot c^{2/3} = c^{9/15} \cdot c^{10/15}$

$\qquad = c^{19/15} = c^{15/15} \cdot c^{4/15}$

$\qquad = c\sqrt[15]{c^4}$

91) $\dfrac{\sqrt{w}}{\sqrt[4]{w}} = \dfrac{w^{1/2}}{w^{1/4}} = \dfrac{w^{2/4}}{w^{1/4}} = w^{1/4} = \sqrt[4]{w}$

93) $\dfrac{\sqrt[5]{t^4}}{\sqrt[3]{t^2}} = \dfrac{t^{4/5}}{t^{2/3}} = \dfrac{t^{12/15}}{t^{10/15}} = t^{2/15} = \sqrt[15]{t^2}$

95) $s = \sqrt[3]{V} = \sqrt[3]{64}$

$\qquad = 4 \text{ inches}$

Section 9.5

1) They have the same index and the same radicand.

3) $5\sqrt{2} + 9\sqrt{2} = 14\sqrt{2}$

5) $7\sqrt[3]{4} + 8\sqrt[3]{4} = 15\sqrt[3]{4}$

7) $6 - \sqrt{13} + 5 - 2\sqrt{13} = 11 - 3\sqrt{13}$

.

9) $15\sqrt[3]{z^2} - 20\sqrt[3]{z^2} = -5\sqrt[3]{z^2}$

Section 9.5: Adding, Subtracting, and Multiplying Radicals

11) $2\sqrt[3]{n^2} + 9\sqrt[5]{n^2} - 11\sqrt[3]{n^2} + \sqrt[5]{n^2}$

$= -9\sqrt[3]{n^2} + 10\sqrt[5]{n^2}$

13) $\sqrt{5c} - 8\sqrt{6c} + \sqrt{5c} + 6\sqrt{6c} = 2\sqrt{5c} - 2\sqrt{6c}$

15) i) Write each radical expression
   in simplest form.
   ii) Combine like radicals.

17) $\sqrt{48} + \sqrt{3} = \sqrt{16 \cdot 3} + \sqrt{3}$ $\boxed{\text{Factor.}}$

$= \boxed{\sqrt{16} \cdot \sqrt{3} + \sqrt{3}}$ Product Rule.

$= 4\sqrt{3} + \sqrt{3}$ $\boxed{\text{Simplify.}}$

$= \boxed{5\sqrt{3}}$ Add like radicals.

19) $6\sqrt{3} - \sqrt{12} = 6\sqrt{3} - 2\sqrt{3} = 4\sqrt{3}$

21) $\sqrt{32} - 3\sqrt{8} = 4\sqrt{2} - 3\left(2\sqrt{2}\right)$

$= 4\sqrt{2} - 6\sqrt{2} = -2\sqrt{2}$

23) $\sqrt{12} + \sqrt{75} - \sqrt{3} = 2\sqrt{3} + 5\sqrt{3} - \sqrt{3} = 6\sqrt{3}$

25) $8\sqrt[3]{9} + \sqrt[3]{72} = 8\sqrt[3]{9} + \sqrt[3]{8} \cdot \sqrt[3]{9}$

$= 8\sqrt[3]{9} + 2\sqrt[3]{9} = 10\sqrt[3]{9}$

27) $\sqrt[3]{6} - \sqrt[3]{48} = \sqrt[3]{6} - \left(2\sqrt[3]{6}\right) = -\sqrt[3]{6}$

29) $6q\sqrt{q} + 7\sqrt{q^3} = 6q\sqrt{q} + 7q\sqrt{q}$

$= 13q\sqrt{q}$

31) $4d^2\sqrt{d} - 24\sqrt{d^5}$

$= 4d^2\sqrt{d} - 24d^2\sqrt{d} = -20d^2\sqrt{d}$

33) $9t^3\sqrt[3]{t} - 5\sqrt[3]{t^{10}} = 9t^3\sqrt[3]{t} - 5t^3\sqrt[3]{t}$

$= 4t^3\sqrt[3]{t}$

35) $5a\sqrt[4]{a^7} + \sqrt[4]{a^{11}}$

$= 5a\left(a\sqrt[4]{a^3}\right) + \left(a^2\sqrt[4]{a^3}\right)$

$= 5a^2\sqrt[4]{a^3} + a^2\sqrt[4]{a^3} = 6a^2\sqrt[4]{a^3}$

37) $2\sqrt{8p} - 6\sqrt{2p} = 2\left(2\sqrt{2p}\right) - 6\sqrt{2p}$

$= 4\sqrt{2p} - 6\sqrt{2p} = -2\sqrt{2p}$

39) $7\sqrt[3]{81a^5} + 4a\sqrt[3]{3a^2}$

$= 7\left(3a\sqrt[3]{3a^2}\right) + 4a\sqrt[3]{3a^2}$

$= 21a\sqrt[3]{3a^2} + 4a\sqrt[3]{3a^2} = 25a\sqrt[3]{3a^2}$

41) $\sqrt{xy^3} + 3y\sqrt{xy} = y\sqrt{xy} + 3y\sqrt{xy}$

$= 4y\sqrt{xy}$

43) $6c^2\sqrt{8d^3} - 9d\sqrt{2c^4d}$

$\qquad = 6c^2\left(2d\sqrt{2d}\right) - 9d\left(c^2\sqrt{2d}\right)$

$\qquad = 12c^2d\sqrt{2d} - 9c^2d\sqrt{2d} = 3c^2d\sqrt{2d}$

45) $18a^5\sqrt[3]{7a^2b} + 2a^3\sqrt[3]{7a^8b}$

$\qquad = 18a^5\sqrt[3]{7a^2b} + 2a^3\left(a^2\sqrt[3]{7a^2b}\right)$

$\qquad = 18a^5\sqrt[3]{7a^2b} + 2a^5\sqrt[3]{7a^2b}$

$\qquad = 20a^5\sqrt[3]{7a^2b}$

47) $15cd\sqrt[4]{9cd} - \sqrt[4]{9c^5d^5}$

$\qquad = 15cd\sqrt[4]{9cd} - \left(\sqrt[4]{c^4d^4}\cdot\sqrt[4]{9cd}\right)$

$\qquad = 15cd\sqrt[4]{9cd} - cd\sqrt[4]{9cd}$

$\qquad = 14cd\sqrt[4]{9cd}$

49) $\sqrt[3]{a^9b} - \sqrt[3]{b^7} = a^3\sqrt[3]{b} - b^2\sqrt[3]{b}$

$\qquad = \sqrt[3]{b}\left(a^3 - b^2\right)$

51) $3(x+5) = 3x + 15$

53) $7\left(\sqrt{6}+2\right) = 7\sqrt{6} + 14$

55) $\sqrt{10}\left(\sqrt{3}-1\right) = \sqrt{30} - \sqrt{10}$

57) $-6\left(\sqrt{32}+\sqrt{2}\right) = -6\left(4\sqrt{2}+\sqrt{2}\right)$

$\qquad = -6\left(5\sqrt{2}\right) = -30\sqrt{2}$

59) $4\left(\sqrt{45}-\sqrt{20}\right) = 4\left(3\sqrt{5}-2\sqrt{5}\right)$

$\qquad = 4\left(\sqrt{5}\right) = 4\sqrt{5}$

61) $\sqrt{5}\left(\sqrt{24}-\sqrt{54}\right) = \sqrt{5}\left(2\sqrt{6}-3\sqrt{6}\right)$

$\qquad = \sqrt{5}\left(-\sqrt{6}\right)$

$\qquad = -\sqrt{30}$

63) $\sqrt[4]{3}\left(5-\sqrt[4]{27}\right) = \sqrt[4]{3}\cdot5 - \sqrt[4]{3}\cdot\sqrt[4]{27}$

$\qquad = 5\sqrt[4]{3} - 3$

65) $\sqrt{t}\left(\sqrt{t}-\sqrt{81u}\right) = \sqrt{t}\left(\sqrt{t}-9\sqrt{u}\right)$

$\qquad = t - 9\sqrt{tu}$

67) $\sqrt{ab}\left(\sqrt{5a}+\sqrt{27b}\right)$

$\qquad = \sqrt{ab}\left(\sqrt{5a}+3\sqrt{3b}\right)$

$\qquad = \sqrt{5a^2b} + 3\sqrt{3ab^2} = a\sqrt{5b} + 3b\sqrt{3a}$

69) $\sqrt[3]{c^2}\left(\sqrt[3]{c^2}+\sqrt[3]{125cd}\right)$

$\qquad = \sqrt[3]{c^2}\cdot\sqrt[3]{c^2} + \sqrt[3]{c^2}\cdot\sqrt[3]{125cd}$

$\qquad = \sqrt[3]{c^4} + \sqrt[3]{125c^3d}$

$\qquad = c\sqrt[3]{c} + 5c\sqrt[3]{d}$

71) Both are examples of multiplication of two binomials. They can be multiplied using FOIL.

Section 9.5: Adding, Subtracting, and Multiplying Radicals

73) $(a+b)(a-b)=a^2-b^2$

75) $(p+7)(p+6)=p^2+6p+7p+42$
$$= p^2+13p+42$$

77)

$(6+\sqrt{7})(2+\sqrt{7})$

$= \boxed{6\cdot2+6\sqrt{7}+2\sqrt{7}+\sqrt{7}\cdot\sqrt{7}}$ Use FOIL.

$= 12+6\sqrt{7}+2\sqrt{7}+7$ $\boxed{\text{Multiply.}}$

$= \boxed{19+8\sqrt{7}}$ Combine like terms.

79) $(\sqrt{2}+8)(\sqrt{2}-3)=2-3\sqrt{2}+8\sqrt{2}-24=-22+5\sqrt{2}$

81)

$(\sqrt{5}-4\sqrt{3})(2\sqrt{5}-\sqrt{3})$

$= 2(5)-\sqrt{15}-8\sqrt{15}+4(3)=10-9\sqrt{15}+12=22-9\sqrt{15}$

83) $(5+2\sqrt{3})(\sqrt{7}+\sqrt{2})=5\sqrt{7}+5\sqrt{2}+2\sqrt{21}+2\sqrt{6}$

85) $(\sqrt[3]{25}-3)(\sqrt[3]{5}-\sqrt[3]{6})$
$$= \sqrt[3]{25}\,\sqrt[3]{5}-\sqrt[3]{25}\,\sqrt[3]{6}-3\sqrt[3]{5}+3\sqrt[3]{6}$$
$$= \sqrt[3]{125}-\sqrt[3]{150}-3\sqrt[3]{5}+3\sqrt[3]{6}$$
$$= 5-\sqrt[3]{150}-3\sqrt[3]{5}+3\sqrt[3]{6}$$

87) $(\sqrt{6p}-2\sqrt{q})(8\sqrt{q}+5\sqrt{6p})$
$$= 8\sqrt{6pq}+5(6p)-16q-10\sqrt{6pq}$$
$$= -2\sqrt{6pq}+30p-16q$$

89) $(\sqrt{3}+1)^2=(\sqrt{3})^2+2(\sqrt{3})(1)+(1)^2$
$$= 3+2\sqrt{3}+1=4+2\sqrt{3}$$

91)

$(\sqrt{11}-\sqrt{5})^2$

$=(\sqrt{11})^2-2(\sqrt{11})(\sqrt{5})+(\sqrt{5})^2=11-2\sqrt{55}+5=16-2\sqrt{55}$

93)

$(\sqrt{h}+\sqrt{7})^2$

$=(\sqrt{h})^2+2(\sqrt{h})(\sqrt{7})+(\sqrt{7})^2=h+2\sqrt{7h}+7$

95)

$(\sqrt{x}-\sqrt{y})^2$

$=(\sqrt{x})^2-2(\sqrt{x})(\sqrt{y})+(\sqrt{y})^2=x-2\sqrt{xy}+y$

97) $(c+9)(c-9)=c^2-(9)^2=c^2-81$

99)

$(6-\sqrt{5})(6+\sqrt{5})=6^2-(\sqrt{5})^2=36-5=31$

101) $(4\sqrt{3}+\sqrt{2})(4\sqrt{3}-\sqrt{2})$
$$= (4\sqrt{3})^2-(\sqrt{2})^2$$
$$= 16(3)-2=48-2=46$$

103) $\left(\sqrt[3]{2}-3\right)\left(\sqrt[3]{2}+3\right)=\left(\sqrt[3]{2}\right)^{2}-3^{2}$

$$=\sqrt[3]{4}-9$$

105) $\left(\sqrt{c}+\sqrt{d}\right)\left(\sqrt{c}-\sqrt{d}\right)$

$$=\left(\sqrt{c}\right)^{2}-\left(\sqrt{d}\right)^{2}=c-d$$

107) $\left(8\sqrt{f}-\sqrt{g}\right)\left(8\sqrt{f}+\sqrt{g}\right)$

$$=\left(8\sqrt{f}\right)^{2}-\left(\sqrt{g}\right)^{2}=64f-g$$

109) $\left(1+2\sqrt[3]{5}\right)\left(1-2\sqrt[3]{5}+4\sqrt[3]{25}\right)$

$$=1-2\sqrt[3]{5}+4\sqrt[3]{25}+2\sqrt[3]{5}$$

$$-4\sqrt[3]{25}+8\sqrt[3]{125}$$

$$=1+8\sqrt[3]{125}=1+8\left(5\right)=1+40=41$$

111)

$$f\left(\sqrt{7}+2\right)=\left(\sqrt{7}+2\right)^{2}$$

$$=\left(\sqrt{7}\right)^{2}+2\left(\sqrt{7}\right)\left(2\right)+\left(2\right)^{2}$$

$$=7+4\sqrt{7}+4=11+4\sqrt{7}$$

113)

$$f\left(1-2\sqrt{3}\right)=\left(1-2\sqrt{3}\right)^{2}$$

$$=\left(1\right)^{2}-2\left(1\right)\left(2\sqrt{3}\right)+\left(2\sqrt{3}\right)^{2}$$

$$=1-4\sqrt{3}+12=13-4\sqrt{3}$$

Section 9.6

1) Eliminate the radical of the denominator.

3) $\dfrac{1}{\sqrt{5}}=\dfrac{1}{\sqrt{5}}\cdot\dfrac{\sqrt{5}}{\sqrt{5}}=\dfrac{\sqrt{5}}{5}$

5) $\dfrac{9}{\sqrt{6}}=\dfrac{9}{\sqrt{6}}\cdot\dfrac{\sqrt{6}}{\sqrt{6}}=\dfrac{9\sqrt{6}}{6}=\dfrac{3\sqrt{6}}{2}$

7) $-\dfrac{20}{\sqrt{8}}=-\dfrac{20}{2\sqrt{2}}=-\dfrac{10}{\sqrt{2}}=-\dfrac{10}{\sqrt{2}}\cdot\dfrac{\sqrt{2}}{\sqrt{2}}$

$$=-\dfrac{10\sqrt{2}}{2}=-5\sqrt{2}$$

9) $\dfrac{\sqrt{3}}{\sqrt{28}}=\dfrac{\sqrt{3}}{2\sqrt{7}}=\dfrac{\sqrt{3}}{2\sqrt{7}}\cdot\dfrac{\sqrt{7}}{\sqrt{7}}$

$$=\dfrac{\sqrt{21}}{2\left(7\right)}=\dfrac{\sqrt{21}}{14}$$

11) $\sqrt{\dfrac{20}{60}}=\sqrt{\dfrac{1}{3}}=\dfrac{1}{\sqrt{3}}=\dfrac{1}{\sqrt{3}}\cdot\dfrac{\sqrt{3}}{\sqrt{3}}=\dfrac{\sqrt{3}}{3}$

13) $\sqrt{\dfrac{56}{48}}=\sqrt{\dfrac{7}{6}}=\dfrac{\sqrt{7}}{\sqrt{6}}$

$$=\dfrac{\sqrt{7}}{\sqrt{6}}\cdot\dfrac{\sqrt{6}}{\sqrt{6}}=\dfrac{\sqrt{42}}{6}$$

Section 9.6: Dividing Radicals

15) $\sqrt{\dfrac{10}{7}} \cdot \sqrt{\dfrac{7}{3}} = \sqrt{\dfrac{10}{7} \cdot \dfrac{7}{3}} = \sqrt{\dfrac{10}{3}}$

$\quad = \dfrac{\sqrt{10}}{\sqrt{3}} \cdot \dfrac{\sqrt{3}}{\sqrt{3}} = \dfrac{\sqrt{30}}{3}$

17) $\sqrt{\dfrac{6}{5}} \cdot \sqrt{\dfrac{1}{8}} = \sqrt{\dfrac{6}{5} \cdot \dfrac{1}{8}} = \sqrt{\dfrac{3}{20}} = \dfrac{\sqrt{3}}{\sqrt{20}}$

$\quad = \dfrac{\sqrt{3}}{2\sqrt{5}} \cdot \dfrac{\sqrt{5}}{\sqrt{5}} = \dfrac{\sqrt{15}}{2(5)} = \dfrac{\sqrt{15}}{10}$

19) $\dfrac{8}{\sqrt{y}} = \dfrac{8}{\sqrt{y}} \cdot \dfrac{\sqrt{y}}{\sqrt{y}} = \dfrac{8\sqrt{y}}{y}$

21) $\dfrac{\sqrt{5}}{\sqrt{t}} = \dfrac{\sqrt{5}}{\sqrt{t}} \cdot \dfrac{\sqrt{t}}{\sqrt{t}} = \dfrac{\sqrt{5t}}{t}$

23) $\sqrt{\dfrac{64v^7}{5w}} = \dfrac{8v^3 \sqrt{v}}{\sqrt{5w}} = \dfrac{8v^3 \sqrt{v}}{\sqrt{5w}} \cdot \dfrac{\sqrt{5w}}{\sqrt{5w}}$

$\quad = \dfrac{8v^3 \sqrt{5vw}}{5w}$

25) $\sqrt{\dfrac{a^3 b^3}{3ab^4}} = \sqrt{\dfrac{a^2}{3b}} = \dfrac{a}{\sqrt{3b}}$

$\quad = \dfrac{a}{\sqrt{3b}} \cdot \dfrac{\sqrt{3b}}{\sqrt{3b}} = \dfrac{a\sqrt{3b}}{3b}$

27) $-\dfrac{\sqrt{75}}{\sqrt{b^3}} = -\dfrac{5\sqrt{3}}{b\sqrt{b}} = -\dfrac{5\sqrt{3}}{b\sqrt{b}} \cdot \dfrac{\sqrt{b}}{\sqrt{b}}$

$\quad = -\dfrac{5\sqrt{3b}}{b(b)} = -\dfrac{5\sqrt{3b}}{b^2}$

29) $\dfrac{\sqrt{13}}{\sqrt{j^5}} = \dfrac{\sqrt{13}}{j^2 \sqrt{j}} = \dfrac{\sqrt{13}}{j^2 \sqrt{j}} \cdot \dfrac{\sqrt{j}}{\sqrt{j}}$

$\quad = \dfrac{\sqrt{13j}}{j^2(j)} = \dfrac{\sqrt{13j}}{j^3}$

31) 2^2 or 4

33) 3

35) c^2

37) 2^3 or 8

39) m

41) $\dfrac{4}{\sqrt[3]{3}} = \dfrac{4}{\sqrt[3]{3}} \cdot \dfrac{\sqrt[3]{3^2}}{\sqrt[3]{3^2}} = \dfrac{4\sqrt[3]{9}}{\sqrt[3]{3^3}} = \dfrac{4\sqrt[3]{9}}{3}$

43) $\dfrac{12}{\sqrt[3]{2}} = \dfrac{12}{\sqrt[3]{2}} \cdot \dfrac{\sqrt[3]{2^2}}{\sqrt[3]{2^2}} = \dfrac{12\sqrt[3]{4}}{\sqrt[3]{2^3}}$

$\quad = \dfrac{12\sqrt[3]{4}}{2} = 6\sqrt[3]{4}$

45) $\dfrac{9}{\sqrt[3]{25}} = \dfrac{9}{\sqrt[3]{5^2}} = \dfrac{9}{\sqrt[3]{5^2}} \cdot \dfrac{\sqrt[3]{5}}{\sqrt[3]{5}}$

$= \dfrac{9\sqrt[3]{5}}{\sqrt[3]{5^3}} = \dfrac{9\sqrt[3]{5}}{5}$

47) $\sqrt[4]{\dfrac{5}{9}} = \dfrac{\sqrt[4]{5}}{\sqrt[4]{3^2}} = \dfrac{\sqrt[4]{5}}{\sqrt[4]{3^2}} \cdot \dfrac{\sqrt[4]{3^2}}{\sqrt[4]{3^2}}$

$= \dfrac{\sqrt[4]{5} \cdot \sqrt[4]{9}}{\sqrt[4]{3^4}} = \dfrac{\sqrt[4]{45}}{3}$

49) $\sqrt[5]{\dfrac{3}{8}} = \dfrac{\sqrt[5]{3}}{\sqrt[5]{2^3}} = \dfrac{\sqrt[5]{3}}{\sqrt[5]{2^3}} \cdot \dfrac{\sqrt[5]{2^2}}{\sqrt[5]{2^2}}$

$= \dfrac{\sqrt[5]{3} \cdot \sqrt[5]{4}}{\sqrt[5]{2^5}} = \dfrac{\sqrt[5]{12}}{2}$

51) $\dfrac{10}{\sqrt[3]{z}} = \dfrac{10}{\sqrt[3]{z}} \cdot \dfrac{\sqrt[3]{z^2}}{\sqrt[3]{z^2}} = \dfrac{10\sqrt[3]{z^2}}{\sqrt[3]{z^3}} = \dfrac{10\sqrt[3]{z^2}}{z}$

53) $\sqrt[3]{\dfrac{3}{n^2}} = \dfrac{\sqrt[3]{3}}{\sqrt[3]{n^2}} \cdot \dfrac{\sqrt[3]{n}}{\sqrt[3]{n}} = \dfrac{\sqrt[3]{3n}}{\sqrt[3]{n^3}} = \dfrac{\sqrt[3]{3n}}{n}$

55) $\dfrac{\sqrt[3]{7}}{\sqrt[3]{2k^2}} = \dfrac{\sqrt[3]{7}}{\sqrt[3]{2k^2}} \cdot \dfrac{\sqrt[3]{2^2 k}}{\sqrt[3]{2^2 k}} = \dfrac{\sqrt[3]{7} \cdot \sqrt[3]{4k}}{\sqrt[3]{2^3 k^3}}$

$= \dfrac{\sqrt[3]{28k}}{2k}$

57) $\dfrac{9}{\sqrt[5]{a^3}} = \dfrac{9}{\sqrt[5]{a^3}} \cdot \dfrac{\sqrt[5]{a^2}}{\sqrt[5]{a^2}} = \dfrac{9\sqrt[5]{a^2}}{\sqrt[5]{a^5}} = \dfrac{9\sqrt[5]{a^2}}{a}$

59) $\sqrt[4]{\dfrac{5}{2m}} = \dfrac{\sqrt[4]{5}}{\sqrt[4]{2m}} \cdot \dfrac{\sqrt[4]{2^3 m^3}}{\sqrt[4]{2^3 m^3}} = \dfrac{\sqrt[4]{5} \cdot \sqrt[4]{8m^3}}{\sqrt[4]{2^4 m^4}}$

$= \dfrac{\sqrt[4]{40m^3}}{2m}$

61) Change the sign between the
    two terms.

63) $(5 + \sqrt{2})(5 - \sqrt{2}) = (5)^2 - (\sqrt{2})^2$

$= 25 - 2 = 23$

65) $(\sqrt{2} + \sqrt{6})(\sqrt{2} - \sqrt{6})$

$= (\sqrt{2})^2 - (\sqrt{6})^2 = 2 - 6 = -4$

67) $(\sqrt{t} - 8)(\sqrt{t} + 8) = (\sqrt{t})^2 - (8)^2$

$= t - 64$

69) $\dfrac{6}{4 - \sqrt{5}} = \dfrac{6}{4 - \sqrt{5}} \cdot \dfrac{4 + \sqrt{5}}{4 + \sqrt{5}}$

$\boxed{\text{Multiply by the conjugate.}}$

$= \dfrac{6(4 + \sqrt{5})}{(4)^2 - (\sqrt{5})^2}$ $\boxed{(a+b)(a-b) = a^2 - b^2}$

$= \boxed{\dfrac{24 + 6\sqrt{5}}{16 - 5}}$ Multiply terms in numerator,

square terms in denominator

$= \boxed{\dfrac{24 + 6\sqrt{5}}{11}}$ Simplify.

Section 9.6: Dividing Radicals

71) $\dfrac{3}{2+\sqrt{3}} = \dfrac{3}{2+\sqrt{3}} \cdot \dfrac{2-\sqrt{3}}{2-\sqrt{3}}$

$= \dfrac{3(2-\sqrt{3})}{(2)^2 - (\sqrt{3})^2} = \dfrac{3(2-\sqrt{3})}{4-3}$

$= 6 - 3\sqrt{3}$

79) $\dfrac{\sqrt{m}}{\sqrt{m}+\sqrt{n}} = \dfrac{\sqrt{m}}{\sqrt{m}+\sqrt{n}} \cdot \dfrac{\sqrt{m}-\sqrt{n}}{\sqrt{m}-\sqrt{n}}$

$= \dfrac{\sqrt{m}(\sqrt{m}-\sqrt{n})}{(\sqrt{m})^2 - (\sqrt{m})^2} = \dfrac{m-\sqrt{mn}}{m-n}$

73) $\dfrac{10}{9-\sqrt{2}} = \dfrac{10}{9-\sqrt{2}} \cdot \dfrac{9+\sqrt{2}}{9+\sqrt{2}}$

$= \dfrac{10(9+\sqrt{2})}{(9)^2 - (\sqrt{2})^2} = \dfrac{10(9+\sqrt{2})}{81-2}$

$= \dfrac{90+10\sqrt{2}}{79}$

81) $\dfrac{b-25}{\sqrt{b}-5} = \dfrac{b-25}{\sqrt{b}-5} \cdot \dfrac{\sqrt{b}+5}{\sqrt{b}+5}$

$= \dfrac{(b-25)(\sqrt{b}+5)}{(\sqrt{b})^2 - (5)^2}$

$= \dfrac{(b-25)(\sqrt{b}+5)}{b-25} = \sqrt{b}+5$

75) $\dfrac{\sqrt{8}}{\sqrt{3}+\sqrt{2}} = \dfrac{2\sqrt{2}}{\sqrt{3}+\sqrt{2}} \cdot \dfrac{\sqrt{3}-\sqrt{2}}{\sqrt{3}-\sqrt{2}}$

$= \dfrac{2\sqrt{2}(\sqrt{3}-\sqrt{2})}{(\sqrt{3})^2 - (\sqrt{2})^2}$

$= \dfrac{2\sqrt{6}-2(2)}{3-2} = 2\sqrt{6}-4$

83) $\dfrac{\sqrt{x}+\sqrt{y}}{\sqrt{x}-\sqrt{y}} = \dfrac{\sqrt{x}+\sqrt{y}}{\sqrt{x}-\sqrt{y}} \cdot \dfrac{\sqrt{x}+\sqrt{y}}{\sqrt{x}+\sqrt{y}}$

$= \dfrac{(\sqrt{x}+\sqrt{y})^2}{(\sqrt{x})^2 - (\sqrt{y})^2}$

$= \dfrac{x+2\sqrt{xy}+y}{x-y}$

77) $\dfrac{\sqrt{3}-\sqrt{5}}{\sqrt{10}-\sqrt{3}} = \dfrac{\sqrt{3}-\sqrt{5}}{\sqrt{10}-\sqrt{3}} \cdot \dfrac{\sqrt{10}+\sqrt{3}}{\sqrt{10}+\sqrt{3}}$

$= \dfrac{(\sqrt{3}-\sqrt{5})(\sqrt{10}+\sqrt{3})}{(\sqrt{10})^2 - (\sqrt{3})^2}$

$= \dfrac{\sqrt{30}+3-\sqrt{50}-\sqrt{15}}{10-3}$

$= \dfrac{\sqrt{30}+3-5\sqrt{2}-\sqrt{15}}{7}$

85) $\dfrac{\sqrt{5}}{3} = \dfrac{\sqrt{5}}{3} \cdot \dfrac{\sqrt{5}}{\sqrt{5}} = \dfrac{5}{3\sqrt{5}}$

87) $\dfrac{\sqrt{x}}{\sqrt{7}} = \dfrac{\sqrt{x}}{\sqrt{7}} \cdot \dfrac{\sqrt{x}}{\sqrt{x}} = \dfrac{x}{\sqrt{7x}}$

89) $\dfrac{2+\sqrt{3}}{6} = \dfrac{2+\sqrt{3}}{6} \cdot \dfrac{2-\sqrt{3}}{2-\sqrt{3}}$

$= \dfrac{4-3}{6\left(2-\sqrt{3}\right)} = \dfrac{1}{12-6\sqrt{3}}$

91) $\dfrac{\sqrt{x}-2}{x-4} = \dfrac{\sqrt{x}-2}{x-4} \cdot \dfrac{\sqrt{x}+2}{\sqrt{x}+2}$

$= \dfrac{x-4}{\left(x-4\right)\left(\sqrt{x}+2\right)} = \dfrac{1}{\sqrt{x}+2}$

93) $\dfrac{4-\sqrt{c+11}}{c-5} = \dfrac{4-\sqrt{c+11}}{c-5} \cdot \dfrac{4+\sqrt{c+11}}{4+\sqrt{c+11}}$

$= \dfrac{16-\left(c+11\right)}{\left(c-5\right)\left(4+\sqrt{c+11}\right)}$

$= \dfrac{16-c-11}{\left(c-5\right)\left(4+\sqrt{c+11}\right)}$

$= \dfrac{5-c}{\left(c-5\right)\left(4+\sqrt{c+11}\right)}$

$= -\dfrac{1}{4+\sqrt{c+11}}$

95) No, because when we multiply the numerator and denominator by the conjugate of the denominator, we are multiplying the original expression by 1.

97) $\dfrac{5+10\sqrt{3}}{5} = \dfrac{5\left(1+2\sqrt{3}\right)}{5} = 1+2\sqrt{3}$

99) $\dfrac{30-18\sqrt{5}}{4} = \dfrac{6\left(5-3\sqrt{5}\right)}{4}$

$= \dfrac{3\left(5-3\sqrt{5}\right)}{2} = \dfrac{15-9\sqrt{5}}{2}$

101) $\dfrac{\sqrt{45}+6}{9} = \dfrac{3\sqrt{5}+6}{9} = \dfrac{3\left(\sqrt{5}+2\right)}{9}$

$= \dfrac{\sqrt{5}+2}{3}$

103) $\dfrac{-10-\sqrt{50}}{5} = \dfrac{-10-5\sqrt{2}}{5}$

$= \dfrac{5\left(-2-\sqrt{2}\right)}{5} = -2-\sqrt{2}$

105) a) $r(A) = \sqrt{\dfrac{A}{\pi}}$

$r(8\pi) = \sqrt{\dfrac{8\pi}{\pi}} = \sqrt{8} = 2\sqrt{2}$

When the area of the circle is 8π in^2, the radius is $2\sqrt{2}$ in.

b) $r(A)=\sqrt{\dfrac{A}{\pi}}$

$r(7)=\sqrt{\dfrac{7}{\pi}}=\dfrac{\sqrt{7}}{\sqrt{\pi}}\dfrac{\sqrt{\pi}}{\sqrt{\pi}}=\dfrac{\sqrt{7\pi}}{\pi}$

When the area of the circle is 7 in^2,

the radius is $\dfrac{\sqrt{7\pi}}{\pi}$ in.

(This is approximately 1.5 in.)

c) $r(A)=\sqrt{\dfrac{A}{\pi}}=\dfrac{\sqrt{A}}{\sqrt{\pi}}\cdot\dfrac{\sqrt{\pi}}{\sqrt{\pi}}=\dfrac{\sqrt{A\pi}}{\pi}$

13) $k^{-3/5}\cdot k^{3/10}=k^{-6/10+3/10}=k^{-3/10}=\dfrac{1}{k^{3/10}}$

15) $\left(\dfrac{27a^{-8}}{b^9}\right)^{2/3}=\dfrac{(27)^{2/3}\left(a^{-8}\right)^{2/3}}{\left(b^9\right)^{2/3}}$

$=\dfrac{\left(27^{1/3}\right)^2 a^{-16/3}}{b^{18/3}}=\dfrac{9a^{-16/3}}{b^6}$

$=\dfrac{9}{a^{16/3}b^6}$

Chapter 9: Putting It All Together

1) $\sqrt[4]{81}=3$

3) $-\sqrt[6]{64}=-2$

5) $\sqrt{-169}$ is not a real number.

7) $(144)^{1/2}=\sqrt{144}=12$

9) $-1000^{2/3}=-\left(1000^{1/3}\right)^2=-\left(\sqrt[3]{1000}\right)^2$
$=-(10)^2=-100$

11) $125^{-1/3}=\left(\dfrac{1}{125}\right)^{1/3}=\sqrt[3]{\dfrac{1}{125}}=\dfrac{1}{5}$

17) $\sqrt{24}=\sqrt{4\cdot6}=\sqrt{4}\cdot\sqrt{6}=2\sqrt{6}$

19) $\sqrt[3]{72}=\sqrt[3]{8}\cdot\sqrt[3]{9}=2\sqrt[3]{9}$

21) $\sqrt[4]{243}=\sqrt[4]{81\cdot3}=\sqrt[4]{81}\cdot\sqrt[4]{3}=3\sqrt[4]{3}$

23) $\sqrt[3]{96m^7n^{15}}=\sqrt[3]{96}\cdot\sqrt[3]{m^7}\cdot\sqrt[3]{n^{15}}$
$=\sqrt[3]{8\cdot12}\cdot\sqrt[3]{m^6\cdot m}\cdot\sqrt[3]{n^{15}}$
$=2\sqrt[3]{12}\cdot m^2\cdot\sqrt[3]{mn^5}$
$=2m^2n^5\sqrt[3]{12m}$

25) $\sqrt[3]{12}\cdot\sqrt[3]{2}=\sqrt[3]{24}=\sqrt[3]{8\cdot3}=2\sqrt[3]{3}$

27) $(6+\sqrt{7})(2+\sqrt{7})$
$=12+6\sqrt{7}+2\sqrt{7}+7=19+8\sqrt{7}$

29) $\dfrac{18}{\sqrt{6}}=\dfrac{18}{\sqrt{6}}\cdot\dfrac{\sqrt{6}}{\sqrt{6}}=\dfrac{18\sqrt{6}}{6}=3\sqrt{6}$

31) $3\sqrt{75m^3n}+m\sqrt{12mn}$

$=3\left(5m\sqrt{3mn}\right)+m\left(2\sqrt{3mn}\right)$

$=15m\sqrt{3mn}+2m\sqrt{3mn}=17m\sqrt{3mn}$

33) $\dfrac{\sqrt{60t^8u^3}}{\sqrt{5t^2u}}=\sqrt{\dfrac{60t^8u^3}{5t^2u}}=\sqrt{12t^6u^2}$

$=\sqrt{12}\cdot\sqrt{t^6}\cdot\sqrt{u^2}$

$=2\sqrt{3}\cdot t^3\cdot u=2t^3u\sqrt{3}$

35) $\left(2\sqrt{3}+10\right)^2$

$=\left(2\sqrt{3}\right)^2+2\left(2\sqrt{3}\right)(10)+(10)^2$

$=4(3)+40\sqrt{3}+100=12+40\sqrt{3}+100$

$=112+40\sqrt{3}$

37) $\dfrac{\sqrt{2}}{4+\sqrt{10}}=\dfrac{\sqrt{2}}{4+\sqrt{10}}\cdot\dfrac{4-\sqrt{10}}{4-\sqrt{10}}$

$=\dfrac{\sqrt{2}\left(4-\sqrt{10}\right)}{(4)^2-\left(\sqrt{10}\right)^2}=\dfrac{4\sqrt{2}-\sqrt{20}}{16-10}$

$=\dfrac{4\sqrt{2}-\sqrt{4\cdot5}}{6}=\dfrac{4\sqrt{2}-2\sqrt{5}}{6}$

$=\dfrac{2\left(2\sqrt{2}-\sqrt{5}\right)}{6}=\dfrac{2\sqrt{2}-\sqrt{5}}{3}$

39)

$\sqrt[3]{\dfrac{b^2}{9c}}=\dfrac{\sqrt[3]{b^2}}{\sqrt[3]{9c}}=\dfrac{\sqrt[3]{b^2}}{\sqrt[3]{9c}}\cdot\dfrac{\sqrt[3]{3c^2}}{\sqrt[3]{3c^2}}=\dfrac{\sqrt[3]{3b^2c^2}}{\sqrt[3]{27c^3}}=\dfrac{\sqrt[3]{3b^2c^2}}{3c}$

41) $x-2\geq0$

$x\geq2$

$[2,\infty)$

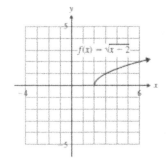

43) $(-\infty,\infty)$

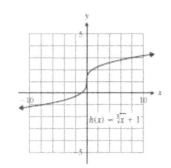

45) $-x\geq0$

$x\leq0$

$(-\infty,0]$

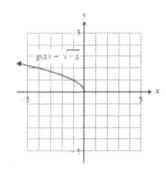

Section 9.7

1) Sometimes these are extraneous solutions.

3) $\sqrt{q} = 7$

$\left(\sqrt{q}\right)^2 = 7^2$

$q = 49$

Check $\sqrt{49} = 7$ $\{49\}$

5) $\sqrt{w} - \dfrac{2}{3} = 0$

$\sqrt{w} = \dfrac{2}{3}$

$\left(\sqrt{w}\right)^2 = \left(\dfrac{2}{3}\right)^2$

$w = \dfrac{4}{9}$ $\left\{\dfrac{4}{9}\right\}$

Check is left to the student.

7) $\sqrt{a} + 5 = 3$

$\sqrt{a} = -2$

$\left(\sqrt{a}\right)^2 = (-2)^2$

$a = 4$

Check $\sqrt{4} + 5 = 3$

$2 + 5 \neq 3$ $\varnothing$

9) $\sqrt{b-11} - 3 = 0$

$\sqrt{b-11} = 3$

$\left(\sqrt{b-11}\right)^2 = (3)^2$

$b - 11 = 9$

$b = 20$

Check is left to the student. $\{20\}$

11) $\sqrt{4g-1} + 7 = 1$

$\sqrt{4g-1} = -6$

$\left(\sqrt{4g-1}\right)^2 = (-6)^2$

$4g - 1 = 36$

$4g = 37$

$g = \dfrac{37}{4}$

$\dfrac{37}{4}$ is an extraneous solution. $\varnothing$

13) $\sqrt{3f+2} + 9 = 11$

$\sqrt{3f+2} = 2$

$\left(\sqrt{3f+2}\right)^2 = (2)^2$

$3f + 2 = 4$

$3f = 2$

$f = \dfrac{2}{3}$

Check is left to the student. $\left\{\dfrac{2}{3}\right\}$

15) $m = \sqrt{m^2 - 3m + 6}$

$m^2 = \left(\sqrt{m^2 - 3m + 6}\right)^2$

$m^2 = m^2 - 3m + 6$

$0 = -3m + 6$

$3m = 6$

$m = 2$

Check is left to the student. $\{2\}$

17) $\sqrt{9r^2 - 2r + 10} = 3r$

$\left(\sqrt{9r^2 - 2r + 10}\right)^2 = (3r)^2$

$9r^2 - 2r + 10 = 9r^2$

$-2r + 10 = 0$

$-2r = -10$

$r = 5$

Check is left to the student. $\{5\}$

19) $(n+5)^2 = n^2 + 2(n)(5) + 5^2$

$= n^2 + 10n + 25$

21) $(c-6)^2 = c^2 - 2(c)(6) + 6^2$

$= c^2 - 12c + 36$

23) $\qquad p + 6 = \sqrt{12 + p}$

$(p+6)^2 = \left(\sqrt{12 + p}\right)^2$

$p^2 + 12p + 36 = 12 + p$

$p^2 + 11p + 24 = 0$

$(p+8)(p+3) = 0$

$p + 8 = 0 \text{ or } p + 3 = 0$

$p = -8 \qquad p = -3$

-8 is an extraneous solution. $\{-3\}$

25) $6 + \sqrt{c^2 + 3c - 9} = c$

$\sqrt{c^2 + 3c - 9} = c - 6$

$\left(\sqrt{c^2 + 3c - 9}\right)^2 = (c-6)^2$

$c^2 + 3c - 9 = c^2 - 12c + 36$

$3c - 9 = -12c + 36$

$15c = 45, c = 3$

3 is an extraneous solution. $\varnothing$

27) $w - \sqrt{10w + 6} = -3$

$w + 3 = \sqrt{10w + 6}$

$(w+3)^2 = \left(\sqrt{10w + 6}\right)^2$

$w^2 + 6w + 9 = 10w + 6$

$w^2 - 4w + 3 = 0$

$(w-3)(w-1) = 0$

$w - 3 = 0 \text{ or } w - 1 = 0$

$w = 3 \qquad w = 1$

Check is left to the student. $\{1,3\}$

29)

$$3v = 8 + \sqrt{3v + 4}$$

$$3v - 8 = \sqrt{3v + 4}$$

$$(3v - 8)^2 = \left(\sqrt{3v + 4}\right)^2$$

$$9v^2 - 48v + 64 = 3v + 4$$

$$9v^2 - 51v + 60 = 0 \qquad \text{Divide by 3.}$$

$$3v^2 - 17v + 20 = 0$$

$$(3v - 5)(v - 4) = 0$$

$$3v - 5 = 0 \quad \text{or} \quad v - 4 = 0$$

$$3v = 5 \qquad\qquad v = 4$$

$$v = \frac{5}{3}$$

$\dfrac{5}{3}$ is an extraneous solution. $\{4\}$

31) $m + 4 = 5\sqrt{m}$

$$(m + 4)^2 = \left(5\sqrt{m}\right)^2$$

$$m^2 + 8m + 16 = 25m$$

$$m^2 + 8m - 25m + 16 = 0$$

$$m^2 - 17m + 16 = 0$$

$$(m - 16)(m - 1) = 0$$

$$m - 16 = 0 \quad \text{or} \quad m - 1 = 0$$

$$m = 16 \qquad\qquad m = 1$$

Check is left to the student. $\{1, 16\}$

33) $y + 2\sqrt{6 - y} = 3$

$$2\sqrt{6 - y} = 3 - y$$

$$\left(2\sqrt{6 - y}\right)^2 = (3 - y)^2$$

$$4(6 - y) = 9 - 6y + y^2$$

$$24 - 4y = y^2 - 6y + 9$$

$$0 = y^2 - 6y + 4y - 24 + 9$$

$$y^2 - 2y - 15 = 0$$

$$(y - 5)(y + 3) = 0$$

$$y - 5 = 0 \qquad\qquad y + 3 = 0$$

$$y = 5 \qquad\qquad y = -3$$

5 is an extraneous solution. $\{-3\}$

35) $\sqrt{r^2 - 8r - 19} = r - 9$

$$\left(\sqrt{r^2 - 8r - 19}\right)^2 = (r - 9)^2$$

$$r^2 - 8r - 19 = r^2 - 18r + 81$$

$$-8r - 19 = -18r + 81$$

$$10r = 100$$

$$r = 10$$

Check is left to the student. $\{10\}$

37) $5\sqrt{1 - 5h} = 4\sqrt{1 - 8h}$

$$\left(5\sqrt{1 - 5h}\right)^2 = \left(4\sqrt{1 - 8h}\right)^2$$

$$25(1 - 5h) = 16(1 - 8h)$$

$$25 - 125h = 16 - 128h$$

$$3h = -9$$

$$h = -3$$

Check is left to the student. $\{-3\}$

39) $3\sqrt{3x+6} = 2\sqrt{9x-9}$

$\left(3\sqrt{3x+6}\right)^2 = \left(2\sqrt{9x-9}\right)^2$

$9(3x+6) = 4(9x-9)$

$27x+54 = 36x-36$

$90 = 9x$

$10 = x$

Check is left to the student. $\{10\}$

41) $\sqrt{m} = 3\sqrt{7}$

$\left(\sqrt{m}\right)^2 = \left(3\sqrt{7}\right)^2$

$m = 9(7)$

$m = 63$

Check is left to the student. $\{63\}$

43) $\sqrt{2w-1} + 2\sqrt{w+4} = 0$

$\left(\sqrt{2w-1}\right)^2 = \left(-2\sqrt{w+4}\right)^2$

$2w-1 = 4w+16$

$2w-1-4w = 16$

$-2w-1 = 16$

$-2w = 17$

$\dfrac{-17}{2}$ is an extraneous solution. $\varnothing$

45) $\left(\sqrt{x}+5\right)^2 = \left(\sqrt{x}\right)^2 + 2\left(\sqrt{x}\right)(5) + (5)^2$

$= x + 10\sqrt{x} + 25$

47) $\left(9 - \sqrt{a+4}\right)^2$

$= (9)^2 - 2(9)\left(\sqrt{a+4}\right) + \left(\sqrt{a+4}\right)^2$

$= 81 - 18\sqrt{a+4} + a + 4$

$= 85 - 18\sqrt{a+4} + a$

49) $\left(2\sqrt{3n-1}+7\right)^2$

$= \left(2\sqrt{3n-1}\right)^2 + 2\left(2\sqrt{3n-1}\right)(7) + 7^2$

$= 4(3n-1) + 28\sqrt{3n-1} + 49$

$= 12n - 4 + 28\sqrt{3n-1} + 49$

$= 12n + 28\sqrt{3n-1} + 45$

51) $\left(\sqrt{2y-1}\right)^2 = \left(2 + \sqrt{y-4}\right)^2$

$2y-1 = 4 + 4\sqrt{(y-4)} + \left(\sqrt{y-4}\right)^2$

$2y-1 = 4 + 4\sqrt{y-4} + y - 4$

$2y-1 = y + 4\sqrt{y-4}$

$y-1 = 4\sqrt{y-4}$

Section 9.7: Solving Radical Equations

$$(y-1)^2 = \left(4\sqrt{y-4}\right)^2$$
$$y^2 - 2y + 1 = 16y - 64$$
$$y^2 - 2y - 16y + 1 + 64 = 0$$
$$y^2 - 18y + 65 = 0$$
$$(y-13)(y-5) = 0$$
$$y - 13 = 0 \quad \text{or} \quad y - 5 = 0$$
$$y = 13 \qquad y = 5$$

Check is left to the student. $\{5, 13\}$

53) $\qquad 1 + \sqrt{3s-2} = \sqrt{2s+5}$
$$\left(1+\sqrt{3s-2}\right)^2 = \left(\sqrt{2s+5}\right)^2$$
$$1 + 2\sqrt{3s-2} + 3s - 2 = 2s + 5$$
$$3s - 1 + 2\sqrt{3s-2} = 2s + 5$$
$$2\sqrt{3s-2} = 6 - s$$
$$\left(2\sqrt{3s-2}\right)^2 = (6-s)^2$$
$$4(3s-2) = 36 - 12s + s^2$$
$$12s - 8 = s^2 - 12s + 36$$
$$0 = s^2 - 24s + 44$$
$$0 = (s-2)(s-22)$$

$$s - 2 = 0 \quad \text{or} \quad s - 22 = 0$$
$$s = 2 \qquad s = 22$$

22 is an extraneous solution. $\{2\}$

55) $\sqrt{5a+19} - \sqrt{a+12} = 1$
$$\sqrt{5a+19} = 1 + \sqrt{a+12}$$
$$\left(\sqrt{5a+19}\right)^2 = \left(1+\sqrt{a+12}\right)^2$$
$$5a + 19 = 1 + 2\sqrt{a+12} + a + 12$$
$$5a + 19 = a + 13 + 2\sqrt{a+12}$$
$$4a + 6 = 2\sqrt{a+12}$$
$$2a + 3 = \sqrt{a+12}$$
$$(2a+3)^2 = \left(\sqrt{a+12}\right)^2$$
$$4a^2 + 12a + 9 = a + 12$$
$$4a^2 + 11a - 3 = 0$$
$$(4a-1)(a+3) = 0$$

$$4a - 1 = 0 \quad \text{or} \quad a + 3 = 0$$
$$4a = 1$$
$$a = \frac{1}{4} \qquad a = -3$$

-3 is an extraneous solution. $\left\{\frac{1}{4}\right\}$

57) $\sqrt{3k+1} - \sqrt{k-1} = 2$

$\sqrt{3k+1} = 2 + \sqrt{k-1}$

$\left(\sqrt{3k+1}\right)^2 = \left(2 + \sqrt{k-1}\right)^2$

$3k+1 = 4 + 4\sqrt{k-1} + k - 1$

$3k+1 = k + 3 + 4\sqrt{k-1}$

$2k - 2 = 4\sqrt{k-1}$

$k - 1 = 2\sqrt{k-1}$

$(k-1)^2 = \left(2\sqrt{k-1}\right)^2$

$k^2 - 2k + 1 = 4(k-1)$

$k^2 - 2k + 1 = 4k - 4$

$k^2 - 6k + 5 = 0$

$(k-1)(k-5) = 0$

$k - 1 = 0$ or $k - 5 = 0$

$k = 1 \qquad k = 5$

Check is left to the student. $\{1,5\}$

59) $\qquad \sqrt{3x+4} - 5 = \sqrt{3x-11}$

$\left(\sqrt{3x+4} - 5\right)^2 = \left(\sqrt{3x-11}\right)^2$

$3x+4 - 10\sqrt{3x+4} + 25 = 3x - 11$

$3x + 29 - 10\sqrt{3x+4} = 3x - 11$

$-10\sqrt{3x+4} = -40$

$\sqrt{3x+4} = 4$

$\left(\sqrt{3x+4}\right)^2 = (4)^2$

$3x + 4 = 16$

$3x = 12$

$x = 4$

4 is an extraneous solution. $\varnothing$

61) $\sqrt{3v+3} - \sqrt{v-2} = 3$

$\sqrt{3v+3} = \sqrt{v-2} + 3$

$\left(\sqrt{3v+3}\right)^2 = \left(\sqrt{v-2} + 3\right)^2$

$3v+3 = v - 2 + 6\sqrt{v-2} + 9$

$3v+3 = v + 6\sqrt{v-2} + 7$

$2v - 4 = 6\sqrt{v-2}$

$v - 2 = 3\sqrt{v-2}$

$(v-2)^2 = \left(3\sqrt{v-2}\right)^2$

$v^2 - 4v + 4 = 9(v-2)$

$v^2 - 4v + 4 = 9v - 18$

$v^2 - 13v + 22 = 0$

$(v-2)(v-11) = 0$

$v - 2 = 0$ or $v - 11 = 0$

$v = 2 \qquad v = 11$

Check is left to the student. $\{2,11\}$

63) Raise both sides of the equation to the third power.

65) $\quad \sqrt[3]{y} = 5$

$\left(\sqrt[3]{y}\right)^3 = 5^3$

$y = 125$

Check is left to the student. $\{125\}$

67) $\quad \sqrt[3]{m} = -4$

$\left(\sqrt[3]{m}\right)^3 = (-4)^3$

$m = -64$

Check is left to the student. $\{-64\}$

Section 9.7: Solving Radical Equations

69) $\sqrt[3]{2x-5}+3=1$

$\quad\quad \sqrt[3]{2x-5}=-2$

$\quad\quad \left(\sqrt[3]{2x-5}\right)^3=(-2)^3$

$\quad\quad\quad 2x-5=-8$

$\quad\quad\quad\quad 2x=-3$

$\quad\quad\quad\quad\quad x=-\dfrac{3}{2}$

Check is left to the student. $\left\{-\dfrac{3}{2}\right\}$

71) $\sqrt[3]{6j-2}=\sqrt[3]{j-7}$

$\quad \left(\sqrt[3]{6j-2}\right)^3=\left(\sqrt[3]{j-7}\right)^3$

$\quad\quad\quad 6j-2=j-7$

$\quad\quad\quad\quad 5j=-5$

$\quad\quad\quad\quad\quad j=-1$

Check is left to the student. $\{-1\}$

73) $\sqrt[3]{3y-1}-\sqrt[3]{2y-3}=0$

$\quad \left(\sqrt[3]{3y-1}\right)^3=\left(\sqrt[3]{2y-3}\right)^3$

$\quad\quad\quad 3y-1=2y-3$

$\quad\quad\quad\quad y=-2$

Check is left to the student. $\{-2\}$

75) $\quad\quad \sqrt[3]{2n^2}=\sqrt[3]{7n+4}$

$\quad\quad \left(\sqrt[3]{2n^2}\right)^3=\left(\sqrt[3]{7n+4}\right)^3$

$\quad\quad\quad 2n^2=7n+4$

$\quad\quad 2n^2-7n-4=0$

$\quad\quad (2n+1)(n-4)=0$

$\quad\quad 2n+1=0 \quad\quad n-4=0$

$\quad\quad n=-\dfrac{1}{2} \quad\quad\quad n=4$

Check is left to the student. $\left\{-\dfrac{1}{2},4\right\}$

77) $\quad p^{1/2}=6$

$\quad\quad \left(p^{1/2}\right)^2=(6)^2$

$\quad\quad\quad p=36$

Check is left to the student $\quad\{36\}$

79) $\quad 7=(2z-3)^{1/2}$

$\quad\quad (7)^2=\left((2z-3)^{1/2}\right)^2$

$\quad\quad\quad 49=2z-3$

$\quad\quad\quad 52=2z$

$\quad\quad\quad 26=z$

Check is left to the student. $\{26\}$

81) $\quad (y+4)^{1/3}=3$

$\quad\quad \left((y+4)^{1/3}\right)^3=(3)^3$

$\quad\quad\quad y+4=27$

$\quad\quad\quad\quad y=23$

Check is left to the student. $\{23\}$

83) $\sqrt[4]{n+7} = 2$

$\left(\sqrt[4]{n+7}\right)^4 = (2)^4$

$n + 7 = 16$

$n = 9$

Check is left to the student. $\{9\}$

85) $\sqrt{13 + \sqrt{r}} = \sqrt{r+7}$

$\left(\sqrt{13 + \sqrt{r}}\right)^2 = \left(\sqrt{r+7}\right)^2$

$13 + \sqrt{r} = r + 7$

$\sqrt{r} = r - 6$

$\left(\sqrt{r}\right)^2 = (r-6)^2$

$r = r^2 - 12r + 36$

$0 = r^2 - 13r + 36$

$0 = (r-9)(r-4)$

$r - 9 = 0$ or $r - 4 = 0$

$r = 9$ $r = 4$

4 is an extraneous solution. $\{9\}$

87) $\sqrt{y + \sqrt{y+5}} = \sqrt{y+2}$

$\left(\sqrt{y + \sqrt{y+5}}\right)^2 = \left(\sqrt{y+2}\right)^2$

$y + \sqrt{y+5} = y + 2$

$\sqrt{y+5} = 2$

$\left(\sqrt{y+5}\right)^2 = (2)^2$

$y + 5 = 4$

$y = -1$

Check is left to the student. $\{-1\}$

89) $v = \sqrt{\dfrac{2E}{m}}$

$v^2 = \left(\sqrt{\dfrac{2E}{m}}\right)^2$

$v^2 = \dfrac{2E}{m}$

$mv^2 = 2E$

$\dfrac{mv^2}{2} = E$

91) $c = \sqrt{a^2 + b^2}$

$c^2 = \left(\sqrt{a^2 + b^2}\right)^2$

$c^2 = a^2 + b^2$

$c^2 - a^2 = b^2$

93) $T = \sqrt[4]{\dfrac{E}{\sigma}}$

$T^4 = \left(\sqrt[4]{\dfrac{E}{\sigma}}\right)^4$

$T^4 = \dfrac{E}{\sigma}$

$\sigma T^4 = E$

$\sigma = \dfrac{E}{T^4}$

95) a) Let $T = -17$

$V_s = 20\sqrt{-17 + 273}$

$= 20\sqrt{256}$

$= 20(16)$

$= 320$ 320 m/s

Section 9.7: Solving Radical Equations

b) Let $T = 16$

$V_s = 20\sqrt{16 + 273}$

$ = 20\sqrt{289}$

$ = 20(17)$

$ = 340 \qquad\qquad$ 340 m/s

c) The speed of sound increases.

d) $\qquad V_s = 20\sqrt{T + 273}$

$\qquad V_s^2 = \left(20\sqrt{T + 273}\right)^2$

$\qquad V_s^2 = 400(T + 273)$

$\qquad \dfrac{V_s^2}{400} = T + 273$

$\dfrac{V_s^2}{400} - 273 = T$

97) a) Let $V = 28\pi$ and $h = 7$,
solve for r.

$$r = \sqrt{\dfrac{28\pi}{\pi(7)}} = \sqrt{\dfrac{28}{7}} = \sqrt{4} = 2$$

2 in.

b) $r = \sqrt{\dfrac{V}{\pi h}}$

$r^2 = \left(\sqrt{\dfrac{V}{\pi h}}\right)^2$

$r^2 = \dfrac{V}{\pi h}$

$\pi r^2 h = V$

99) a) $c = \sqrt{(32)(14,400)}$

$c \approx 678.82$ ft/sec

$c \approx \dfrac{678.82(3600)}{5280}$

$c \approx 463$ mph

b) distance $= (\text{rate})(\text{time})$

$60 \text{ miles} = (678.82 \text{ ft/sec})(t)$

$t = \dfrac{(60)(5280)}{678.82}$

$t \approx 466.69 \text{ seconds}$

$\approx \dfrac{466.69}{60} \approx 7.78$

≈ 8 minutes

101) $\quad D = 1.2\sqrt{h}$

$\qquad D^2 = \left(1.2\sqrt{h}\right)^2$

$\qquad D^2 = 1.44h$

$\qquad \dfrac{D^2}{1.44} = h$

$\qquad \dfrac{(4.8)^2}{1.44} = h$

$\qquad h = 16 \text{ ft}$

103) $W = 35.74 - 35.75V^{4/25}$

$W - 35.74 = -35.75V^{4/25}$

$\dfrac{W - 35.74}{-35.75} = V^{4/25}$

$V = \left(\dfrac{W - 35.74}{-35.75}\right)^{25/4}$

$V = \left(\dfrac{-10 - 35.74}{-35.75}\right)^{25/4}$

$V = \left(\dfrac{-45.74}{-35.75}\right)^{25/4}$

$V \approx 5\,\text{mph}$

Section 9.8

1) False

3) True

5) $\sqrt{-81} = \sqrt{-1}\cdot\sqrt{81} = i\cdot 9 = 9i$

7) $\sqrt{-25} = \sqrt{-1}\cdot\sqrt{25} = i\cdot 5 = 5i$

9) $\sqrt{-6} = \sqrt{-1}\cdot\sqrt{6} = i\sqrt{6}$

11) $\sqrt{-27} = \sqrt{-1}\cdot\sqrt{27} = i\cdot 3\sqrt{3} = 3i\sqrt{3}$

13) $\sqrt{-60} = \sqrt{-1}\cdot\sqrt{60} = i\cdot 2\sqrt{15} = 2i\sqrt{15}$

15) Write each radical in terms of i before multiplying.

$\sqrt{-5}\cdot\sqrt{-10} = i\sqrt{5}\cdot i\sqrt{10} = i^2\sqrt{50}$
$= -1\sqrt{25}\cdot\sqrt{2} = -5\sqrt{2}$

17) $\sqrt{-1}\cdot\sqrt{-5} = (i)(i\sqrt{5}) = i^2\sqrt{5} = -\sqrt{5}$

19) $\sqrt{-12}\cdot\sqrt{-3} = (i\sqrt{12})(i\sqrt{3})$
$= i^2\sqrt{36} = -1(6) = -6$

21) $\dfrac{\sqrt{-60}}{\sqrt{-15}} = \dfrac{i\sqrt{60}}{i\sqrt{15}} = \sqrt{\dfrac{60}{15}} = \sqrt{4} = 2$

23) $\left(\sqrt{-13}\right)^2 = \left(i\sqrt{13}\right)^2 = i^2(13)$
$= -1(13) = -13$

25) Add the real parts and add the imaginary parts.

27) -1

29) $(-4+9i)+(7+2i) = 3+11i$

31) $(13-8i)-(9+i) = 4-9i$

Section 9.8: Complex Numbers

33) $\left(-\dfrac{3}{4}-\dfrac{1}{6}i\right)-\left(-\dfrac{1}{2}+\dfrac{2}{3}i\right)$

$=\left(-\dfrac{3}{4}+\dfrac{1}{2}\right)+\left(-\dfrac{1}{6}i-\dfrac{2}{3}i\right)$

$=\left(-\dfrac{3}{4}+\dfrac{2}{4}\right)+\left(-\dfrac{1}{6}i-\dfrac{4}{6}i\right)=-\dfrac{1}{4}-\dfrac{5}{6}i$

35) $16i-(3+10i)+(3+i)$

$=16i-3-10i+3+i=7i$

37) $3(8-5i)=24-15i$

39) $\dfrac{2}{3}(-9+2i)=-6+\dfrac{4}{3}i$

41) $6i(5+6i)=30i+36i^2$

$\qquad=30i+36(-1)=-36+30i$

43) $(2+5i)(1+6i)=2+12i+5i+30i^2$

$\qquad=2+17i+30(-1)$

$\qquad=2+17i-30$

$\qquad=-28+17i$

45) $(-1+3i)(4-6i)=-4+6i+12i-18i^2$

$\qquad=-4+18i-18(-1)$

$\qquad=-4+18i+18$

$\qquad=14+18i$

47) $(5-3i)(9-3i)=45-15i-27i+9i^2$

$\qquad=45-42i+9(-1)$

$\qquad=45-42i-9$

$\qquad=36-42i$

49) $\left(\dfrac{3}{4}+\dfrac{3}{4}i\right)\left(\dfrac{2}{5}+\dfrac{1}{5}i\right)$

$=\dfrac{3}{10}+\dfrac{3}{20}i+\dfrac{3}{10}i+\dfrac{3}{20}i^2$

$=\dfrac{3}{10}+\dfrac{9}{20}i+\dfrac{3}{20}(-1)$

$=\dfrac{3}{10}+\dfrac{9}{20}i-\dfrac{3}{20}=\dfrac{3}{20}+\dfrac{9}{20}i$

51) $(11+4i)(11-4i)$

$\qquad=121-16i^2$

$\qquad=121-16(-1)=121+16=137$

53) $(-3-7i)(-3+7i)$

$\qquad=9-49i^2=9-49(-1)$

$\qquad=9+49=58$

55) $(-6+4i)(-6-4i)$

$\qquad=36-16i^2$

$\qquad=36-16(-1)=36+16=52$

57) Answers may vary.

59) $\dfrac{4}{2-3i} = \dfrac{4}{2-3i} \cdot \dfrac{2+3i}{2+3i} = \dfrac{8+12i}{2^2+3^2}$

$\quad\ = \dfrac{8+12i}{4+9} = \dfrac{8+12i}{13} = \dfrac{8}{13} + \dfrac{12}{13}i$

61) $\dfrac{8i}{4+i} = \dfrac{8i}{4+i} \cdot \dfrac{4-i}{4-i} = \dfrac{32i-8i^2}{4^2+1^2}$

$\quad\ = \dfrac{32i-8(-1)}{16+1} = \dfrac{8}{17} + \dfrac{32}{17}i$

63) $\dfrac{2i}{-3+7i} = \dfrac{2i}{-3+7i} \cdot \dfrac{-3-7i}{-3-7i}$

$\quad\ = \dfrac{-6i-14i^2}{(-3)^2+7^2} = \dfrac{-6i-14(-1)}{9+49}$

$\quad\ = \dfrac{-6i+14}{58} = \dfrac{14}{58} - \dfrac{6}{58}i$

$\quad\ = \dfrac{7}{29} - \dfrac{3}{29}i$

65) $\dfrac{3-8i}{-6+7i} = \dfrac{3-8i}{-6+7i} \cdot \dfrac{-6-7i}{-6-7i}$

$\quad\ = \dfrac{-18-21i+48i+56i^2}{(-6)^2+7^2}$

$\quad\ = \dfrac{-18+27i+56(-1)}{36+49}$

$\quad\ = \dfrac{-74+27i}{85} = -\dfrac{74}{85} + \dfrac{27}{85}i$

67) $\dfrac{2+3i}{5-6i} = \dfrac{2+3i}{5-6i} \cdot \dfrac{5+6i}{5+6i}$

$\quad\ = \dfrac{10+12i+15i+18i^2}{5^2+6^2}$

$\quad\ = \dfrac{10+27i+18(-1)}{25+36}$

$\quad\ = \dfrac{-8+27i}{61} = -\dfrac{8}{61} + \dfrac{27}{61}i$

69) $\dfrac{9}{i} = \dfrac{9}{i} \cdot \dfrac{-i}{-i} = \dfrac{-9i}{1^2} = -9i$

71) $i^{24} = \boxed{\left(i^2\right)^{12}}$ Rewrite i^{24} in terms of i^2

$\quad$ using the power rule.

$\quad\ = (-1)^{12}$ $\boxed{i^2 = -1}$

$\quad\ = \boxed{1}$ Simplify.

73) $i^{24} = \left(i^2\right)^{12}$

$\qquad\quad = (-1)^{12}$

$\qquad\quad = 1$

75) $i^{28} = \left(i^2\right)^{14}$

$\qquad\quad = (-1)^{14}$

$\qquad\quad = 1$

Section 9.8: Complex Numbers

77) $i^9 = i^8 \cdot i$

$\quad = \left(i^2\right)^4 \cdot i$

$\quad = (-1)^4 \cdot i$

$\quad = i$

79) $i^{35} = i^{34} \cdot i$

$\quad = \left(i^2\right)^{17} \cdot i$

$\quad = (-1)^{17} \cdot i$

$\quad = -i$

81) $i^{23} = i^{22} \cdot i$

$\quad = \left(i^2\right)^{11} \cdot i$

$\quad = (-1)^{11} \cdot i$

$\quad = -i$

83) $i^{42} = \left(i^2\right)^{21}$

$\quad = (-1)^{21}$

$\quad = -1$

85) $(2i)^5 = 2^5 \cdot i^4 \cdot i$

$\quad = 32\left(i^2\right)^2 \cdot i$

$\quad = 32(1) \cdot i$

$\quad = 32i$

87) $(-i)^{14} = (-1)^{14} \cdot i^{14}$

$\quad = 1\left(i^2\right)^7$

$\quad = 1(-1)^7$

$\quad = -1$

89)

$(-2+5i)^3 = (-2+5i)^2(-2+5i)$

$\quad = \left((-2)^2 + 2(-2)5i + (5i)^2\right)(-2+5i)$

$\quad = \left(4 - 20i + 25i^2\right)(-2+5i)$

$\quad = (4 - 20i - 25)(-2+5i)$

$= (-21 - 20i)(-2+5i)$

$= 42 - 105i + 40i - 100i^2$

$= 42 - 65i + 100$

$= 142 - 65i$

91) $1 + \sqrt{-8} = 1 + \sqrt{8(-1)}$

$\quad = 1 + \sqrt{8}\,\sqrt{-1}$

$\quad = 1 + 2i\sqrt{2}$

93) $8 - \sqrt{-45} = 8 - \sqrt{45(-1)}$

$\quad = 8 - \sqrt{45}\,\sqrt{-1}$

$\quad = 8 - 3i\sqrt{5}$

95) $\dfrac{-12+\sqrt{-32}}{4} = \dfrac{-12+\sqrt{32(-1)}}{4}$

$= \dfrac{-12+\sqrt{32}\,\sqrt{-1}}{4}$

$= -\dfrac{12}{4} + \dfrac{4i\sqrt{2}}{4}$

$= -3+i\sqrt{2}$

97) $Z = Z_1 + Z_2$

$= 3+2j+7+4j$

$= 10+6j$

99) $Z = Z_1 + Z_2$

$= 5-2j+11+6j$

$= 16+4j$

Chapter 9 Review

1) $\sqrt{\dfrac{169}{4}} = \dfrac{13}{2}$

3) $-\sqrt{81} = -9$

5) $\sqrt[3]{-1} = -1$

7) $\sqrt[6]{-64}$ is not real

9) 13

11) $|p|$

13) h

15) a) $f(x) = \sqrt{5x+3}$

$f(4) = \sqrt{5(4)+3}$

$= \sqrt{23}$

b) $f(x) = \sqrt{5x+3}$

$f(p) = \sqrt{5(p)+3}$

$= \sqrt{5p+3}$

c) $5x+3 \geq 0$

$5x \geq -3$

$x \geq -\dfrac{3}{5} \quad \left[-\dfrac{3}{5}, \infty\right)$

17)

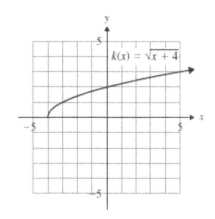

293

Chapter 9 Review

19) The denominator of the fractional exponent becomes the index on the radical. The numerator is the power to which we raise the radical expression. $8^{2/3} = \left(\sqrt[3]{8}\right)^2$

21) $36^{1/2} = \sqrt{36} = 6$

23) $\left(\dfrac{27}{125}\right)^{1/3} = \sqrt[3]{\dfrac{27}{125}} = \dfrac{3}{5}$

25) $32^{3/5} = \left(\sqrt[5]{32}\right)^3 = (2)^3 = 8$

27) $81^{-1/2} = \left(\dfrac{1}{81}\right)^{1/2} = \sqrt{\dfrac{1}{81}} = \dfrac{1}{9}$

29) $81^{-3/4} = \left(\dfrac{1}{81}\right)^{3/4} = \left(\sqrt[4]{\dfrac{1}{81}}\right)^3$
$= \left(\dfrac{1}{3}\right)^3 = \dfrac{1}{27}$

31) $\left(\dfrac{27}{1000}\right)^{-2/3} = \left(\dfrac{1000}{27}\right)^{2/3} = \left(\sqrt[3]{\dfrac{1000}{27}}\right)^2$
$= \left(\dfrac{10}{3}\right)^2 = \dfrac{100}{9}$

33) $3^{6/7} \cdot 3^{8/7} = 3^{6/7+8/7} = 3^{14/7} = 3^2 = 9$

35) $\left(8^{1/5}\right)^{10} = 8^{\frac{1}{5}\cdot 10} = 8^2 = 64$

37) $\dfrac{7^2}{7^{5/3}\cdot 7^{1/3}} = \dfrac{7^2}{7^{5/3+1/3}} = \dfrac{7^2}{7^{6/3}} = \dfrac{7^2}{7^2} = 1$

39) $\left(64a^4b^{12}\right)^{5/6} = 64^{5/6}\cdot\left(a^4\right)^{5/6}\cdot\left(b^{12}\right)^{5/6}$
$= \left(\sqrt[6]{64}\right)^5\cdot a^{4\cdot\frac{5}{6}}\cdot b^{12\cdot\frac{5}{6}}$
$= 2^5\cdot a^{10/3}\cdot b^{10}$
$= 32a^{10/3}b^{10}$

41) $\left(\dfrac{81c^{-5}d^9}{16c^{-1}d^2}\right)^{-1/4} = \left(\dfrac{81d^7}{16c^4}\right)^{-1/4}$
$= \left(\dfrac{16c^4}{81d^7}\right)^{1/4} = \dfrac{16^{1/4}\cdot c^{4\cdot\frac{1}{4}}}{81^{1/4}d^{7\cdot\frac{1}{4}}} = \dfrac{2c}{3d^{7/4}}$

43) $\sqrt[12]{27^4} = \left(27^4\right)^{1/12} = 27^{1/3} = \sqrt[3]{27} = 3$

45) $\sqrt[3]{7^3} = 7^{3/3} = 7$

47) $\sqrt[4]{k^{28}} = k^{28/4} = k^7$

49) $\sqrt{w^6} = w^{6/2} = w^3$

51) $\sqrt{1000} = \sqrt{100\cdot 10} = \sqrt{100}\cdot\sqrt{10}$
$= 10\sqrt{10}$

53) $\sqrt{\dfrac{18}{49}} = \dfrac{\sqrt{18}}{\sqrt{49}} = \dfrac{\sqrt{9 \cdot 2}}{7} = \dfrac{3\sqrt{2}}{7}$

55) $\sqrt{k^{12}} = k^{12/2} = k^6$

57) $\sqrt{x^9} = \sqrt{x^8} \cdot \sqrt{x} = x^{8/2} \cdot \sqrt{x} = x^4\sqrt{x}$

59)

$\sqrt{45t^2} = \sqrt{45} \cdot \sqrt{t^2} = \sqrt{9} \cdot \sqrt{5} \cdot t^{2/2}$
$\qquad = 3\sqrt{5} \cdot t = 3t\sqrt{5}$

61) $\sqrt{72x^7y^{13}}$
$\quad = \sqrt{72} \cdot \sqrt{x^7} \cdot \sqrt{y^{13}}$
$\quad = 6\sqrt{2} \cdot x^3\sqrt{x} \cdot y^6\sqrt{y} = 6x^3y^6\sqrt{2xy}$

63) $\sqrt{5} \cdot \sqrt{3} = \sqrt{5 \cdot 3} = \sqrt{15}$

65) $\sqrt{2} \cdot \sqrt{12} = \sqrt{2 \cdot 12} = \sqrt{24}$
$\qquad = \sqrt{4} \cdot \sqrt{6} = 2\sqrt{6}$

67) $\sqrt{11x^5} \cdot \sqrt{11x^8} = \sqrt{121x^{13}}$
$\qquad = 11 \cdot \sqrt{x^{12}} \cdot \sqrt{x}$
$\qquad = 11x^6\sqrt{x}$

69) $\dfrac{\sqrt{200k^{21}}}{\sqrt{2k^5}} = \sqrt{\dfrac{200k^{21}}{2k^5}} = \sqrt{100k^{21-5}}$
$\qquad\qquad = \sqrt{100k^{16}} = 10k^8$

71) $\sqrt[3]{16} = \sqrt[3]{8} \cdot \sqrt[3]{2} = 2\sqrt[3]{2}$

73) $\sqrt[4]{48} = \sqrt[4]{16} \cdot \sqrt[4]{3} = 2\sqrt[4]{3}$

75) $\sqrt[4]{z^{24}} = z^{24/4} = z^6$

77) $\sqrt[3]{a^{20}} = \sqrt[3]{a^{18}} \cdot \sqrt[3]{a^2} = a^{18/3} \cdot \sqrt[3]{a^2}$
$\qquad = a^6\sqrt[3]{a^2}$

79) $\sqrt[3]{16z^{15}} = \sqrt[3]{16} \cdot \sqrt[3]{z^{15}} = 2\sqrt[3]{2} \cdot z^{15/3}$
$\qquad\qquad = 2\sqrt[3]{2} \cdot z^5 = 2z^5\sqrt[3]{2}$

81) $\sqrt[4]{\dfrac{h^{12}}{81}} = \dfrac{\sqrt[4]{h^{12}}}{\sqrt[4]{81}} = \dfrac{h^{12/4}}{3} = \dfrac{h^3}{3}$

83) $\sqrt[3]{3} \cdot \sqrt[3]{7} = \sqrt[3]{3 \cdot 7} = \sqrt[3]{21}$

85) $\sqrt[4]{4t^7} \cdot \sqrt[4]{8t^{10}} = \sqrt[4]{32t^{17}} = \sqrt[4]{32} \cdot \sqrt[4]{t^{17}}$
$\qquad\qquad = 2\sqrt[4]{2} \cdot t^4\sqrt[4]{t} = 2t^4\sqrt[4]{2t}$

87) $\sqrt[3]{n} \cdot \sqrt{n} = n^{1/3} \cdot n^{1/2} = n^{2/6+3/6}$
$\qquad\qquad = n^{5/6} = \sqrt[6]{n^5}$

89) $8\sqrt{5} + 3\sqrt{5} = 11\sqrt{5}$

91) $\sqrt{80} - \sqrt{48} + \sqrt{20}$
$= 4\sqrt{5} - 4\sqrt{3} + 2\sqrt{5} = 6\sqrt{5} - 4\sqrt{3}$

93) $3p\sqrt{p} - 7\sqrt{p^3}$
$= 3p\sqrt{p} - 7\left(p\sqrt{p}\right)$
$= 3p\sqrt{p} - 7p\sqrt{p} = -4p\sqrt{p}$

95) $10d^2\sqrt{8d} - 32d\sqrt{2d^3}$
$= 10d^2\left(2\sqrt{2d}\right) - 32d\left(d\sqrt{2d}\right)$
$= 20d^2\sqrt{2d} - 32d^2\sqrt{2d}$
$= -12d^2\sqrt{2d}$

97) $3\sqrt{k}\left(\sqrt{20k} + \sqrt{2}\right) = 3\sqrt{k}\left(2\sqrt{5k} + \sqrt{2}\right)$
$= 6\sqrt{5k^2} + 3\sqrt{2k} = 6k\sqrt{5} + 3\sqrt{2k}$

99) $\left(\sqrt{2r} + 5\sqrt{s}\right)\left(3\sqrt{s} + 4\sqrt{2r}\right)$
$= 3\sqrt{2rs} + 4(2r) + 15s + 20\sqrt{2rs}$
$= 23\sqrt{2rs} + 8r + 15s$

101) $\left(1 + \sqrt{y+1}\right)^2$
$= 1^2 + 2(1)\left(\sqrt{y+1}\right) + \left(\sqrt{y+1}\right)^2$
$= 1 + 2\sqrt{y+1} + y + 1 = 2 + 2\sqrt{y+1} + y$

103) $\dfrac{14}{\sqrt{3}} = \dfrac{14}{\sqrt{3}} \cdot \dfrac{\sqrt{3}}{\sqrt{3}} = \dfrac{14\sqrt{3}}{3}$

105) $\dfrac{\sqrt{18k}}{\sqrt{n}} = \dfrac{3\sqrt{2k}}{\sqrt{n}} = \dfrac{3\sqrt{2k}}{\sqrt{n}} \cdot \dfrac{\sqrt{n}}{\sqrt{n}} = \dfrac{3\sqrt{2kn}}{n}$

107) $\dfrac{7}{\sqrt[3]{2}} = \dfrac{7}{\sqrt[3]{2}} \cdot \dfrac{\sqrt[3]{2^2}}{\sqrt[3]{2^2}} = \dfrac{7\sqrt[3]{2^2}}{\sqrt[3]{2^3}} = \dfrac{7\sqrt[3]{4}}{2}$

109) $\dfrac{\sqrt[3]{x^2}}{\sqrt[3]{y}} = \dfrac{\sqrt[3]{x^2}}{\sqrt[3]{y}} \cdot \dfrac{\sqrt[3]{y^2}}{\sqrt[3]{y^2}} = \dfrac{\sqrt[3]{x^2 y^2}}{\sqrt[3]{y^3}}$
$= \dfrac{\sqrt[3]{x^2 y^2}}{y}$

111) $\dfrac{2}{3+\sqrt{3}} = \dfrac{2}{3+\sqrt{3}} \cdot \dfrac{3-\sqrt{3}}{3-\sqrt{3}} = \dfrac{2\left(3-\sqrt{3}\right)}{(3)^2 - \left(\sqrt{3}\right)^2}$
$= \dfrac{2\left(3-\sqrt{3}\right)}{9-3} = \dfrac{2\left(3-\sqrt{3}\right)}{6} = \dfrac{3-\sqrt{3}}{3}$

113) $\dfrac{8 - 24\sqrt{2}}{8} = \dfrac{8\left(1 - 3\sqrt{2}\right)}{8} = 1 - 3\sqrt{2}$

115)

$$\sqrt{x+8} = 3$$
$$\left(\sqrt{x+8}\right)^2 = 3^2$$
$$x+8 = 9$$
$$x = 1$$

Check $\sqrt{1+8} = 1$

$$\sqrt{9} = 3 \quad \{1\}$$

117)

$$\sqrt{3j+4} = -\sqrt{4j-1}$$
$$\left(\sqrt{3j+4}\right)^2 = \left(-\sqrt{4j-1}\right)^2$$
$$3j+4 = 4j-1$$
$$5 = j$$

Check $\sqrt{3(5)+4} = -\sqrt{4(5)-1}$

$$\sqrt{19} \neq -\sqrt{19} \qquad \varnothing$$

119)

$$a = \sqrt{a+8} - 6$$
$$a+6 = \sqrt{a+8}$$
$$\left(a+6\right)^2 = \left(\sqrt{a+8}\right)^2$$
$$a^2 + 12a + 36 = a+8$$
$$a^2 + 11a + 28 = 0$$
$$\left(a+7\right)\left(a+4\right) = 0$$
$$a+7 = 0 \text{ or } a+4 = 0$$
$$a = -7 \qquad a = -4 \quad \{-4\}$$

-7 is an extraneous solution.

121)

$$\sqrt{4a+1} - \sqrt{a-2} = 3$$
$$\sqrt{4a+1} = 3 + \sqrt{a-2}$$
$$\left(\sqrt{4a+1}\right)^2 = \left(3+\sqrt{a-2}\right)^2$$

$$4a+1 = 9 + 6\sqrt{a-2} + a - 2$$
$$4a+1 = 7 + a + 6\sqrt{a-2}$$
$$3a-6 = 6\sqrt{a-2}$$
$$a-2 = 2\sqrt{a-2}$$
$$\left(a-2\right)^2 = \left(2\sqrt{a-2}\right)^2$$
$$a^2 - 4a + 4 = 4\left(a-2\right)$$
$$a^2 - 4a + 4 = 4a - 8$$
$$a^2 - 8a + 12 = 0$$
$$\left(a-6\right)\left(a-2\right) = 0$$
$$a-6 = 0 \text{ or } a-2 = 0$$
$$a = 6 \qquad a = 2 \quad \{2,6\}$$

Check is left to the student.

123)

$$r = \sqrt{\frac{3V}{\pi h}}$$
$$r^2 = \left(\sqrt{\frac{3V}{\pi h}}\right)^2$$
$$r^2 = \frac{3V}{\pi h}$$
$$\pi r^2 h = 3V$$
$$\frac{1}{3}\pi r^2 h = V$$

125) $\sqrt{-49} = i\sqrt{49} = 7i$

127) $\sqrt{-2} \cdot \sqrt{-8} = i\sqrt{2} \cdot i\sqrt{8}$
$$= i^2\sqrt{16} = -1\cdot 4 = -4$$

129) $(2+i)+(10-4i)=12-3i$

143) $\dfrac{8}{i}=\dfrac{8}{i}\cdot\dfrac{-i}{-i}=\dfrac{-8i}{(1)^2}=-8i$

131) $\left(\dfrac{4}{5}-\dfrac{1}{3}i\right)-\left(\dfrac{1}{2}+i\right)$

$=\left(\dfrac{4}{5}-\dfrac{1}{2}\right)+\left(-\dfrac{1}{3}i-i\right)$

$=\left(\dfrac{8}{10}-\dfrac{5}{10}\right)+\left(-\dfrac{1}{3}i-\dfrac{3}{3}i\right)$

$=\dfrac{3}{10}-\dfrac{4}{3}i$

145) $\dfrac{9-4i}{6-i}=\dfrac{9-4i}{6-i}\cdot\dfrac{6+i}{6+i}$

$=\dfrac{54+9i-24i-4i^2}{(6)^2+(1)^2}$

$=\dfrac{54-15i-4(-1)}{36+1}$

$=\dfrac{54-15i+4}{37}=\dfrac{58-15i}{37}$

$=\dfrac{58}{37}-\dfrac{15}{37}i$

133) $5(-6+7i)=-30+35i$

135) $3i(-7+12i)=-21i+36i^2$

$=-21i+36(-1)$

$=-36-21i$

147) $i^{10}=\left(i^2\right)^5$

$=(-1)^5$

$=-1$

$(4-6i)(3-6i)$

$=12-24i-18i+36i^2$

$=12-42i+36(-1)$

137) $=12-42i-36=-24-42i$

149)

$i^{33}=i^{32}\cdot i$

$=\left(i^2\right)^{16}\cdot i$

$=(-1)^{16}\cdot i$

$=i$

139) $(2-7i)(2+7i)=(2)^2+(7)^2$

$=4+49=53$

141) $\dfrac{6}{2+5i}=\dfrac{6}{2+5i}\cdot\dfrac{2-5i}{2-5i}$

$=\dfrac{12-30i}{(2)^2+(5)^2}=\dfrac{12-30i}{4+25}$

$=\dfrac{12-30i}{29}=\dfrac{12}{29}-\dfrac{30}{29}i$

Chapter 9 Test

1) $\sqrt{144}=12$

3) not real

5) $\sqrt[5]{(-19)^5} = -19$

7) $x - 2 \geq 0$

$x \geq 2 \quad [2, \infty)$

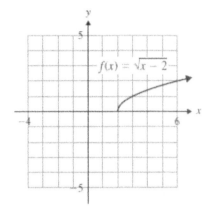

9) $27^{4/3} = \left(27^{1/3}\right)^4 = \left(\sqrt[3]{27}\right)^4 = 3^4 = 81$

11) $\left(\dfrac{8}{125}\right)^{-2/3} = \left(\dfrac{125}{8}\right)^{2/3} = \left(\sqrt[3]{\dfrac{125}{8}}\right)^2$

$= \left(\dfrac{5}{2}\right)^2 = \dfrac{25}{4}$

13) $\dfrac{35a^{1/6}}{14a^{5/6}} = \dfrac{5}{2}a^{1/6 - 5/6} = \dfrac{5}{2}a^{-4/6} = \dfrac{5}{2a^{2/3}}$

15) $\sqrt{75} = \sqrt{25} \cdot \sqrt{3} = 5\sqrt{3}$

17) $\sqrt{\dfrac{24}{2}} = \sqrt{12} = \sqrt{4} \cdot \sqrt{3} = 2\sqrt{3}$

19) $\sqrt[4]{p^{24}} = p^{24/4} = p^6$

21) $\sqrt{63m^5 n^8} = \sqrt{63} \cdot \sqrt{m^5} \cdot \sqrt{n^8}$

$= 3\sqrt{7} \cdot m^2 \sqrt{m} \cdot n^4 = 3m^2 n^4 \sqrt{7m}$

23)

$\sqrt[3]{\dfrac{a^{14}b^7}{27}} = \dfrac{\sqrt[3]{a^{14}} \cdot \sqrt[3]{b^7}}{\sqrt[3]{27}}$

$= \dfrac{a^4 \sqrt[3]{a^2} \cdot b^2 \sqrt[3]{b}}{3} = \dfrac{a^4 b^2 \sqrt[3]{a^2 b}}{3}$

25) $\sqrt[3]{z^4} \cdot \sqrt[3]{z^6} = \sqrt[3]{z^{10}} = \sqrt[3]{z^9} \cdot \sqrt[3]{z} = z^3 \sqrt[3]{z}$

27) $9\sqrt{7} - 3\sqrt{7} = 6\sqrt{7}$

$2h^3 \sqrt[4]{h} - 16\sqrt[4]{h^{13}}$

$= 2h^3 \sqrt[4]{h} - 16\left(\sqrt[4]{h^{12}} \cdot \sqrt[4]{h}\right)$

29) $= 2h^3 \sqrt[4]{h} - 16h^3 \sqrt[4]{h} = -14h^3 \sqrt[4]{h}$

31) $\left(3 - 2\sqrt{5}\right)\left(\sqrt{2} + 1\right)$

$= 3\sqrt{2} + 3 - 2\sqrt{10} - 2\sqrt{5}$

Chapter 9 Test

33) $\left(\sqrt{2p+1}+2\right)^2$

$$=\left(\sqrt{2p+1}\right)^2+2\left(\sqrt{2p+1}\right)(2)+2^2$$
$$=2p+1+4\sqrt{2p+1}+4$$
$$=2p+5+4\sqrt{2p+1}$$

35) $\dfrac{2}{\sqrt{5}}=\dfrac{2}{\sqrt{5}}\cdot\dfrac{\sqrt{5}}{\sqrt{5}}=\dfrac{2\sqrt{5}}{5}$

37) $\dfrac{\sqrt{6}}{\sqrt{a}}=\dfrac{\sqrt{6}}{\sqrt{a}}\cdot\dfrac{\sqrt{a}}{\sqrt{a}}=\dfrac{\sqrt{6a}}{a}$

39) $\dfrac{2-\sqrt{48}}{2}=\dfrac{2-4\sqrt{3}}{2}=\dfrac{2\left(1-2\sqrt{3}\right)}{2}$
$$=1-2\sqrt{3}$$

41)
$$z=\sqrt{1-4z}-5$$
$$z+5=\sqrt{1-4z}$$
$$(z+5)^2=\left(\sqrt{1-4z}\right)^2$$
$$z^2+10z+25=1-4z$$
$$z^2+14z+24=0$$
$$(z+2)(z+12)=0$$
$$z+2=0 \text{ or } z+12=0$$
$$z=-2 \qquad z=-12 \qquad \{-2\}$$

-12 is an extraneous solution.

43) $\sqrt{3k+1}-\sqrt{2k-1}=1$
$$\sqrt{3k+1}=1+\sqrt{2k-1}$$
$$\left(\sqrt{3k+1}\right)^2=\left(1+\sqrt{2k-1}\right)^2$$
$$3k+1=1+2\sqrt{2k-1}+2k-1$$
$$3k+1=2k+2\sqrt{2k-1}$$
$$k+1=2\sqrt{2k-1}$$
$$(k+1)^2=\left(2\sqrt{2k-1}\right)^2$$
$$k^2+2k+1=4(2k-1)$$
$$k^2+2k+1=8k-4$$
$$k^2-6k+5=0$$
$$(k-5)(k-1)=0$$
$$k-5=0 \text{ or } k-1=0$$
$$k=5 \qquad k=1 \qquad \{1,5\}$$

Check is left to the student.

45) $\sqrt{-64}=i\sqrt{64}=8i$

47) $i^{19}=i^{18}\cdot i$
$$=\left(i^2\right)^9\cdot i$$
$$=(-1)^9\cdot i$$
$$=-i$$

49)

$$(2-7i)(-1+3i)$$
$$=-2+6i+7i-21i^2$$
$$=-2+13i-21(-1)$$
$$=-2+13i+21=19+13i$$

50)

$$\frac{8+i}{2-3i}=\frac{8+i}{2-3i}\cdot\frac{2+3i}{2+3i}$$

$$=\frac{16+24i+2i+3i^2}{(2)^2+(3)^2}$$

$$=\frac{16+26i+3(-1)}{4+9}$$

$$=\frac{16+26i-3}{13}=\frac{13+26i}{13}$$

$$=\frac{13}{13}+\frac{26}{13}i=1+2i$$

Cumulative Review: Chapters 1-9

1) $4x-3y+9-\dfrac{2}{3}x+y-1$

$$=\frac{12}{3}x-\frac{2}{3}x-3y+y+9-1$$

$$=\frac{10}{3}x-2y+8$$

3) $3(2c-1)+7=9c+5(c+2)$
$$6c-3+7=9c+5c+10$$
$$6c+4=14c+10$$
$$-6=8c$$
$$-\frac{6}{8}=c$$
$$-\frac{3}{4}=c$$
$$\left\{-\frac{3}{4}\right\}$$

5)

$$m=\frac{-2-3}{1-5}=\frac{-5}{-4}=\frac{5}{4}$$

$$y-y_1=m(x-x_1)$$

$$y-3=\frac{5}{4}(x-5)$$

$$y-3=\frac{5}{4}x-\frac{25}{4}$$

$$y=\frac{5}{4}x-\frac{13}{4}$$

7) $(5p^2-2)(3p^2-4p-1)$
$$=15p^4-20p^3-5p^2-6p^2+8p+2$$
$$=15p^4-20p^3-11p^2+8p+2$$

9) $4w^2+5w-6=(4w-3)(w+2)$

11)

$$6y^2-4=5y$$
$$6y^2-5y-4=0$$
$$6y^2-8y+3y-4=0$$
$$2y(3y-4)+1(3y-4)=0$$
$$(2y+1)(3y-4)=0$$

$$2y+1=0 \ \text{ or } \ 3y-4=0$$

$$y=-\frac{1}{2} \qquad y=\frac{4}{3} \qquad \left\{-\frac{1}{2},\frac{4}{3}\right\}$$

13) $l=\text{length}$
$$w=l-5$$
$$84=lw$$
$$84=l(l-5)$$
$$84=l^2-5l$$
$$0=l^2-5l-84$$
$$0=(l-12)(l+7)$$

$l - 12 = 0$ or $l + 7 = 0$

$l = 12$ $l = -7$

$w = 12 - 5$ Length cannot be negative.

$w = 7$

length $= 12$ in., width $= 7$ in.

15) $\dfrac{10m^2}{9n} \cdot \dfrac{6n^2}{35m^5} = \dfrac{\overset{2}{\cancel{10m^2}}}{\underset{3}{\cancel{9n}}} \cdot \dfrac{\overset{2n}{\cancel{6n^2}}}{\underset{7m^3}{\cancel{35m^5}}} = \dfrac{4n}{21m^3}$

17)

$|6g + 1| \geq 11$

$6g + 1 \geq 11$ or $6g + 1 \leq -11$

$6g \geq 10$ $6g \leq -12$

$g \geq \dfrac{10}{6}$ $g \leq -2$

$g \geq \dfrac{5}{3}$

$(-\infty, -2] \cup \left[\dfrac{5}{3}, \infty\right)$

19) a) $\sqrt{500} = \sqrt{100} \cdot \sqrt{5} = 10\sqrt{5}$

b) $\sqrt[3]{56} = \sqrt[3]{8} \cdot \sqrt[3]{7} = 2\sqrt[3]{7}$

c) $\sqrt{p^{10}q^7} = \sqrt{p^{10}} \cdot \sqrt{q^7} = p^5q^3\sqrt{q}$

d) $\sqrt[4]{32a^{15}} = \sqrt[4]{32} \cdot \sqrt[4]{a^{15}} = 2\sqrt[4]{2} \cdot a^3\sqrt[4]{a^3}$

$= 2a^3\sqrt[4]{2a^3}$

21)

$2\sqrt{3}(5 - \sqrt{3}) = 10\sqrt{3} - 2(3)$

$= 10\sqrt{3} - 6$

23) a) $\sqrt{2b - 1} + 7 = 6$

$\sqrt{2b - 1} = -1$ $\varnothing$

b) $\sqrt{3z + 10} = 2 - \sqrt{z + 4}$

$\left(\sqrt{3z + 10}\right)^2 = \left(2 - \sqrt{z + 4}\right)^2$

$3z + 10 = 4 - 4\sqrt{z + 4} + z + 4$

$3z + 10 = 8 + z - 4\sqrt{z + 4}$

$2z + 2 = -4\sqrt{z + 4}$

$z + 1 = -2\sqrt{z + 4}$

$(z + 1)^2 = \left(-2\sqrt{z + 4}\right)^2$

$z^2 + 2z + 1 = 4(z + 4)$

$z^2 + 2z + 1 = 4z + 16$

$z^2 - 2z - 15 = 0$

$(z - 5)(z + 3) = 0$

$z - 5 = 0$ or $z + 3 = 0$

$z = 5$ or $z = -3$ $\{-3\}$

5 is an extraneous solution.

25) a) $(-3 + 4i) + (5 + 3i) = 2 + 7i$

b) $(3 + 6i)(-2 + 7i)$

$= -6 + 21i - 12i + 42i^2$

$= -6 + 9i + 42(-1)$

$= -6 + 9i - 42 = -48 + 9i$

c) $\dfrac{2 - i}{-4 + 3i} = \dfrac{2 - i}{-4 + 3i} \cdot \dfrac{-4 - 3i}{-4 - 3i}$

$= \dfrac{-8 - 6i + 4i + 3i^2}{(-4)^2 + (3)^2}$

$= \dfrac{-8 - 2i + 3(-1)}{16 + 9}$

$= \dfrac{-8 - 2i - 3}{25} = \dfrac{-11 - 2i}{25}$

$= -\dfrac{11}{25} - \dfrac{2}{25}i$

Chapter 10: Quadratic Equations and Functions

Section 10.1

1) Factoring:
$$y^2 - 16 = 0$$
$$(y+4)(y-4) = 0$$

$$y + 4 = 0 \text{ or } y - 4 = 0$$
$$y = -4 \qquad y = 4$$

Square Root Property:

$$y^2 - 16 = 0$$
$$y^2 = 16$$
$$y = \pm\sqrt{16}$$
$$y = \pm 4 \qquad \{-4, 4\}$$

3) $b^2 = 36$
$$b = \pm\sqrt{36}$$
$$b = \pm 6 \qquad \{-6, 6\}$$

5) $r^2 - 27 = 0$
$$r = \pm\sqrt{27}$$
$$= \pm 3\sqrt{3} \qquad \{3\sqrt{3}, -3\sqrt{3}\}$$

7) $n^2 = \dfrac{4}{9}$
$$n = \pm\sqrt{\dfrac{4}{9}} = \pm\dfrac{2}{3} \qquad \left\{-\dfrac{2}{3}, \dfrac{2}{3}\right\}$$

9) $q^2 = -4$
$$= \pm 2\sqrt{-1}$$
$$= \pm 2i \qquad \{-2i, 2i\}$$

11) $z^2 + 3 = 0$
$$z^2 = -3$$
$$z = \pm i\sqrt{3} \qquad \{-i\sqrt{3}, i\sqrt{3}\}$$

13) $z^2 + 5 = 19$
$$z^2 = 14$$
$$z = \pm\sqrt{14} \qquad \{-\sqrt{14}, \sqrt{14}\}$$

15) $2d^2 + 5 = 55$
$$2d^2 = 50$$
$$d^2 = 25$$
$$d = \pm 5 \qquad \{-5, 5\}$$

17) $5f^2 + 39 = -21$
$$5f^2 = -60$$
$$f^2 = -12$$
$$f = \pm i2\sqrt{3} \qquad \{-2i\sqrt{3}, 2i\sqrt{3}\}$$

Section 10.1: The Square Root Property and Completing the Square

19) $(r+10)^2 = 4$

$r+10 = \pm 2$

$r = -12, -8$ $\{-12, -8\}$

21) $(q-7)^2 = 1$

$q-7 = \pm 1$

$q = 6, 8$ $\{6, 8\}$

23) $(p+4)^2 - 18 = 0$

$(p+4)^2 = 18$

$p+4 = \pm 3\sqrt{2}$

$p = -4 - 3\sqrt{2}, -4 + 3\sqrt{2}$

$\left\{ -4 - 3\sqrt{2}, -4 + 3\sqrt{2} \right\}$

25) $(c+3)^2 - 4 = -29$

$(c+3)^2 = -25$

$c+3 = \pm 5i$

$c = -3 + 5i, -3 - 5i$

$\{-3 - 5i, -3 + 5i\}$

27) $1 = 15 + (k-2)^2$

$(k-2)^2 = -14$

$k-2 = \pm i\sqrt{14}$

$k = 2 + i\sqrt{14}, 2 - i\sqrt{14}$

$\left\{ 2 - i\sqrt{14}, 2 + i\sqrt{14} \right\}$

29) $20 = (2w+1)^2$

$2w+1 = \pm 2\sqrt{5}$

$w = \dfrac{-1 - 2\sqrt{5}}{2}, \dfrac{-1 + 2\sqrt{5}}{2}$

$\left\{ \dfrac{-1 - 2\sqrt{5}}{2}, \dfrac{-1 + 2\sqrt{5}}{2} \right\}$

31) $8 = (3q-10)^2 - 6$

$14 = (3q-10)^2$

$\pm\sqrt{14} = 3q - 10$

$10 \pm \sqrt{14} = 3q$

$\dfrac{10 \pm \sqrt{14}}{3} = q$

$\left\{ \dfrac{10 - \sqrt{14}}{3}, \dfrac{10 + \sqrt{14}}{3} \right\}$

33) $36 + (4p-5)^2 = 6$

$(4p-5)^2 = -30$

$4p-5 = \pm i\sqrt{30}$

$p = \dfrac{5 - i\sqrt{30}}{4}, \dfrac{5 + i\sqrt{30}}{4}$

$\left\{ \dfrac{5 - i\sqrt{30}}{4}, \dfrac{5 + i\sqrt{30}}{4} \right\}$

35) $(6g+11)^2 + 50 = 1$

$(6g+11)^2 = -49$

$6g+11 = \pm 7i$

$g = \dfrac{-11-7i}{6}, \dfrac{-11+7i}{6}$

$\left\{\dfrac{-11-7i}{6}, \dfrac{-11+7i}{6}\right\}$

37) $\left(\dfrac{3}{4}n-8\right)^2 = 4$

$\dfrac{3}{4}n-8 = \pm\sqrt{4}$

$\dfrac{3}{4}n-8 = \pm 2$

$\dfrac{3}{4}n-8 = 2$ or $\dfrac{3}{4}n-8 = -2$

$\dfrac{3}{4}n = 10 \qquad\qquad \dfrac{3}{4}n = 6$

$n = \dfrac{40}{3} \qquad\qquad\quad n = 8$

$\left\{8, \dfrac{40}{3}\right\}$

39) $(5y-2)^2 + 6 = 22$

$(5y-2)^2 = 16$

$5y-2 = \pm 4$

$y = -\dfrac{2}{5}, \dfrac{6}{5} \qquad \left\{-\dfrac{2}{5}, \dfrac{6}{5}\right\}$

41) $d = \sqrt{(x_2-x_1)^2 + (y_2-y_1)^2}$

$d = \sqrt{(3-7)^2 + (-2-1)^2}$

$= \sqrt{(-4)^2 + (-3)^2}$

$= \sqrt{25}$

$= 5$

43) $d = \sqrt{(x_2-x_1)^2 + (y_2-y_1)^2}$

$d = \sqrt{[-2-(-5)]^2 + [-8-(-6)]^2}$

$= \sqrt{(3)^2 + (-2)^2}$

$\sqrt{13}$

45) $d = \sqrt{(x_2-x_1)^2 + (y_2-y_1)^2}$

$d = \sqrt{(0-0)^2 + (7-13)^2}$

$= \sqrt{(-6)^2}$

$= 6$

47) $d = \sqrt{(x_2-x_1)^2 + (y_2-y_1)^2}$

$d = \sqrt{[2-(-4)]^2 + (6-11)^2}$

$d = \sqrt{6^2 + (-5)^2}$

$d = \sqrt{36+25} = \sqrt{61}$

Section 10.1: The Square Root Property and Completing the Square

49) $d = \sqrt{(x_2 - x_1)^2 + (y_2 - y_1)^2}$

$d = \sqrt{(5-3)^2 + [-7-(-3)]^2}$

$d = \sqrt{2^2 + (-4)^2}$

$d = \sqrt{4+16} = \sqrt{20} = 2\sqrt{5}$

51) It is a trinomial whose factored
form is the square of a binomial.

$x^2 - 6x + 9$

53)

$w^2 + 8w$

$\dfrac{1}{2}(8) = 4$ Find half of the coefficient of w.

$4^2 = 16$ Square the result.

$w^2 + 8w + 16$ Add the constant to the expression.

The perfect square trinomial is $w^2 + 8w + 16$.

The factored form of the trinomial is $(w+4)^2$.

55) 1) $\dfrac{1}{2}(12) = 6$ 2) $6^2 = 36$

$a^2 + 12a + 36;$ $(a+6)^2$

57) 1) $\dfrac{1}{2}(-18) = -9$ 2) $(-9)^2 = 81$

$c^2 - 18c + 81;$ $(c-9)^2$

59) 1) $\dfrac{1}{2}(3) = \dfrac{3}{2}$ 2) $\left(\dfrac{3}{2}\right)^2 = \dfrac{9}{4}$

$r^2 + 3r + \dfrac{9}{4};$ $\left(r + \dfrac{3}{2}\right)^2$

61) 1) $\dfrac{1}{2}(-9) = -\dfrac{9}{2}$ 2) $\left(-\dfrac{9}{2}\right)^2 = \dfrac{81}{4}$

$b^2 - 9b + \dfrac{81}{4};$ $\left(b - \dfrac{9}{2}\right)^2$

63) 1) $\dfrac{1}{2}\left(\dfrac{1}{3}\right) = \dfrac{1}{6}$ 2) $\left(\dfrac{1}{6}\right)^2 = \dfrac{1}{36}$

$x^2 + \dfrac{1}{3}x + \dfrac{1}{36};$ $\left(x + \dfrac{1}{6}\right)^2$

65) To solve $2p^2 - 7p = 8$ bycompleting the
square,divide both sides of the equation
by 2.

67) $x^2 + 6x + 8 = 0$

$x^2 + 6x = -8$

$x^2 + 6x + 9 = -8 + 9$

$(x+3)^2 = 1$

$x+3 = \pm\sqrt{1}$

$x+3 = \pm 1$

$x+3 = 1$ or $x+3 = -1$

$x = -2$ $x = -4$ $\{-4, -2\}$

69) $k^2 - 8k + 15 = 0$

$\quad k^2 - 8k = -15$

$\quad k^2 - 8k + 16 = -15 + 16$

$\quad (k-4)^2 = 1$

$\quad k - 4 = \pm\sqrt{1}$

$\quad k - 4 = \pm 1$

$\quad k - 4 = 1 \ \text{ or } \ k - 4 = -1$

$\quad k = 5 \qquad\qquad k = 3 \qquad \{3, 5\}$

71) $\quad s^2 + 10 = -10s$

$\quad s^2 + 10s = -10$

$\quad s^2 + 10s + 25 = -10 + 25$

$\quad (s+5)^2 = 15$

$\quad s + 5 = \pm\sqrt{15}$

$\quad s = -5 \pm \sqrt{15}$

$\quad \left\{ -5 - \sqrt{15}, -5 + \sqrt{15} \right\}$

73) $\quad t^2 = 2t - 9$

$\quad t^2 - 2t = -9$

$\quad t^2 - 2t + 1 = -9 + 1$

$\quad (t-1)^2 = -8$

$\quad t - 1 = \pm\sqrt{-8}$

$\quad t = 1 \pm 2i\sqrt{2}$

$\quad \left\{ 1 - 2i\sqrt{2}, 1 + 2i\sqrt{2} \right\}$

75) $v^2 + 4v + 8 = 0$

$\quad v^2 + 4v = -8$

$\quad v^2 + 4v + 4 = -8 + 4$

$\quad (v+2)^2 = -4$

$\quad v + 2 = \pm\sqrt{-4}$

$\quad v = -2 \pm 2i$

$\quad \{-2 - 2i, -2 + 2i\}$

77) $m^2 + 3m - 40 = 0$

$\quad m^2 + 3m = 40$

$\quad m^2 + 3m + \dfrac{9}{4} = 40 + \dfrac{9}{4}$

$\quad \left(m + \dfrac{3}{2} \right)^2 = \dfrac{169}{4}$

$\quad m + \dfrac{3}{2} = \pm\sqrt{\dfrac{169}{4}}$

$\quad m + \dfrac{3}{2} = \pm\dfrac{13}{2}$

$\quad m + \dfrac{3}{2} = \dfrac{13}{2} \ \text{ or } \ m + \dfrac{3}{2} = -\dfrac{13}{2}$

$\quad m = \dfrac{10}{2} \qquad\qquad m = -\dfrac{16}{2}$

$\quad m = 5 \qquad\qquad\quad m = -8$

$\quad \{-8, 5\}$

79) $x^2 - 7x + 12 = 0$

$$x^2 - 7x = -12$$

$$x^2 - 7x + \frac{49}{4} = -12 + \frac{49}{4}$$

$$\left(x - \frac{7}{2}\right)^2 = \frac{1}{4}$$

$$x - \frac{7}{2} = \pm\sqrt{\frac{1}{4}} = \pm\frac{1}{2}$$

$$x - \frac{7}{2} = \frac{1}{2} \quad \text{or} \quad x - \frac{7}{2} = -\frac{1}{2}$$

$$x = \frac{8}{2} \qquad\qquad x = \frac{6}{2}$$

$$x = 4 \qquad\qquad x = 3 \qquad \{3, 4\}$$

81) $r^2 - r = 3$

$$r^2 - r + \frac{1}{4} = 3 + \frac{1}{4}$$

$$\left(r - \frac{1}{2}\right)^2 = \frac{13}{4}$$

$$r - \frac{1}{2} = \pm\sqrt{\frac{13}{4}}$$

$$r = \frac{1}{2} \pm \frac{\sqrt{13}}{2}$$

$$\left\{\frac{1}{2} - \frac{\sqrt{13}}{2}, \frac{1}{2} + \frac{\sqrt{13}}{2}\right\}$$

83) $c^2 + 5c + 7 = 0$

$$c^2 + 5c + \frac{25}{4} = -7 + \frac{25}{4}$$

$$\left(c + \frac{5}{2}\right)^2 = -\frac{3}{4}$$

$$c + \frac{5}{2} = \pm i\frac{\sqrt{3}}{2}$$

$$c = -\frac{5}{2} \pm i\frac{\sqrt{3}}{2}$$

$$\left\{-\frac{5}{2} - \frac{\sqrt{3}}{2}i, -\frac{5}{2} + \frac{\sqrt{3}}{2}i\right\}$$

85) $3k^2 - 6k + 12 = 0$

$$k^2 - 2k + 1 + 3 = 0$$

$$(k-1)^2 = -3$$

$$k - 1 = \sqrt{-3}$$

$$k = 1 \pm i\sqrt{3}$$

$$\left\{1 - i\sqrt{3}, 1 + i\sqrt{3}\right\}$$

87) $4r^2 + 24r = 8$

$$r^2 + 6r = 2$$

$$r^2 + 6r + 9 = 11$$

$$(r+3)^2 = 11$$

$$r + 3 = \pm\sqrt{11}$$

$$r = -3 \pm \sqrt{11}$$

$$\left\{-3 - \sqrt{11}, -3 + \sqrt{11}\right\}$$

89) $10d = 2d^2 + 12$

$5d = d^2 + 6$

$d^2 - 5d = -6$

$d^2 - 5d + \dfrac{25}{4} = -6 + \dfrac{25}{4}$

$\left(d - \dfrac{5}{2}\right)^2 = \dfrac{1}{4}$

$d - \dfrac{5}{2} = \pm\dfrac{1}{2}$

$d = 2,3 \qquad \{2,3\}$

$a - \dfrac{7}{8} = \pm\dfrac{1}{8}$

$a = \dfrac{7}{8} \pm \dfrac{1}{8} \qquad \left\{\dfrac{3}{4}, 1\right\}$

95) $(y+5)(y-3) = 5$

$y^2 - 3y + 5y - 15 = 5$

$y^2 + 2y = 20$

$y^2 + 2y + 1 = 20 + 1$

$(y+1)^2 = 21$

91) $2n^2 + 8 = 5n$

$2n^2 - 5n = -8$

$n^2 - \dfrac{5}{2}n + \dfrac{25}{16} = -4 + \dfrac{25}{16}$

$\left(n - \dfrac{5}{4}\right)^2 = -\dfrac{39}{16}$

$n - \dfrac{5}{4} = \pm\dfrac{\sqrt{39}}{4}i$

$n = \dfrac{5}{4} \pm \dfrac{\sqrt{39}}{4}i$

$\left\{\dfrac{5}{4} - \dfrac{\sqrt{39}}{4}i, \dfrac{5}{4} + \dfrac{\sqrt{39}}{4}i\right\}$

$y + 1 = \pm\sqrt{21}$

$y = -1 \pm \sqrt{21}$ $\left\{-1 - \sqrt{21}, -1 + \sqrt{21}\right\}$

97) $(2m+1)(m-3) = -7$

$2m - 6m + m - 3 = -7$

$2m - 5m = -4$

$m - \dfrac{5}{2}m = -2$

$m - \dfrac{5}{2}m + \dfrac{25}{16} = -2 + \dfrac{25}{16}$

$\left(m - \dfrac{5}{4}\right)^2 = -\dfrac{7}{16}$

$m - \dfrac{5}{4} = \pm\dfrac{\sqrt{7}}{4}i$

$m = \dfrac{5}{4} \pm \dfrac{\sqrt{7}}{4}i$ $\left\{\dfrac{5}{4} - \dfrac{\sqrt{7}}{4}i, \dfrac{5}{4} + \dfrac{\sqrt{7}}{4}i\right\}$

93) $4a^2 - 7a + 3 = 0$

$4a^2 - 7a = -3$

$a^2 - \dfrac{7}{4}a + \dfrac{49}{64} = -\dfrac{3}{4} + \dfrac{49}{64}$

$\left(a - \dfrac{7}{8}\right)^2 = \dfrac{1}{64}$

Section 10.2: The Quadratic Formula

99) $a^2 + b^2 = c^2$

$a^2 + 8^2 = 10^2$

$a^2 = 100 - 64$

$a^2 = 36$

$a = 6$

101) $a^2 + b^2 = c^2$

$5^2 + 2^2 = c^2$

$c^2 = 29$

$c = \sqrt{29}$

103)

$\qquad$ Width $= w = 4$

$\qquad$ Length $= l$

$\qquad$ Diagonal $= d = 2\sqrt{13}$

$l^2 = d^2 - w^2$

$ = \left(2\sqrt{13}\right)^2 - 4^2$

$ = 52 - 16$

$ = 36$

$d = 6$ in.

105) $\quad l = 13 =$ Ladder length

$b = 5 =$ Distance from the wall

$h =$ Height of the wall

$l^2 = b^2 + h^2$

$13^2 = 5^2 + h^2$

$h^2 = 169 - 25$

$ = 144$

$h = 12$ ft

107) $\quad f(x) = 49 = (x+3)^2$

$x + 3 = \pm 7$

$x = -3 \pm 7$

$x = -10$ or $x = 4$

109) $\quad x =$ Width of the garden

$x + 8 =$ Length of the garden

$153 =$ Area of the garden

$x(x+8) = 153$

$x^2 + 8x = 153$

$x^2 + 8x + 16 = 153 + 16$

$(x+4)^2 = 169$

$x + 4 = \pm 13$

$x = 9$ or $x = -17$

Width of the garden: 9 ft;

Length: 17 ft

Section 10.2

1) The fraction bar should also be under $-b$:

$$x = \frac{-b \pm \sqrt{b^2 - 4ac}}{2a}$$

3) You cannot divide only the -2 by 2.

$$\frac{-2 \pm 6\sqrt{11}}{2} = \frac{2\left(-1 \pm 3\sqrt{11}\right)}{2}$$

$$= -1 \pm 3\sqrt{11}$$

5) $x^2 + 4x + 3 = 0$

$\quad a = 1, \ b = 4 \text{ and } c = 3$

$x = \dfrac{-4 \pm \sqrt{4^2 - 4(1)(3)}}{2(1)}$

$= \dfrac{-4 \pm \sqrt{16 - 12}}{2}$

$= \dfrac{-4 \pm \sqrt{4}}{2} = \dfrac{-4 \pm 2}{2}$

$\dfrac{-4 + 2}{2} = \dfrac{-2}{2} = -1, \ \dfrac{-4 - 2}{2} = \dfrac{-6}{2} = -3$

$\{-3, \ -1\}$

7) $3t^2 + t - 10 = 0$

$\quad a = 3, \ b = 1 \text{ and } c = -10$

$t = \dfrac{-1 \pm \sqrt{1^2 - 4(3)(-10)}}{2(3)}$

$= \dfrac{-1 \pm \sqrt{1 + 120}}{6}$

$= \dfrac{-1 \pm \sqrt{121}}{6} = \dfrac{-1 \pm 11}{6}$

$\dfrac{-1 + 11}{6} = \dfrac{10}{6} = \dfrac{5}{3},$

$\dfrac{-1 - 11}{6} = \dfrac{-12}{6} = -2 \qquad \left\{ -2, \dfrac{5}{3} \right\}$

9) $k^2 + 2 = 5k$

$k^2 - 5k + 2 = 0$

$\quad a = 1, \ b = -5 \text{ and } c = 2$

$k = \dfrac{-(-5) \pm \sqrt{(-5)^2 - 4(1)(2)}}{2(1)}$

$= \dfrac{5 \pm \sqrt{25 - 8}}{2} = \dfrac{5 \pm \sqrt{17}}{2}$

$\left\{ \dfrac{5 - \sqrt{17}}{2}, \dfrac{5 + \sqrt{17}}{2} \right\}$

11. $\qquad y^2 = 8y - 25$

$y^2 - 8y + 25 = 0$

$\quad a = 1, \ b = -8 \text{ and } c = 25$

$y = \dfrac{-(-8) \pm \sqrt{(-8)^2 - 4(1)(25)}}{2(1)}$

$= \dfrac{8 \pm \sqrt{64 - 100}}{2} = \dfrac{8 \pm \sqrt{-36}}{2}$

$= \dfrac{8 \pm 6i}{2} = \dfrac{8}{2} \pm \dfrac{6}{2}i = 4 \pm 3i$

$\{4 - 3i, \ 4 + 3i\}$

13. $\qquad 3 - 2w = -5w^2$

$5w^2 - 2w + 3 = 0$

$\quad a = 5, \ b = -2 \text{ and } c = 3$

$w = \dfrac{-(-2) \pm \sqrt{(-2)^2 - 4(5)(3)}}{2(5)}$

$= \dfrac{2 \pm \sqrt{4 - 60}}{10} = \dfrac{2 \pm \sqrt{-56}}{10}$

$= \dfrac{2 \pm 2i\sqrt{14}}{10} = \dfrac{2}{10} \pm \dfrac{2i\sqrt{14}}{10} = \dfrac{1}{5} \pm \dfrac{\sqrt{14}}{5}i$

$\left\{ \dfrac{1}{5} - \dfrac{\sqrt{14}}{5}i, \ \dfrac{1}{5} + \dfrac{\sqrt{14}}{5}i \right\}$

Section 10.2: The Quadratic Formula

15) $r^2 + 7r = 0$

$a = 1,\ b = 7$ and $c = 0$

$r = \dfrac{-7 \pm \sqrt{7^2 - 4(1)(0)}}{2(1)}$

$= \dfrac{-7 \pm \sqrt{49}}{2} = \dfrac{-7 \pm 7}{2}$

$\dfrac{-7 + 7}{2} = \dfrac{0}{2} = 0,$

$\dfrac{-7 - 7}{2} = \dfrac{-14}{2} = -7 \qquad \{-7,\ 0\}$

17) $\quad 3v(v + 3) = 7v + 4$

$\quad 3v^2 + 9v = 7v + 4$

$\quad 3v^2 + 2v - 4 = 0$

$a = 3,\ b = 2$ and $c = -4$

$v = \dfrac{-2 \pm \sqrt{2^2 - 4(3)(-4)}}{2(3)}$

$= \dfrac{-2 \pm \sqrt{4 + 48}}{6} = \dfrac{-2 \pm \sqrt{52}}{6}$

$= \dfrac{-2 \pm 2\sqrt{13}}{6} = \dfrac{2\left(-1 \pm \sqrt{13}\right)}{6}$

$= \dfrac{-1 \pm \sqrt{13}}{3};\ \left\{\dfrac{-1 - \sqrt{13}}{3},\ \dfrac{-1 + \sqrt{13}}{3}\right\}$

19) $(2c - 5)(c - 5) = -3$

$\quad 2c^2 - 15c + 25 = -3$

$\quad 2c^2 - 15c + 28 = 0$

$a = 2,\ b = -15$ and $c = 28$

$c = \dfrac{-(-15) \pm \sqrt{(-15)^2 - 4(2)(28)}}{2(2)}$

$= \dfrac{15 \pm \sqrt{225 - 224}}{4} = \dfrac{15 \pm \sqrt{1}}{4}$

$= \dfrac{15 \pm 1}{4}$

$\dfrac{15 + 1}{4} = \dfrac{16}{4} = 4,$

$\dfrac{15 - 1}{4} = \dfrac{14}{4} = \dfrac{7}{2} \qquad \left\{\dfrac{7}{2},\ 4\right\}$

21) $\quad \dfrac{1}{6}u^2 + \dfrac{4}{3}u = \dfrac{5}{2}$

$\quad 6\left(\dfrac{1}{6}u^2 + \dfrac{4}{3}u\right) = 6\left(\dfrac{5}{2}\right)$

$\quad u^2 + 8u = 15$

$\quad u^2 + 8u - 15 = 0$

$a = 1,\ b = 8$ and $c = -15$

$u = \dfrac{-8 \pm \sqrt{8^2 - 4(1)(-15)}}{2(1)}$

$= \dfrac{-8 \pm \sqrt{64 + 60}}{2} = \dfrac{-8 \pm \sqrt{124}}{2}$

$= \dfrac{-8 \pm 2\sqrt{31}}{2} = \dfrac{2\left(-4 \pm \sqrt{31}\right)}{2}$

$= -4 \pm \sqrt{31}$

$\left\{-4 - \sqrt{31},\ -4 + \sqrt{31}\right\}$

23) $2(p+10)=(p+10)(p-2)$

$2p+20=p^2+8p-20$

$0=p^2+6p-40$

$a=1,\, b=6 \text{ and } c=-40$

$p=\dfrac{-6\pm\sqrt{6^2-4(1)(-40)}}{2(1)}$

$=\dfrac{-6\pm\sqrt{36+160}}{2}=\dfrac{-6\pm\sqrt{196}}{2}$

$=\dfrac{-6\pm14}{2}$

$\dfrac{-6+14}{2}=\dfrac{8}{2}=4,$

$\dfrac{-6-14}{2}=\dfrac{-20}{2}=-10 \qquad \{-10,\,4\}$

25) $4g^2+9=0$

$a=4,\, b=0 \text{ and } c=9$

$g=\dfrac{-0\pm\sqrt{0^2-4(4)(9)}}{2(4)}$

$=\dfrac{\pm\sqrt{-144}}{8}=\dfrac{\pm12i}{8}=\pm\dfrac{3}{2}i$

$\left\{-\dfrac{3}{2}i,\,\dfrac{3}{2}i\right\}$

27) $\qquad x(x+6)=-34$

$x^2+6x+34=0$

$a=1,\, b=6 \text{ and } c=34$

$x=\dfrac{-6\pm\sqrt{6^2-4(1)(34)}}{2(1)}$

$=\dfrac{-6\pm\sqrt{36-136}}{2}=\dfrac{-6\pm\sqrt{-100}}{2}$

$=\dfrac{-6\pm10i}{2}=-\dfrac{6}{2}\pm\dfrac{10i}{2}=-3\pm5i$

$\{-3-5i,\,-3+5i\}$

29) $(2s+3)(s-1)=s^2-s+6$

$2s^2+s-3=s^2-s+6$

$s^2+2s-9=0$

$a=1,\, b=2 \text{ and } c=-9$

$s=\dfrac{-2\pm\sqrt{2^2-4(1)(-9)}}{2(1)}$

$=\dfrac{-2\pm\sqrt{4+36}}{2}=\dfrac{-2\pm\sqrt{40}}{2}$

$=\dfrac{-2\pm2\sqrt{10}}{2}=\dfrac{2\left(-1\pm\sqrt{10}\right)}{2}$

$=-1\pm\sqrt{10}$

$\left\{-1-\sqrt{10},\,-1+\sqrt{10}\right\}$

31) $\qquad 3(3-4y)=-4y^2$

$9-12y=-4y^2$

$4y^2-12y+9=0$

$a=4,\, b=-12 \text{ and } c=9$

$y=\dfrac{-(-12)\pm\sqrt{(-12)^2-4(4)(9)}}{2(4)}$

$=\dfrac{12\pm\sqrt{144-144}}{8}=\dfrac{12\pm\sqrt{0}}{8}$

$=\dfrac{12}{8}=\dfrac{3}{2} \qquad\qquad \left\{\dfrac{3}{2}\right\}$

Section 10.2: The Quadratic Formula

33) $-\dfrac{1}{6} = \dfrac{2}{3}p^2 + \dfrac{1}{2}p$

$6\left(-\dfrac{1}{6}\right) = 6\left(\dfrac{2}{3}p^2 + \dfrac{1}{2}p\right)$

$-1 = 4p^2 + 3p$

$0 = 4p^2 + 3p + 1$

$a = 4, b = 3 \text{ and } c = 1$

$p = \dfrac{-3 \pm \sqrt{3^2 - 4(4)(1)}}{2(4)}$

$= \dfrac{-3 \pm \sqrt{9 - 16}}{8} = \dfrac{-3 \pm \sqrt{-7}}{8}$

$= \dfrac{-3 \pm i\sqrt{7}}{8} = -\dfrac{3}{8} \pm \dfrac{\sqrt{7}}{8}i$

$\left\{-\dfrac{3}{8} - \dfrac{\sqrt{7}}{8}i, \ -\dfrac{3}{8} + \dfrac{\sqrt{7}}{8}i\right\}$

35) $4q^2 + 6 = 20q$

$\dfrac{4q^2}{2} + \dfrac{6}{2} = \dfrac{20q}{2}$

$2q^2 + 3 = 10q$

$2q^2 - 10q + 3 = 0$

$a = 2, b = -10 \text{ and } c = 3$

$q = \dfrac{-(-10) \pm \sqrt{(-10)^2 - 4(2)(3)}}{2(2)}$

$= \dfrac{10 \pm \sqrt{100 - 24}}{4} = \dfrac{10 \pm \sqrt{76}}{4}$

$= \dfrac{10 \pm 2\sqrt{19}}{4} = \dfrac{2(5 \pm \sqrt{19})}{4}$

$= \dfrac{5 \pm \sqrt{19}}{2}; \quad \left\{\dfrac{5 - \sqrt{19}}{2}, \dfrac{5 + \sqrt{19}}{2}\right\}$

37) $0 = x^2 + 6x - 2$

$a = 1 \quad b = 6 \quad c = -2$

$x = \dfrac{-6 \pm \sqrt{6^2 - 4(1)(-2)}}{2(1)}$

$x = \dfrac{-6 \pm \sqrt{36 + 8}}{2} \ x = \dfrac{-6 \pm \sqrt{44}}{2}$

$x = \dfrac{-6 \pm \sqrt{4 \cdot 11}}{2} = \dfrac{-6 \pm 2\sqrt{11}}{2} = -3 \pm \sqrt{11}$

$x = -3 - \sqrt{11} \text{ or } x = -3 + \sqrt{11}$

39) $12 = 2t^2 - t + 7$

$0 = 2t^2 - t - 5$

$a = 2 \quad b = -1 \quad c = -5$

$t = \dfrac{-(-1) \pm \sqrt{(-1)^2 - 4(2)(-5)}}{2(2)}$

$t = \dfrac{1 \pm \sqrt{1 + 40}}{4} = \dfrac{1 \pm \sqrt{41}}{4}$

$t = \dfrac{1 + \sqrt{41}}{4} \text{ or } t = \dfrac{1 - \sqrt{41}}{4}$

41) $g(x) = f(x)$

$$2x + 3 = 5x^2 + 21x - 1$$

$$0 = 5x^2 + 19x - 4$$

$$a = 5 \quad b = 19 \quad c = -4$$

$$x = \frac{-19 \pm \sqrt{19^2 - 4(5)(-4)}}{2(5)}$$

$$= \frac{-19 \pm \sqrt{361 + 80}}{2(5)} = \frac{-19 \pm \sqrt{441}}{10}$$

$$= \frac{-19 \pm 21}{10}$$

$$\frac{-19 + 21}{10} = \frac{2}{10} = \frac{1}{5}$$

$$\frac{-19 - 21}{10} = -\frac{40}{10} = -4$$

$$x = -4 \text{ or } x = \frac{1}{5}$$

43) There is one rational solution.

45) $a = 10, \ b = -9 \text{ and } c = 3$

$$b^2 - 4ac = (-9)^2 - 4(10)(3)$$

$$= 81 - 120 = -39$$

two nonreal complex solutions

47) $4y^2 - 49 = -28y$

$$4y^2 + 28y - 49 = 0$$

$$a = 4, \ b = 28 \text{ and } c = 49$$

$$b^2 - 4ac = 28^2 - 4(4)(49)$$

$$= 784 - 784 = 0$$

one rational solution

49) $-5 = u(u + 6)$

$$-5 = u^2 + 6u$$

$$0 = u^2 + 6u + 5$$

$$a = 1, \ b = 6 \text{ and } c = 5$$

$$b^2 - 4ac = 6^2 - 4(1)(5) = 36 - 20 = 16$$

two rational solutions

51) $a = 2, \ b = -4 \text{ and } c = -5$

$$b^2 - 4ac = (-4)^2 - 4(2)(-5)$$

$$= 16 + 40 = 56$$

two irrational solutions

53) $a = 1 \text{ and } c = 16$

$$b^2 - 4ac = 0$$

$$b^2 - 4(1)(16) = 0$$

$$b^2 - 64 = 0$$

$$b^2 = 64$$

$$b = \pm\sqrt{64} = \pm 8$$

-8 or 8

55) $a = 4 \text{ and } b = -12$

$$b^2 - 4ac = 0$$

$$(-12)^2 - 4(4)c = 0$$

$$144 - 16c = 0$$

$$144 = 16c$$

$$9 = c$$

57) $b = 12$ and $c = 9$

$$b^2 - 4ac = 0$$

$$(12)^2 - 4a(9) = 0$$

$$144 - 36a = 0$$

$$144 = 36a$$

$$4 = a$$

59) $x = $ length of one leg

$2x + 1 = $ length of other leg

$\sqrt{29} = $ length of hypotenuse

$$x^2 + (2x+1)^2 = (\sqrt{29})^2$$

$$x^2 + 4x^2 + 4x + 1 = 29$$

$$5x^2 + 4x - 28 = 0$$

$$(5x + 14)(x - 2) = 0$$

$5x + 14 = 0$ or $x - 2 = 0$

$5x = -14$ $\boxed{x = 2}$

$x = -\dfrac{14}{5}$

$2x + 1 = 2(2) + 1 = 5$

The lengths of the legs are 2 in. and 5 in.

61) a) Let $h = 8$ and solve for t.

$$8 = -16t^2 + 24t + 24$$

$$0 = -16t^2 + 24t + 16$$

$$0 = 2t^2 - 3t - 2$$

$$0 = (2t + 1)(t - 2)$$

$2t + 1 = 0$ or $t - 2 = 0$

$2t = -1$ $\boxed{t = 2}$

$t = -\dfrac{1}{2}$

The ball reaches 8 feet after 2 sec.

b) Let $h = 0$ and solve for t.

$$0 = -16t^2 + 24t + 24; \ 0 = 2t^2 - 3t - 3$$

$$a = 2, \ b = -3 \text{ and } c = -3$$

$$t = \frac{-(-3) \pm \sqrt{(-3)^2 - 4(2)(-3)}}{2(2)}$$

$$= \frac{3 \pm \sqrt{9 + 24}}{4} = \frac{3 \pm \sqrt{33}}{4}$$

Reject $t = \dfrac{3 - \sqrt{33}}{4}$ because it is negative.

The ball will hit the ground after

$\dfrac{3 + \sqrt{33}}{4}$ sec ≈ 2.2 sec.

Chapter 10: Putting It All Together

1) $z^2 - 50 = 0$

$$z^2 = 50$$

$$z = \pm\sqrt{50} = \pm 5\sqrt{2}$$

$$\left\{-5\sqrt{2},\ 5\sqrt{2}\right\}$$

3) $\qquad a(a+1) = 20$

$$a^2 + a = 20$$

$$a^2 + a - 20 = 0$$

$$(a+5)(a-4) = 0$$

$$a + 5 = 0 \ \text{ or } \ a - 4 = 0$$

$$a = -5 \qquad a = 4 \qquad \{-5,\ 4\}$$

5) $u^2 + 7u + 9 = 0$

$$a = 1,\ b = 7 \text{ and } c = 9$$

$$u = \frac{-7 \pm \sqrt{7^2 - 4(1)(9)}}{2(1)}$$

$$= \frac{-7 \pm \sqrt{49 - 36}}{2} = \frac{-7 \pm \sqrt{13}}{2}$$

$$\left\{\frac{-7 - \sqrt{13}}{2},\ \frac{-7 + \sqrt{13}}{2}\right\}$$

7) $\qquad 2k(2k+7) = 3(k+1)$

$$4k^2 + 14k = 3k + 3$$

$$4k^2 + 11k - 3 = 0$$

$$(4k-1)(k+3) = 0$$

$$4k - 1 = 0 \ \text{ or } \ k + 3 = 0$$

$$4k = 1 \qquad k = -3$$

$$k = \frac{1}{4} \qquad\qquad \left\{-3,\ \frac{1}{4}\right\}$$

9) $m^2 + 14m + 60 = 0$

$$m^2 + 14m = -60$$

$$m^2 + 14m + 49 = -60 + 49$$

$$(m+7)^2 = -11$$

$$m + 7 = \pm\sqrt{-11}$$

$$m + 7 = \pm i\sqrt{11}$$

$$m = -7 \pm i\sqrt{11}$$

$$\left\{-7 - i\sqrt{11},\ -7 + i\sqrt{11}\right\}$$

11) $10 + (3b-1)^2 = 4$

$$(3b-1)^2 = -6$$

$$3b - 1 = \pm\sqrt{-6}$$

$$3b - 1 = \pm i\sqrt{6}$$

$$3b = 1 \pm i\sqrt{6}$$

$$b = \frac{1 \pm i\sqrt{6}}{3}$$

$$\left\{\frac{1}{3} - \frac{\sqrt{6}}{3}i,\ \frac{1}{3} + \frac{\sqrt{6}}{3}i\right\}$$

13) $\qquad 1 = \dfrac{x^2}{12} - \dfrac{x}{3}$

$$12(1) = 12\left(\frac{x^2}{12} - \frac{x}{3}\right)$$

$$12 = x^2 - 4x$$

$$0 = x^2 - 4x - 12$$

$$0 = (x-6)(x+2)$$

$$x - 6 = 0 \ \text{ or } \ x + 2 = 0$$

$$x = 6 \qquad x = -2 \qquad \{-2,\ 6\}$$

15) $r^2 - 4r = 3$

$r^2 - 4r + 4 = 3 + 4$

$(r - 2)^2 = 7$

$r - 2 = \pm\sqrt{7}$

$r = 2 \pm \sqrt{7}$

$\left\{ 2 - \sqrt{7}, \ 2 + \sqrt{7} \right\}$

$$v = \frac{-(-2) \pm \sqrt{(-2)^2 - 4(3)(1)}}{2(3)}$$

$$= \frac{2 \pm \sqrt{4 - 12}}{6} = \frac{2 \pm \sqrt{-8}}{6}$$

$$= \frac{2 \pm 2i\sqrt{2}}{6} = \frac{1 \pm i\sqrt{2}}{3}$$

$$\left\{ \frac{1}{3} - \frac{\sqrt{2}}{3} i, \ \frac{1}{3} + \frac{\sqrt{2}}{3} i \right\}$$

17) $p(p + 8) = 3(p^2 + 2) + p$

$p^2 + 8p = 3p^2 + 6 + p$

$0 = 2p^2 - 7p + 6$

$0 = (2p - 3)(p - 2)$

$2p - 3 = 0$ or $p - 2 = 0$

$2p = 3 \qquad\qquad p = 2$

$p = \dfrac{3}{2} \qquad\qquad \left\{ \dfrac{3}{2}, 2 \right\}$

23) $(c - 5)^2 + 16 = 0$

$(c - 5)^2 = -16$

$c - 5 = \pm\sqrt{-16}$

$c - 5 = \pm 4i$

$c = 5 \pm 4i$

$\{ 5 - 4i, \ 5 + 4i \}$

19) $\dfrac{10}{z} = 1 + \dfrac{21}{z^2}$

$z^2 \left(\dfrac{10}{z} \right) = z^2 \left(1 + \dfrac{21}{z^2} \right)$

$10z = z^2 + 21$

$0 = z^2 - 10z + 21$

$0 = (z - 7)(z - 3)$

$z - 7 = 0$ or $z - 3 = 0$

$z = 7 \qquad z = 3 \qquad \{ 3, 7 \}$

25) $3g = g^2$

$0 = g^2 - 3g$

$0 = g(g - 3)$

$g - 3 = 0$ or $g = 0$

$g = 3 \qquad\qquad \{ 0, 3 \}$

21) $(3v + 4)(v - 2) = -9$

$3v^2 - 2v - 8 = -9$

$3v^2 - 2v + 1 = 0$

$a = 3, \ b = -2 \text{ and } c = 1$

27)
$$4m^3 = 9m$$
$$4m^3 - 9m = 0$$
$$m(4m^2 - 9) = 0$$
$$m(2m+3)(2m-3) = 0$$
$$2m+3 = 0 \text{ or } 2m-3 = 0 \text{ or } m = 0$$
$$2m = -3 \qquad\qquad 2m = 3$$
$$m = -\frac{3}{2} \qquad\qquad m = \frac{3}{2}$$
$$\left\{ -\frac{3}{2}, 0, \frac{3}{2} \right\}$$

$$q = \frac{-5 \pm \sqrt{5^2 - 4(2)(8)}}{2(2)}$$
$$= \frac{-5 \pm \sqrt{25-64}}{4} = \frac{-5 \pm \sqrt{-39}}{4}$$
$$= \frac{-5 \pm i\sqrt{39}}{4}$$
$$\left\{ -\frac{5}{4} - \frac{\sqrt{39}}{4}i, \ -\frac{5}{4} + \frac{\sqrt{39}}{4}i \right\}$$

29)
$$\frac{1}{3}q^2 + \frac{5}{6}q + \frac{4}{3} = 0$$
$$6\left(\frac{1}{3}q^2 + \frac{5}{6}q + \frac{4}{3} \right) = 6(0)$$
$$2q^2 + 5q + 8 = 0$$
$$a = 2, \ b = 5 \text{ and } c = 8$$

Section 10.3

1)
$$t - \frac{48}{t} = 8$$
$$t\left(t - \frac{48}{t} \right) = t(8)$$
$$t^2 - 48 = 8t$$
$$t^2 - 8t - 48 = 0$$
$$(t-12)(t+4) = 0$$

$$t - 12 = 0 \text{ or } t + 4 = 0$$
$$t = 12 \qquad t = -4 \qquad \{-4, 12\}$$

3)
$$\frac{2}{x} + \frac{6}{x-2} = -\frac{5}{2}$$
$$2x(x-2)\left(\frac{2}{x} + \frac{6}{x-2} \right) = 2x(x-2)\left(-\frac{5}{2} \right)$$
$$4(x-2) + 12x = -5x(x-2)$$
$$4x - 8 + 12x = -5x^2 + 10x$$
$$5x^2 + 6x - 8 = 0$$
$$(5x-4)(x+2) = 0$$

$$5x - 4 = 0 \text{ or } x + 2 = 0$$
$$5x = 4 \qquad\qquad x = -2$$
$$x = \frac{4}{5} \qquad\qquad \left\{ -2, \frac{4}{5} \right\}$$

Section 10.3: Equations in Quadratic Form

5) $$1 = \frac{2}{c} + \frac{1}{c-5}$$

$$c(c-5)(1) = c(c-5)\left(\frac{2}{c} + \frac{1}{c-5}\right)$$

$$c^2 - 5c = 2(c-5) + c$$

$$c^2 - 5c = 2c - 10 + c$$

$$c^2 - 5c = 3c - 10$$

$$c^2 - 8c = -10$$

$$c^2 - 8c + 16 = -10 + 16$$

$$(c-4)^2 = 6$$

$$c - 4 = \pm\sqrt{6}$$

$$c = 4 \pm \sqrt{6}$$

$$\left\{4 - \sqrt{6},\ 4 + \sqrt{6}\right\}$$

7) $$\frac{3}{2v+2} + \frac{1}{v} = \frac{3}{2}$$

$$\frac{3}{2(v+1)} + \frac{1}{v} = \frac{3}{2}$$

$$2v(v+1)\left(\frac{3}{2(v+1)} + \frac{1}{v}\right) = 2v(v+1)\left(\frac{3}{2}\right)$$

$$3v + 2v + 2 = 3v^2 + 3v$$

$$0 = 3v^2 - 2v - 2$$

$$v = \frac{-(-2) \pm \sqrt{(-2)^2 - 4(3)(-2)}}{2(3)}$$

$$= \frac{2 \pm \sqrt{4+24}}{6} = \frac{2 \pm \sqrt{28}}{6}$$

$$= \frac{2 \pm 2\sqrt{7}}{6} = \frac{1 \pm \sqrt{7}}{3}$$

$$\left\{\frac{1 - \sqrt{7}}{3},\ \frac{1 + \sqrt{7}}{3}\right\}$$

9) $$\frac{9}{n^2} = 5 + \frac{4}{n}$$

$$n^2\left(\frac{9}{n^2}\right) = n^2\left(5 + \frac{4}{n}\right)$$

$$9 = 5n^2 + 4n$$

$$0 = 5n^2 + 4n - 9$$

$$0 = (5n+9)(n-1)$$

$$5n + 9 = 0 \quad \text{or} \quad n - 1 = 0$$

$$5n = -9 \qquad\qquad n = 1$$

$$n = -\frac{9}{5} \qquad\qquad \left\{-\frac{9}{5}, 1\right\}$$

11) $$\frac{5}{6r} = 1 - \frac{r}{6r-6}$$

$$6r(r-1)\left(\frac{5}{6r}\right) = 6r(r-1)\left[1 - \frac{r}{6(r-1)}\right]$$

$$5r - 5 = 6r^2 - 6r - r^2$$

$$0 = 5r^2 - 11r + 5$$

$$r = \frac{-(-11) \pm \sqrt{(-11)^2 - 4(5)(5)}}{2(5)}$$

$$= \frac{11 \pm \sqrt{121 - 100}}{10} = \frac{11 \pm \sqrt{21}}{10}$$

$$\left\{\frac{11 - \sqrt{21}}{10},\ \frac{11 + \sqrt{21}}{10}\right\}$$

13)
$$g = \sqrt{g+20}$$
$$g^2 = g+20$$
$$g^2 - g - 20 = 0$$
$$(g-5)(g+4) = 0$$
$$g-5 = 0 \ \text{ or } \ g+4 = 0$$
$$g = 5 \qquad g = -4$$
Only one solution satisfies the original equation. $\{5\}$

15)
$$a = \sqrt{\frac{14a-8}{5}}$$
$$a^2 = \frac{14a-8}{5}$$
$$5a^2 = 14a-8$$
$$5a^2 - 14a + 8 = 0$$
$$(5a-4)(a-2) = 0$$
$$5a-4 = 0 \ \text{ or } \ a-2 = 0$$
$$5a = 4 \qquad a = 2$$
$$a = \frac{4}{5} \qquad\qquad \left\{\frac{4}{5}, 2\right\}$$

17)
$$p - \sqrt{p} = 6$$
$$p - 6 = \sqrt{p}$$
$$(p-6)^2 = \left(\sqrt{p}\right)^2$$
$$p^2 - 12p + 36 = p$$
$$p^2 - 13p + 36 = 0$$
$$(p-9)(p-4) = 0$$
$$p-9 = 0 \ \text{ or } \ p-4 = 0$$
$$p = 9 \qquad p = 4$$

Only one solution satisfies the original equation. $\{9\}$

19)
$$x = 5\sqrt{x} - 4$$
$$x + 4 = 5\sqrt{x}$$
$$(x+4)^2 = \left(5\sqrt{x}\right)^2$$
$$x^2 + 8x + 16 = 25x$$
$$x^2 - 17x + 16 = 0$$
$$(x-16)(x-1) = 0$$
$$x-16 = 0 \ \text{ or } \ x-1 = 0$$
$$x = 16 \qquad x = 1 \qquad \{1,16\}$$

21) $2 + \sqrt{2y-1} = y$
$$\sqrt{2y-1} = y-2$$
$$\left(\sqrt{2y-1}\right)^2 = (y-2)^2$$
$$2y-1 = y^2 - 4y + 4$$
$$0 = y^2 - 6y + 5$$
$$0 = (y-5)(y-1)$$
$$y-5 = 0 \ \text{ or } \ y-1 = 0$$
$$y = 5 \qquad y = 1$$
Only one solution satisfies the original equation. $\{5\}$

23)
$$2 = \sqrt{6k+4} - k$$
$$k + 2 = \sqrt{6k+4}$$
$$(k+2)^2 = \left(\sqrt{6k+4}\right)^2$$
$$k^2 + 4k + 4 = 6k + 4$$
$$k^2 - 2k = 0$$
$$k(k-2) = 0$$
$$k-2 = 0 \ \text{ or } \ k = 0$$
$$k = 2 \qquad\qquad \{0,2\}$$

25) yes   27) yes   29) no

31) yes   33) no

35) $x^4 - 10x^2 + 9 = 0$

$u^2 = x^4; \ u = x^2$

$u^2 - 10u + 9 = 0$

$(u-9)(u-1) = 0$

$u = 1, 9$

$\begin{array}{ll} x^2 = 9 & x^2 = 1 \\ x = \pm\sqrt{9} & x = \pm\sqrt{1} \\ x = \pm 3 & x = \pm 1 \quad \{-3, -1, 1, 3\} \end{array}$

37) $p^4 - 11p^2 + 28 = 0$

$(p^2 - 7)(p^2 - 4) = 0$

$p^2 - 7 = 0 \ \text{ or } \ p^2 - 4 = 0$

$\begin{array}{ll} p^2 = 7 & p^2 = 4 \\ p = \pm\sqrt{7} & p = \pm 2 \end{array}$

$\{-\sqrt{7}, -2, 2, \sqrt{7}\}$

39) $a^4 + 12a^2 = -35$

$a^4 + 12a^2 + 35 = 0$

$(a^2 + 7)(a^2 + 5) = 0$

$a^2 + 7 = 0 \ \text{ or } \ a^2 + 5 = 0$

$\begin{array}{ll} a^2 = -7 & a^2 = -5 \\ a = \pm i\sqrt{7} & a = \pm i\sqrt{5} \end{array}$

$\{-i\sqrt{7}, -i\sqrt{5}, i\sqrt{5}, i\sqrt{7}\}$

41) $b^{2/3} + 3b^{1/3} + 2 = 0; \quad u^2 = b^{2/3}; u = b^{1/3}$

$u^2 + 3u + 2 = 0$

$(u+2)(u+1) = 0$

$\begin{array}{ll} b^{1/3} = -2 & b^{1/3} = -1 \\ \left(\sqrt[3]{b}\right)^3 = (-2)^3 & \left(\sqrt[3]{b}\right)^3 = (-1)^3 \\ b = -8 & b = -1 \quad \{-8, -1\} \end{array}$

43) $t^{2/3} - 6t^{1/3} = 40; \quad u^2 = t^{2/3}; u = t^{1/3}$

$u^2 - 6u - 40 = 0$

$(u+4)(u-10) = 0$

$\begin{array}{ll} t^{1/3} = -4 & t^{1/3} = 10 \\ \left(\sqrt[3]{t}\right)^3 = (-4)^3 & \left(\sqrt[3]{t}\right)^3 = 10^3 \\ t = -64 & t = 1000 \end{array}$

$\{-64, 1000\}$

45) $0 = h - 4h^{1/2} - 21; \quad u^2 = h; \quad u = h^{1/2}$

$0 = u^2 - 4u - 21$

$0 = (u+3)(u-7)$

$-3, 7 = u$

$\begin{array}{ll} h^{1/2} = -3 & h^{1/2} = 7 \\ \left(\sqrt{h}\right)^2 = (-3)^2 & \left(\sqrt{h}\right)^2 = 7^2 \\ h = 9 & h = 49 \quad \{9, 49\} \end{array}$

Only $\{49\}$ solution satisfies the equation.

47) $2a - 5a^{1/2} - 12 = 0; \quad u^2 = a; \quad u = a^{1/2}$

$2u^2 - 5u - 12 = 0$

$(2u + 3)(u - 4) = 0$

$u = -\dfrac{3}{2}, 4$

$2a^{1/2} = -3 \qquad\qquad\qquad a^{1/2} = 4$

$a^{1/2} = -\dfrac{3}{2} \qquad\qquad\qquad \left(\sqrt{a}\right)^2 = 4^2$

$\left(\sqrt{a}\right)^2 = \left(-\dfrac{3}{2}\right)^2 = \dfrac{9}{4} \qquad\qquad a = 16$

Only $\{16\}$ solution satisfies the equation.

49) $\qquad\qquad 9n^4 = -15n^2 - 4$

$9n^4 + 15n^2 + 4 = 0;$

$(3x^2 + 4)(3x^2 + 1) = 0$

$3x^2 + 4 = 0 \quad \text{or} \quad 3x^2 + 1 = 0$

$3x^2 = -4 \qquad\qquad 3x^2 = -1$

$x^2 = -\dfrac{4}{3} \qquad\qquad x^2 = -\dfrac{1}{3}$

$x = \pm\dfrac{2}{\sqrt{3}}i \qquad\qquad x = \pm\dfrac{1}{\sqrt{3}}i$

$x = \pm\dfrac{2\sqrt{3}}{3}i \quad \text{or} \quad x = \pm\dfrac{\sqrt{3}}{3}i$

$\left\{ -\dfrac{2\sqrt{3}}{3}i, -\dfrac{\sqrt{3}}{3}i, \dfrac{\sqrt{3}}{3}i, \dfrac{2\sqrt{3}}{3}i \right\}$

51) $\qquad\qquad z^4 - 2z^2 = 15$

$z^4 - 2z^2 - 15 = 0$

$(z^2 - 5)(z^2 + 3) = 0$

$z^2 - 5 = 0 \quad \text{or} \quad z^2 + 3 = 0$

$z^2 = 5 \qquad\qquad z^2 = -3$

$z = \pm\sqrt{5} \qquad\qquad z = \pm i\sqrt{3}$

$\left\{ -\sqrt{5}, \sqrt{5}, -i\sqrt{3}, i\sqrt{3} \right\}$

53) $w^4 - 6w^2 + 2 = 0$ Let $u = w^2$ and $u^2 = w^4$.

$u^2 - 6u + 2 = 0$

$u = \dfrac{-(-6) \pm \sqrt{(-6)^2 - 4(1)(2)}}{2(1)}$

$= \dfrac{6 \pm \sqrt{28}}{2} = \dfrac{6 \pm 2\sqrt{7}}{2} = 3 \pm \sqrt{7}$

$u = 3 + \sqrt{7} \quad \text{or} \quad u = 3 - \sqrt{7}$

$u = w^2 \qquad\qquad u = w^2$

$w^2 = 3 + \sqrt{7} \qquad\qquad w^2 = 3 - \sqrt{7}$

$w = \pm\sqrt{3 + \sqrt{7}} \qquad w = \pm\sqrt{3 - \sqrt{7}}$

55) $\qquad\qquad 2m^4 + 1 = 7m^2$

$2m^4 - 7m^2 + 1 = 0$

Let $u = m^2$ and $u^2 = m^4$.

$2u^2 - 7u + 1 = 0$

$u = \dfrac{-(-7) \pm \sqrt{(-7)^2 - 4(2)(1)}}{2(2)} = \dfrac{7 \pm \sqrt{41}}{4}$

$$u = \frac{7 + \sqrt{41}}{4} \quad \text{or} \quad u = \frac{7 - \sqrt{41}}{4}$$

$$u = m^2 \qquad\qquad u = m^2$$

$$m^2 = \frac{7 + \sqrt{41}}{4} \qquad m^2 = \frac{7 - \sqrt{41}}{4}$$

$$m = \pm\frac{\sqrt{7 + \sqrt{41}}}{2} \qquad m = \pm\frac{\sqrt{7 - \sqrt{41}}}{2}$$

57) $\quad t^{-2} - 4t^{-1} - 12 = 0$

$$\left(t^{-1} + 2\right)\left(t^{-1} - 6\right) = 0$$

$$t^{-1} + 2 = 0 \quad \text{or} \quad t^{-1} - 6 = 0$$

$$t^{-1} = -2 \qquad\qquad t^{-1} = 6$$

$$t = -\frac{1}{2} \qquad\qquad t = \frac{1}{6} \quad \left\{-\frac{1}{2}, \frac{1}{6}\right\}$$

59) $\qquad\qquad\qquad 4 = 13y^{-1} - 3y^{-2}$

$$3y^{-2} - 13y^{-1} + 4 = 0$$

$$\left(3y^{-1} - 1\right)\left(y^{-1} - 4\right) = 0$$

$$3y^{-1} - 1 = 0 \quad \text{or} \quad y^{-1} - 4 = 0$$

$$3y^{-1} = 1 \qquad\qquad y^{-1} = 4$$

$$y^{-1} = \frac{1}{3} \qquad\qquad y = \frac{1}{4}$$

$$y = 3$$

$$\left\{\frac{1}{4}, 3\right\}$$

61) $(x-2)^2 + 11(x-2) + 24 = 0$

Let $u = x - 2$.

$$u^2 + 11u + 24 = 0$$

$$(u+8)(u+3) = 0$$

$$u + 8 = 0 \quad \text{or} \quad u + 3 = 0$$

$$u = -8 \qquad\qquad u = -3$$

Solve for x using $u = x - 2$.

$$-8 = x - 2 \qquad\qquad -3 = x - 2$$

$$-6 = x \qquad\qquad\qquad -1 = x$$

$$\{-6, -1\}$$

63) $2(3q+4)^2 - 13(3q+4) + 20 = 0$

Let $u = 3q + 4$.

$$2u^2 - 13u + 20 = 0$$

$$(2u - 5)(u - 4) = 0$$

$$2u - 5 = 0 \quad \text{or} \quad u - 4 = 0$$

$$2u = 5$$

$$u = \frac{5}{2} \qquad\qquad u = 4$$

Solve for q using $u = 3q + 4$.

$$\frac{5}{2} = 3q + 4 \qquad\qquad 4 = 3q + 4$$

$$-\frac{3}{2} = 3q \qquad\qquad 0 = 3q$$

$$-\frac{1}{2} = q \qquad\qquad 0 = q$$

$$\left\{-\frac{1}{2}, 0\right\}$$

65) $(5a-3)^2 + 6(5a-3) = -5$

Let $u = 5a-3$.

$$u^2 + 6u + 5 = 0$$
$$(u+1)(u+5) = 0$$
$$u+1 = 0 \quad \text{or} \quad u+5 = 0$$
$$u = -1 \qquad u = -5$$

Solve for a using $u = 5a-3$.

$$-1 = 5a-3 \qquad -5 = 5a-3$$
$$2 = 5a \qquad -2 = 5a$$
$$\frac{2}{5} = a \qquad -\frac{2}{5} = a$$

$$\left\{ -\frac{2}{5}, \frac{2}{5} \right\}$$

67) $3(k+8)^2 + 5(k+8) = 12$

Let $u = k+8$.

$$3u^2 + 5u - 12 = 0$$
$$(3u-4)(u+3) = 0$$
$$3u-4 = 0 \quad \text{or} \quad u+3 = 0$$
$$3u = 4 \qquad u = -3$$
$$u = \frac{4}{3}$$

Solve for k using $u = k+8$.

$$\frac{4}{3} = k+8 \qquad -3 = k+8$$
$$-\frac{20}{3} = k \qquad -11 = k$$

$$\left\{ -11, -\frac{20}{3} \right\}$$

69) $1 - \dfrac{8}{2w+1} = -\dfrac{16}{(2w+1)^2}$

Let $u = 2w+1$.

$$1 - \frac{8}{u} = -\frac{16}{u^2}$$
$$u^2 \left(1 - \frac{8}{u} \right) = u^2 \left(-\frac{16}{u^2} \right)$$
$$u^2 - 8u = -16$$
$$u^2 - 8u + 16 = 0$$
$$(u-4)^2 = 0$$
$$u-4 = 0$$
$$u = 4$$

Solve for w using $u = 2w+1$.

$$4 = 2w+1$$
$$3 = 2w$$
$$\frac{3}{2} = w \qquad \left\{ \frac{3}{2} \right\}$$

71) $1 + \dfrac{2}{h-3} = \dfrac{1}{(h-3)^2}$

Let $u = h-3$

$$1 + \frac{2}{u} = \frac{1}{u^2}$$
$$u^2 \left(1 + \frac{2}{u} \right) = u^2 \left(\frac{1}{u^2} \right)$$
$$u^2 + 2u = 1$$
$$u^2 + 2u + 1 = 1+1$$
$$(u+1)^2 = 2$$
$$u+1 = \pm\sqrt{2}$$
$$u = -1 \pm \sqrt{2}$$

$$-1-\sqrt{2} = h-3 \qquad -1+\sqrt{2} = h-3$$
$$2-\sqrt{2} = h \qquad 2+\sqrt{2} = h$$

Section 10.3: Equations in Quadratic Form

73) $x =$ Time taken by Walter

$x + 3 =$ Time taken by Kevin

$2 =$ Time taken together

$$\frac{1}{x} + \frac{1}{x+3} = \frac{1}{2}$$

$$\frac{(x+3)+x}{x(x+3)} = \frac{1}{2}$$

$$x(x+3) = 2(2x+3)$$

$$x^2 + 3x = 4x + 6$$

$$x^2 - x = 6$$

$$x^2 - x = 6$$

$$x^2 - x + \frac{1}{4} = 6 + \frac{1}{4}$$

$$\left(x - \frac{1}{2}\right)^2 = \frac{25}{4}$$

$$x - \frac{1}{2} = \pm\frac{5}{2}$$

$$x = 3, -2$$

Time taken by Walter: 3 hrs, by Kevin: 6 hrs

75) $x =$ Speed of Boat in Still water

$3 =$ Speed of the current

$$\frac{9}{x+3} + \frac{6}{x-3} = 1$$

$$\frac{9(x-3)+6(x+3)}{x^2-9} = 1$$

$$x^2 - 9 = 15x - 9$$

$$x^2 - 15x = 0$$

$$x(x-15) = 0$$

$$x = 0, 15$$

Speed of boat in still water: 15 mph

77) $t =$ time large drain emptying

$t + 3 =$ time small drain emptying

$2 =$ time both the drains emptying

$$\frac{1}{t} + \frac{1}{t+3} = \frac{1}{2}$$

$$\frac{(t+3)+t}{t(t+3)} = \frac{1}{2}$$

$$2(2t+3) = t^2 + 3t$$

$$4t + 6 = t^2 + 3t$$

$$t^2 - t - 6 = 0$$

$$(t-3)(t+2) = 0$$

$t - 3 = 0 \qquad\qquad t + 2 = 0$

$\boxed{t = 3} \qquad\qquad t = -2$

Reject $t = -2$, because time cannot be negative.

large drain emptys the tank in 3 hr

small drain emptys the tank in 6 hr

79) $\qquad\qquad\qquad s =$ speed to Boulder

$s - 10 =$ speed to home

$$\frac{600}{s} + \frac{600}{s-10} = 22$$

$$\frac{600(s-10)+600s}{s(s-10)} = 22$$

$$1200s - 6000 = 22s(s-10)$$

$$1200s - 6000 = 22s^2 - 220s$$

$$22s^2 - 1420s + 6000 = 0$$

$$11s - 710s + 3000 = 0$$

$$(s-60)(11s-50) = 0$$

$s - 60 = 0$ or $11s - 50 = 0$

$s = 60 \qquad s = \dfrac{50}{11} \approx 4.55$

Reject $s \approx 4.55$ because the speed to home cannot be negative.

speed to Boulder: 60 mph

speed to home: 50 mph

Section 10.4

1) $\qquad A = \pi r^2$

$$\dfrac{A}{\pi} = r^2$$

$$\pm\sqrt{\dfrac{A}{\pi}} = r$$

$$\pm\dfrac{\sqrt{A}}{\sqrt{\pi}} \cdot \dfrac{\sqrt{\pi}}{\sqrt{\pi}} = r$$

$$\dfrac{\pm\sqrt{A\pi}}{\pi} = r$$

3) $\qquad a = \dfrac{v^2}{r}$

$$ar = v^2$$

$$\pm\sqrt{ar} = v$$

5) $\qquad E = \dfrac{I}{d^2}$

$$d^2 E = I$$

$$d^2 = \dfrac{I}{E}$$

$$d = \pm\sqrt{\dfrac{I}{E}}$$

$$d = \pm\dfrac{\sqrt{I}}{\sqrt{E}} \cdot \dfrac{\sqrt{E}}{\sqrt{E}} = \dfrac{\pm\sqrt{IE}}{E}$$

7) $\qquad F = \dfrac{kq_1 q_2}{r^2}$

$$r^2 F = kq_1 q_2$$

$$r^2 = \dfrac{kq_1 q_2}{F}$$

$$r = \pm\sqrt{\dfrac{kq_1 q_2}{F}}$$

$$r = \dfrac{\pm\sqrt{kq_1 q_2}}{\sqrt{F}} \cdot \dfrac{\sqrt{F}}{\sqrt{F}} = \dfrac{\pm\sqrt{kq_1 q_2 F}}{F}$$

Section 10.4: Formulas and Applications

9) $d = \sqrt{\dfrac{4A}{\pi}}$

$d^2 = \dfrac{4A}{\pi}$

$\pi d^2 = 4A$

$\dfrac{\pi d^2}{4} = A$

11) $T_p = 2\pi\sqrt{\dfrac{l}{g}}$

$\dfrac{T_p}{2\pi} = \sqrt{\dfrac{l}{g}}$

$\dfrac{T_p^2}{4\pi^2} = \dfrac{l}{g}$

$\dfrac{g T_p^2}{4\pi^2} = l$

13) $T_p = 2\pi\sqrt{\dfrac{l}{g}}$

$T_p^2 = 4\pi^2\left(\dfrac{l}{g}\right)$

$g T_p^2 = 4\pi^2 l$

$g = \dfrac{4\pi^2 l}{T_p^2}$

15) a) Both are written in the standard form for a quadratic equation, $ax^2 + bx + c = 0$.

b) Use the quadratic formula.

17) $x = \dfrac{-(-5) \pm \sqrt{(-5)^2 - 4rs}}{2r}$

$= \dfrac{5 \pm \sqrt{25 - 4rs}}{2r}$

19) $z = \dfrac{-r \pm \sqrt{r^2 - 4p(-q)}}{2p}$

$= \dfrac{-r \pm \sqrt{r^2 + 4pq}}{2p}$

21) $da^2 - ha = k$

$da^2 - ha - k = 0$

$a = \dfrac{-(-h) \pm \sqrt{(-h)^2 - 4d(-k)}}{2d}$

$= \dfrac{h \pm \sqrt{h^2 + 4dk}}{2d}$

23) $s = \dfrac{1}{2}gt^2 + vt$

$0 = \dfrac{1}{2}gt^2 + vt - s$

$t = \dfrac{-v \pm \sqrt{v^2 + 2\left(\dfrac{1}{2}g\right)s}}{2\left(\dfrac{1}{2}g\right)}$

$t = \dfrac{-v \pm \sqrt{v^2 + gs}}{g}$

25) $x =$ width of sheet metal

$x + 3 =$ length of sheet metal

length of box $= x + 3 - 1 - 1 = x + 1$

width of box $= x - 1 - 1 = x - 2$

height of box $= 1$

Volume $= ($length$)($width$)($height$)$

$$70 = (x+1)(x-2)(1)$$

$$70 = x^2 - x - 2$$

$$0 = x^2 - x - 72$$

$$0 = (x+8)(x-9)$$

$x + 8 = 0$ or $x - 9 = 0$

$x = -8$ $\boxed{x = 9}$

width $= 9$ in.

length $= 9 + 3 = 12$ in.

27)

$x =$ width of non-skid surface

$80 + 2x =$ length of pool plus two strips of non-skid surface

$60 + 2x =$ width of pool plus two strips of non-skid surface

Area of Pool plus strips $-$ Area of Pool

                $=$ Area of Strips

$$(80+2x)(60+2x)-80(60)=576$$

$$4800+280x+4x^2-4800=576$$

$$4x^2+280x-576=0$$

$$x^2+70x-144=0$$

$$(x+72)(x-2)=0$$

$x + 72 = 0$ or $x - 2 = 0$

$x = -72$ $\boxed{x = 2}$

The width of non-skid surface is 2 ft.

29) $x =$ base of the sail

$2x - 1 =$ height of the sail

$$\text{Area} = \frac{1}{2}(\text{base})(\text{height})$$

$$60 = \frac{1}{2}(x)(2x-1)$$

$$120 = 2x^2 - x$$

$$0 = 2x^2 - x - 120$$

$$0 = (2x+15)(x-8)$$

$2x + 15 = 0$ or $x - 8 = 0$

$2x = -15$ $\boxed{x = 8}$

$$x = -\frac{15}{2}$$

base $= 8$ ft.

height $= 2(8) - 1 = 15$ ft.

31) $x =$ height of the ramp

$2x + 4 =$ base of the ramp

$3x - 4 =$ hypotenuse of the ramp

$$a^2 + b^2 = c^2$$

$$x^2 + (2x+4)^2 = (3x-4)^2$$

$$x^2 + 4x^2 + 16x + 16 = 9x^2 - 24x + 16$$

$$5x^2 + 16x + 16 = 9x^2 - 24x + 16$$

$$0 = 4x^2 - 40x$$

$$0 = 4x(x-10)$$

$4x = 0$ or $x - 10 = 0$

$x = 0$ $\boxed{x = 10}$

The height of the ramp is 10 in.

Section 10.4: Formulas and Applications

33) a) $h = 40$

$h = -16t^2 + 60t + 4$

$40 = -16t^2 + 60t + 4$

$0 = -16t^2 + 60t - 36$

$0 = 4t^2 - 15t + 9$

$0 = (4t - 3)(t - 3)$

$4t - 3 = 0$ or $t - 3 = 0$

$\qquad 4t = 3$

$\qquad t = \dfrac{3}{4} \qquad t = 3$

0.75 sec on the way up,
3 sec on the way down

b) $h = 0$

$h = -16t^2 + 60t + 4$

$0 = -16t^2 + 60t + 4$

$0 = 4t^2 - 15t - 1$

$t = \dfrac{-(-15) \pm \sqrt{(-15)^2 - 4(4)(-1)}}{2(4)}$

$= \dfrac{15 \pm \sqrt{225 - 16}}{8} = \dfrac{15 \pm \sqrt{241}}{8}$

Reject $\dfrac{15 - \sqrt{241}}{8}$ as a solution

since this is a negative number.

$\dfrac{15 + \sqrt{241}}{8}$ sec or about 3.8 sec

35) a) $x = 0$

$y = -0.25x^2 + 1.5x + 9.5$

$y = -0.25(0)^2 + 1.5(0) + 9.5$

$y = 9.5$

9.5 million

b) $y = 11.75$

$y = -0.25x^2 + 1.5x + 9.5$

$11.75 = -0.25x^2 + 1.5x + 9.5$

$0 = -0.25x^2 + 1.5x - 2.25$

$0 = 25x^2 - 150x + 225$

$0 = x^2 - 6x + 9$

$0 = (x - 3)^2$

$0 = x - 3$

$3 = x$

11.75 million saw a Broadway play in
1996+3 or in 1999.

37) $D = \dfrac{65}{P}; \ S = 10P + 3$

$D = S$

$\dfrac{65}{P} = 10P + 3$

$65 = P(10P + 3)$

$65 = 10P^2 + 3P$

$0 = 10P^2 + 3P - 65$

$P = \dfrac{-3 \pm \sqrt{3^2 - 4(10)(-65)}}{2(10)}$

$= \dfrac{-3 \pm \sqrt{9 + 2600}}{20} = \dfrac{-3 \pm \sqrt{2609}}{20}$

$= \dfrac{-3 \pm \sqrt{2609}}{20} \approx \dfrac{-3 \pm 51.08}{20}$

$P \approx \dfrac{-3 - 51.08}{20}$ or $P \approx \dfrac{-3 + 51.08}{20}$

$P \approx -2.70 \qquad \boxed{P \approx \$2.40}$

330

39) $P = \dfrac{\sqrt{L^2 - d^2}}{2}$

$= \dfrac{\sqrt{12.5^2 - 12^2}}{2}$

$= \dfrac{\sqrt{12.25}}{2}$

$= \dfrac{3.5}{2}$

$= 1.75$ ft

Section 10.5

1) The graph of $g(x)$ is the same as the graph of $f(x)$, but $g(x)$ is shifted up 6 units.

3) The graph of $h(x)$ is the same as the graph of $f(x)$, but $h(x)$ is shifted left 5 units.

5)

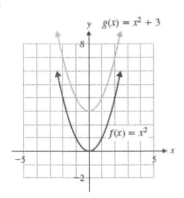

7)

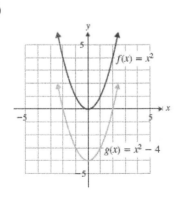

9)

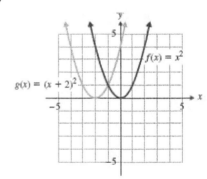

11)

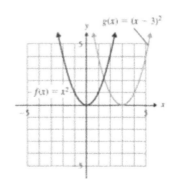

13)

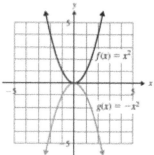

15)

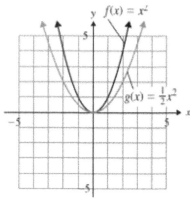

17)

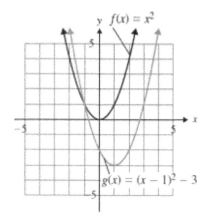

19) a) (h,k)

b) $x = h$

c) a is positive

d) a is negative

e) $a > 1$ or $a < -1$

f) $0 < a < 1$ or $-1 < a < 0$

21)

$V(3, -1); x = 3;$
x-ints: (2, 0), (4, 0);
y-int: (0, 8); domain:
$(-\infty, \infty)$; range: $[-1, \infty)$

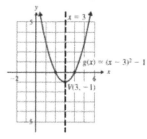

23)

$V(-2, 7); x = -2;$
x-ints: none; y-int:
(0, 11); domain:
$(-\infty, \infty)$; range: $[7, \infty)$

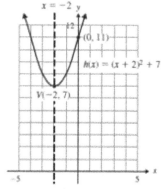

29)

25)

$V(-1, -5); x = -1;$
x-ints: $(-1 - \sqrt{5}, 0),$
$(-1 + \sqrt{5}, 0);$ y-int:
$(0, -4);$ domain: $(-\infty, \infty);$
range: $[-5, \infty)$

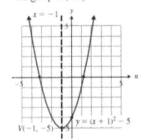

$V(2, -4); x = 2; x$-ints:
none; y-int: $(0, -8);$
domain: $(-\infty, \infty);$
range: $(-\infty, -4]$

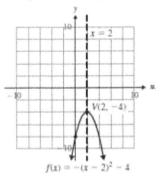

27)

$V(3, 2); x = 3; x$-ints: $(3 - \sqrt{2}, 0), (3 + \sqrt{2}, 0); y$-int:
$(0, -7);$ domain: $(-\infty, \infty);$ range: $(-\infty, 2]$

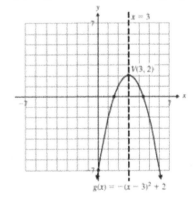

31)

$V(-1, -2); x = -1;$
x-ints: $(-2, 0), (0, 0);$
y-int: $(0, 0);$ domain:
$(-\infty, \infty);$ range: $[-2, \infty)$

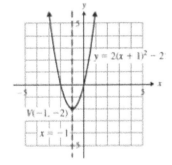

Section 10.5: Quadratic Functions and Their Graphs

33)

$V(0, -1)$; $x = 0$; x-ints: $(-2, 0)$, $(2, 0)$; y-int: $(0, -1)$; domain: $(-\infty, \infty)$; range: $[-1, \infty)$

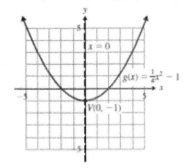

35)

$V(-4, 3)$; $x = -4$; x-ints: $(-7, 0)$, $(-1, 0)$; y-int: $\left(0, -\frac{7}{3}\right)$; domain: $(-\infty, \infty)$; range: $(-\infty, 3]$

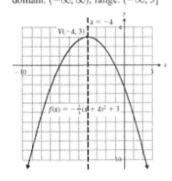

37)

$V(-2, 5)$; $x = -2$; x-int: none; y-int: $(0, 17)$; domain: $(-\infty, \infty)$; range: $[5, \infty)$

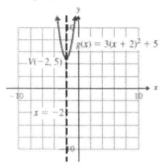

39) a) $h(x)$ b) $f(x)$

    c) $g(x)$ d) $k(x)$

41) $g(x) = (x-8)^2$

43) $g(x) = x^2 + 3.5$

45) $g(x) = (x+4)^2 - 7$

47) $f(x) = (x^2 + 8x) + 11$;

$\left[\dfrac{1}{2} \cdot 8\right]^2 = (4)^2 = 16$;

Add and subtract the number above to the same side of the equation.

$f(x) = (x+4)^2 - 5$

49)

$f(x) = x^2 - 2x - 3$

$f(x) = (x^2 - 2x) - 3; \quad \left[\dfrac{1}{2}(-2)\right]^2 = (-1)^2 = 1$

$f(x) = (x^2 - 2x + 1) - 3 - 1$

$f(x) = (x-1)^2 - 4$

    x-ints: $(-1, 0)$, $(3, 0)$;
    y-int: $(0, -3)$; domain:
    $(-\infty, \infty)$; range: $[-4, \infty)$

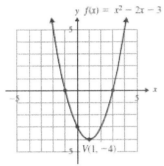

53) $g(x) = x^2 + 4x$

$g(x) = (x^2 + 4x); \quad \left[\dfrac{1}{2}(4)\right]^2 = (2)^2 = 4$

$g(x) = (x^2 + 4x + 4) - 4$

$g(x) = (x+2)^2 - 4$

    x-ints: $(-4, 0)$, $(0, 0)$;
    y-int: $(0, 0)$; domain:
    $(-\infty, \infty)$; range: $[-4, \infty)$

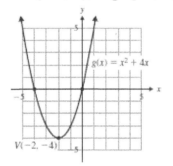

51) $y = x^2 + 6x + 7$

$y = (x^2 + 6x) + 7; \quad \left[\dfrac{1}{2}(6)\right]^2 = (3)^2 = 9$

$y = (x^2 + 6x + 9) + 7 - 9$

$y = (x+3)^2 - 2$

    $(-3 - \sqrt{2}, 0)$,
    $(-3 + \sqrt{2}, 0)$; y-int:
    $(0, 7)$; domain: $(-\infty, \infty)$;
    range: $[-2, \infty)$

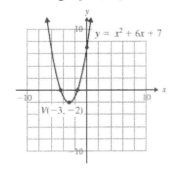

55) $h(x) = -x^2 - 4x + 5;$

$h(x) = -(x^2 + 4x) + 5$

$\left[\dfrac{1}{2}(4)\right]^2 = (2)^2 = 4$

$h(x) = -(x^2 + 4x + 4) + 5 + 4$

$h(x) = -(x+2)^2 + 9$

    x-ints: $(-5, 0)$, $(1, 0)$;
    y-int: $(0, 5)$; domain:
    $(-\infty, \infty)$; range: $(-\infty, 9]$

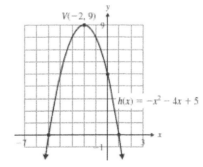

57) $y = -x^2 + 6x - 10$;

$$\left[\frac{1}{2}(-6)\right]^2 = (-3)^2 = 9$$

$$y = -\left(x^2 - 6x + 9\right) - 10 + 9$$

$$y = -(x-3)^2 - 1$$

x-ints: none; y-int: (0, −10);
domain: (−∞, ∞);
range: (−∞, −1]

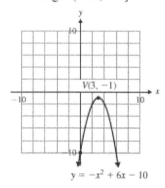

59) $f(x) = 2x^2 - 8x + 4$;

$$f(x) = 2\left(x^2 - 4x\right) + 4;$$

$$\left[\frac{1}{2}(-4)\right]^2 = (-2)^2 = 4$$

$$f(x) = 2\left(x^2 - 4x + 4\right) + 4 - 8$$

$$f(x) = 2(x-2)^2 - 4$$

x-ints: $(2 - \sqrt{2}, 0)$,
$(2 + \sqrt{2}, 0)$; y-int: (0, 4);
domain: (−∞, ∞);
range: [−4, ∞)

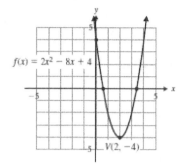

61) $g(x) = -\frac{1}{3}x^2 - 2x - 9$

$$g(x) = -\frac{1}{3}\left(x^2 + 6x\right) - 9;$$

$$\left[\frac{1}{2}(6)\right]^2 = (3)^2 = 9$$

$$g(x) = -\frac{1}{3}\left(x^2 + 6x + 9\right) + 3 - 9$$

$$g(x) = -\frac{1}{3}(x+3)^2 - 6$$

x-ints: none; y-int: (0, −9);
domain: (−∞, ∞); range:
(−∞, −6]

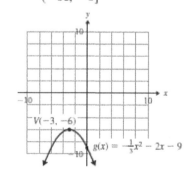

63) $y = x^2 - 3x + 2$

$y = (x^2 - 3x) + 2$

$\left[\dfrac{1}{2}(-3)\right]^2 = \left(-\dfrac{3}{2}\right)^2 = \dfrac{9}{4}$

$y = \left(x^2 - 3x + \dfrac{9}{4}\right) + 2 - \dfrac{9}{4}$

$y = \left(x - \dfrac{3}{2}\right)^2 - \dfrac{1}{4}$; x-ints: (1, 0), (2, 0); y-int: (0, 2);

domain: $(-\infty, \infty)$; range: $\left[-\dfrac{1}{4}, \infty\right]$

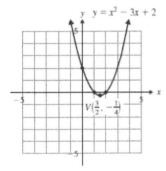

65) $y = x^2 + 2x - 3 \qquad a = 1; b = 2; c = -3$

$V = -\dfrac{b}{2a}, f\left(-\dfrac{b}{2a}\right)$

$x = \dfrac{-b}{2a} = \dfrac{-2}{2 \cdot 1} = -1$

$y = (-1)^2 + 2(-1) - 3 = -4$

$V = (-1, -4)$

$f(0) = -3; \quad 0 = x^2 + 2x - 3$

$\qquad\qquad 0 = (x + 3)(x - 1)$

$\qquad\qquad 1, -3 = x$

Domain: $(-\infty, \infty)$;   Range: $[-4, \infty)$

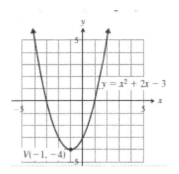

67) $f(x) = -x^2 - 8x - 13$

$a = -1; b = -8; c = -13$

$V = -\dfrac{b}{2a}, f\left(-\dfrac{b}{2a}\right)$

$x = \dfrac{-b}{2a} = \dfrac{-(-8)}{2(-1)} = -4$

$f(-4) = -(-4)^2 - 8(-4) - 13 = 3$

$V = (-4, 3)$

$f(0) = -13; \quad y - \text{intercept: } (0, -13)$

$0 = -x^2 - 8x - 13$

$x = \dfrac{-(-8) \pm \sqrt{(-8)^2 - 4(-1)(-13)}}{2(-1)}$

$x = -4 \pm \sqrt{3}$

x-intercepts: $\left(-4 - \sqrt{3}, 0\right)$ and $\left(-4 + \sqrt{3}, 0\right)$

Domain: $(-\infty, \infty)$;   Range: $(-\infty, 3]$

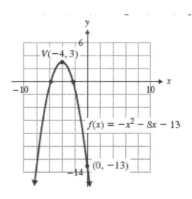

69)

$$g(x) = 2x^2 - 4x + 4 \qquad a = 2; b = -4; c = 4$$

$$V = -\frac{b}{2a}, g\left(-\frac{b}{2a}\right)$$

$$x = \frac{-b}{2a} = \frac{-(-4)}{2 \cdot 2} = 1$$

$$g(1) = 2(1)^2 - 4(1) + 4 = 2$$

$$V = (1, 2)$$

$f(0) = 4; \quad y-\text{intercept: } (0, 2)$

$0 = 2x^2 - 4x + 4$

No real solutions, so no $x-$intercepts

Domain: $(-\infty, \infty); \quad$ Range: $[2, \infty)$

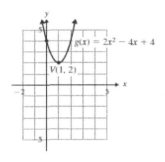

71) $y = -3x^2 + 6x + 1 \quad a = -3; b = 6; c = 1$

$$x = \frac{-b}{2a} = \frac{-6}{2(-3)} = 1$$

$$y = -3(1)^2 + 6(1) + 1 = 4$$

$$V = (1, 4)$$

$f(0) = 1; \quad y-\text{intercept: } (0, 1)$

$0 = -3x^2 + 6x + 1$

$$x = \frac{-6 \pm \sqrt{6^2 - 4(-3)(1)}}{2(-3)} = 1 \pm \frac{2\sqrt{3}}{3}$$

x-intercepts: $\left(1 + \frac{2\sqrt{3}}{3}, 0\right)$ and $\left(1 - \frac{2\sqrt{3}}{3}, 0\right)$

Domain: $(-\infty, \infty); \quad$ Range: $(-\infty, 4]$

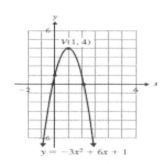

73)

$$f(x) = \frac{1}{2}x^2 - 4x + 5 \qquad a = \frac{1}{2}; b =; -4\ c = 5$$

$$x = \frac{-b}{2a} = \frac{-(-4)}{2\left(\frac{1}{2}\right)} = 4$$

$$f(4) = \frac{1}{2}(4)^2 - 4(4) + 5 = -3$$

$$V = (4, -3)$$

$$f(0) = 5; \quad y-\text{intercept: } (0, 5)$$

$$0 = \frac{1}{2}x^2 - 4x + 5$$

$$x = \frac{-(-4) \pm \sqrt{(-4)^2 - 4\left(\frac{1}{2}\right)(5)}}{2\left(\frac{1}{2}\right)} = 4 \pm \sqrt{6}$$

x-intercepts: $\left(4 + \sqrt{6}, 0\right)$ and $\left(4 - \sqrt{6}, 0\right)$

Domain: $(-\infty, \infty)$; Range: $[-3, \infty)$

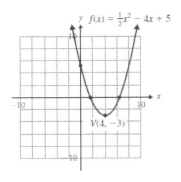

$$a = -\frac{1}{3}; b = -2; c = -5$$

$$x = \frac{-b}{2a} = \frac{-(-2)}{2\left(-\frac{1}{3}\right)} = -3$$

$$h(-3) = -\frac{1}{3}(-3)^2 - 2(-3) - 5 = -2$$

$$V = (-3, -2)$$

$$h(0) = -5; \quad y-\text{intercept: } (0, -5)$$

$$0 = \frac{1}{3}x^2 - 2x - 5$$

No real solution so no x – intercepts.

Domain: $(-\infty, \infty)$; Range: $(-\infty, -2]$

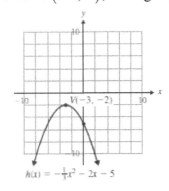

75) $h(x) = -\frac{1}{3}x^2 - 2x - 5$

339

Section 10.6

1) maximum

3) neither

5) minimum

7) If a is positive the graph opens upward, so the y-coordinate of the vertex is the minimum value of the function. If a is negative the graph opens downward, so the y-coordinate of the vertex is the maximum value of the function.

9) a) minimum

b) $h = -\dfrac{6}{2(1)} = -3$

$k = (-3)^2 + 6(-3) + 9$

$\quad = 9 - 18 + 9 = 0 \qquad V(-3,0)$

c) 0

d)

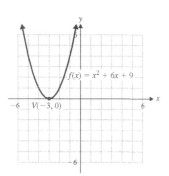

11) a) maximum

b) $h = -\dfrac{4}{2\left(-\dfrac{1}{2}\right)} = \dfrac{4}{1} = 4$

$k = -\dfrac{1}{2}(4)^2 + 4(4) - 6$

$\quad = -\dfrac{1}{2}(16) + 16 - 6$

$\quad = -8 + 16 - 6 = 2 \qquad V(4,2)$

c) 2 				d)

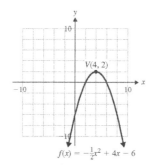

13) a) The max height occurs at the t-coordinate of the vertex.

$t = -\dfrac{320}{2(-16)} = \dfrac{320}{32} = 10$

10 sec

b) Since the object reaches max height $t = 10$, find $h(10)$.

$h(10) = -16(10)^2 + 320(10)$

$\qquad = -16(100) + 3200 = 1600$

1600 ft

c) The object hits the ground when $h(t) = 0$.

$$0 = -16t^2 + 320t$$
$$0 = t^2 - 20t$$
$$0 = t(t - 20)$$
$$t = 0 \text{ or } t - 20 = 0$$
$$t = 20$$

The object will hit the ground after 20 sec.

15) The x-coordinate of the vertex represents the number of months after January that had the greatest number of guests.

$$x = -\frac{120}{2(-10)} = \frac{120}{20} = 6$$

The inn had the greatest number of guests in July.

The number of guests at the inn during July is $N(6)$.

$$N(6) = -10(6)^2 + 120(6) + 120$$
$$= -10(36) + 720 + 120 = 480$$

480 guests stayed at the inn during the month of July.

17) The t-coordinate of the vertex represents the number of years after 1989 in which the greatest number of babies was born to teen mothers.

$$t = -\frac{2.75}{2(-0.721)} \approx 2$$

Greatest number were born in 1991. The number of babies born in (in thousands) to teen mothers is $N(2)$.

$$N(2) = -0.721(2)^2 + 2.75(2) + 528$$
$$= -2.884 + 5.5 + 528 \approx 531$$

Approximately 531,000 babies were born to teen mothers in 1991.

19) $w = $ width of ice rink
$l = $ length of ice rink
$A = $ area of the rink
Maximize: $A = lw$
Constraint: $2l + 2w = 100$
$$2l + 2w = 100$$
$$2l = 100 - 2w$$
$$l = 50 - w$$
$$A = lw$$
$$A(w) = (50 - w)w$$
$$A(w) = -w^2 + 50w$$

Find the w-coordinate of the vertex, the value that maximizes the area.

$$w = -\frac{50}{2(-1)} = 25$$

Find $A(25)$.

$$A(25) = -(25)^2 + 50(25)$$
$$= -625 + 1250 = 625 \text{ ft}^2$$

Section 10.6: Application of Quadratic Functions and Graphing Other Parabolas

21) $w =$ width of dog pen

    $l =$ length of dog pen

    $A =$ area of the dog pen

Maximize: $A = lw$

Since the barn is 1 side of the pen,

the fence is used for only 3 sides.

Constraint: $l + 2w = 48$

$l + 2w = 48$

    $l = 48 - 2w$

     $A = lw$

$A(w) = (48 - 2w)w$

$A(w) = -2w^2 + 48w$

Find the w-coordinate of the vertex,

the value that maximizes the area.

$w = -\dfrac{48}{2(-2)} = 12$

Use $l + 2w = 40$ with $w = 12$ to

find the corresponding length.

$l + 2(12) = 48$

    $l + 24 = 48$

       $l = 24$

The dog pen will be 12 ft $\times$ 24 ft.

23) $x =$ one integer, $y =$ other integer,

    $P =$ product

Maximize: $P = xy$

Constraint: $x + y = 18$

$x + y = 18$

    $y = 18 - x$

     $P = xy$

$P(x) = x(18 - x)$

$P(x) = -x^2 + 18x$

$x = -\dfrac{18}{2(-1)} = 9$

$x + y = 18$

$9 + y = 18$

   $y = 9$

9 and 9

25) $x =$ one integer, $y =$ other integer,

    $P =$ product

Maximize: $P = xy$

Constraint: $x - y = 12$

   $x - y = 12$

  $x - 12 = y$

    $P = xy$

$P(x) = x(x - 12)$

$P(x) = x^2 - 12x$

$x = -\dfrac{-12}{2(1)} = 6$

$x - y = 12$

$6 - y = 12$

 $-y = 6$

  $y = -6$

-6 and 6

27) (h, k)

29) to the left

342

31) $V(-4,1); y=1$

     x-int $x=(0-1)^2-4$

           $x=1-4$

           $x=-3$ $(-3,0)$

     y-int $0=(y-1)^2-4$

           $4=(y-1)^2$

           $\pm\sqrt{4}=y-1$

           $1\pm2=y$

        $y=-1$ or $y=3$ $(0,-1),(0,3)$

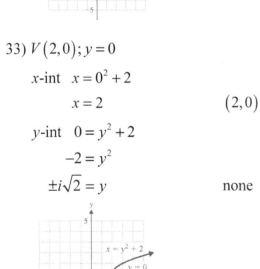

33) $V(2,0); y=0$

     x-int $x=0^2+2$

           $x=2$ $(2,0)$

     y-int $0=y^2+2$

           $-2=y^2$

           $\pm i\sqrt{2}=y$ none

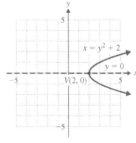

35) $V(5,4); y=4$

     x-int $x=-(0-4)^2+5$

           $x=-(-4)^2+5$

           $x=-16+5$ $(-11,0)$

     y-int $0=-(y-4)^2+5$

           $-5=-(y-4)^2$

           $5=(y-4)^2$

           $\pm\sqrt{5}=y-4$

           $4\pm\sqrt{5}=y$

        $\left(0,4-\sqrt{5}\right),\left(0,4+\sqrt{5}\right)$

37) $V(-9,2); y=2$

     x-int $x=-2(0-2)^2-9$

           $x=-2(4)-9$

           $x=-8-9=-17$ $(-17,0)$

     y-int $0=-2(y-2)^2-9$

           $9=-2(y-2)^2$

           $-\dfrac{9}{2}=(y-2)^2$

           $\pm\dfrac{3}{\sqrt{2}}i=y+1$

           $-1\pm\dfrac{3}{\sqrt{2}}i=y$ none

39) $V(0,-2); y=-2$

x-int $x=\dfrac{1}{4}(0+2)^2$

$x=\dfrac{1}{4}(4)=1$ $(1,0)$

y-int $0=\dfrac{1}{4}(y+2)^2$

$0=(y+2)^2$

$0=y+2$

$-2=y$ $(0,-2)$

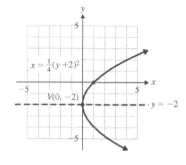

41) $x=y^2-4y+5$

$x=(y^2-4y+4)+5-4$

$x=(y-2)^2+1$

Domain:$[1,\infty)$ Range:$(-\infty,\infty)$

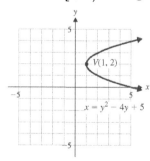

43) $x=-y^2+6y+6$

$x=-(y^2-6y+9)+6+9$

$x=-(y-3)^2+15$

Domain:$(-\infty,15]$ Range:$(-\infty,\infty)$

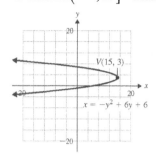

45) $x=\dfrac{1}{3}y^2+\dfrac{8}{3}y-\dfrac{5}{3}$

$3x=y^2+8y-5$

$3x=(y^2+8y+16)-5-16$

$3x=(y+4)^2-21$

$x=\dfrac{1}{3}(y+4)^2-7$

Domain:$[-7,\infty)$ Range:$(-\infty,\infty)$

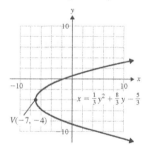

47) $x = -4y^2 - 8y - 10$

$x = -4(y^2 + 2y) - 10$

$x = -4(y^2 + 2y + 1) - 10 + 4$

$x = -4(y+1)^2 - 6$

Domain: $(-\infty, -6]$ Range: $(-\infty, \infty)$

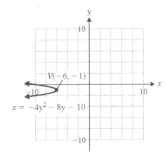

49) $y = -\dfrac{-4}{2(1)} = \dfrac{4}{2} = 2$

$x = 2^2 - 4(2) + 3$

$= 4 - 8 + 3 = -1$ $V(-1, 2)$

x-int $(3, 0)$

$y^2 - 4y + 3 = 0$

$(y-1)(y-3) = 0$

$y = 1, 3$

y-int $(0,1), (0,3)$

Domain: $[-1, \infty)$ Range: $(-\infty, \infty)$

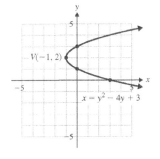

51) $y = -\dfrac{2}{2(-1)} = \dfrac{2}{2} = 1$

$x = -1^2 + 2(1) + 2 = 3$ $V(3, 1)$

x-int $(2, 0)$

$-y^2 + 2y + 2 = 0$

$y^2 - 2y + 1 = 2 + 1$

$(y-1)^2 = 3$

$y - 1 = \pm\sqrt{3}$

$y = 1 \pm \sqrt{3}$

y-int $\left(0, 1-\sqrt{3}\right), \left(0, 1+\sqrt{3}\right)$

Domain: $(-\infty, 4]$ Range: $(-\infty, \infty)$

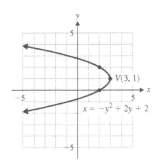

53) $y = -\dfrac{4}{2(-2)} = \dfrac{4}{4} = 1$

$x = -2(1)^2 + 4(1) - 6 = -4$ $V(-4, 1)$

x-int $(-6, 0)$

y-int none

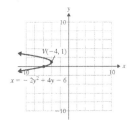

55)

$$y = -\frac{-16}{2(4)} = \frac{16}{8} = 2$$

$$x = 4(2)^2 - 16(2) + 13 = -3 \qquad V(-3, 2)$$

x-int $(13, 0)$

$$4y^2 - 16y + 13 = 0$$

$$y^2 - 4y + 4 = -\frac{13}{4} + \frac{16}{4}$$

$$(y-2)^2 = \frac{3}{4}$$

$$y = 2 \pm \frac{\sqrt{3}}{2}$$

y-int $\left(0, 2 - \frac{\sqrt{3}}{2}\right), \left(0, 2 + \frac{\sqrt{3}}{2}\right)$

Domain: $[-3, \infty)$ Range: $(-\infty, \infty)$

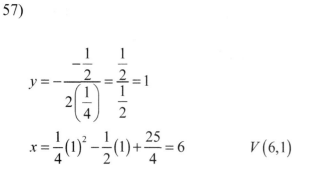

57)

$$y = -\frac{-\frac{1}{2}}{2\left(\frac{1}{4}\right)} = \frac{\frac{1}{2}}{\frac{1}{2}} = 1$$

$$x = \frac{1}{4}(1)^2 - \frac{1}{2}(1) + \frac{25}{4} = 6 \qquad V(6, 1)$$

x-int $\left(\frac{25}{4}, 0\right)$ y-int none

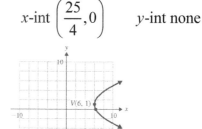

59)

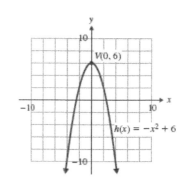

Domain: $(-\infty, \infty)$ Range: $(-\infty, 6]$

61)

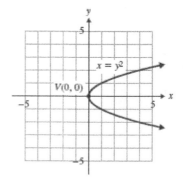

Domain: $[0, \infty)$ Range: $(-\infty, \infty)$

63)

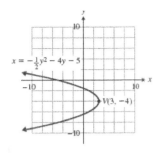

Domain: $(-\infty, 3]$ Range: $(-\infty, \infty)$

65)

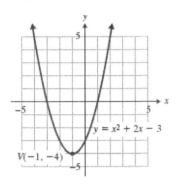

Domain: $(-\infty, \infty)$ Range: $[-4, \infty)$

67)

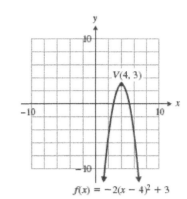

Domain: $(-\infty, \infty)$ Range: $(-\infty, 3]$

Section 10.7

1) The endpoints are included when the inequality symbol is $\leq$ or $\geq$. The endpoints are not included when the symbol is $<$ or $>$.

3) a) $[-5, 1]$ b) $(-\infty, -5) \cup (1, \infty)$

5) a) $[-1, 3]$ b) $(-\infty, -1) \cup (3, \infty)$

7) $x^2 + 6x - 7 \geq 0$

$(x+7)(x-1) \geq 0$

$(x+7)(x-1) = 0$

$x + 7 = 0$ or $x - 1 = 0$

$x = -7$ $x = 1$

Interval A: $(-\infty, -7)$ Positive

Interval B: $(-7, 1)$ Negative

Interval C: $(1, \infty)$ Positive

$(-\infty, -7] \cup [1, \infty)$

9)
$$c^2 + 5c < 36$$
$$c^2 + 5c - 36 < 0$$
$$(c+9)(c-4) < 0$$
$$(c+9)(c-4) = 0$$
$$c + 9 = 0 \text{ or } c - 4 = 0$$
$$c = -9 \qquad c = 4$$

Interval A: $(-\infty, -9)$ Positive

Interval B: $(-9, 4)$ Negative

Interval C: $(4, \infty)$ Positive

$(-9, 4)$

11) $3z^2 + 14z - 24 \le 0$
$$(3z - 4)(z + 6) \le 0$$
$$(3z - 4)(z + 6) = 0$$
$$3z - 4 = 0 \text{ or } z + 6 = 0$$
$$3z = 4 \qquad z = -6$$
$$z = \frac{4}{3}$$

Interval A: $(-\infty, -6)$ Positive

Interval B: $\left(-6, \dfrac{4}{3}\right)$ Negative

Interval C: $\left(\dfrac{4}{3}, \infty\right)$ Positive

$\left[-6, \dfrac{4}{3}\right]$

13)
$$7p^2 - 4 > 12p$$
$$7p^2 - 12p - 4 > 0$$
$$(7p + 2)(p - 2) > 0$$
$$(7p + 2)(p - 2) = 0$$

$$7p + 2 = 0 \text{ or } p - 2 = 0$$
$$7p = -2 \qquad\quad p = 2$$
$$p = -\frac{2}{7}$$

Interval A: $\left(-\infty, -\dfrac{2}{7}\right)$ Positive

Interval B: $\left(-\dfrac{2}{7}, 2\right)$ Negative

Interval C: $(2, \infty)$ Positive

$\left(-\infty, -\dfrac{2}{7}\right) \cup (2, \infty)$

15) $b^2 - 9b > 0$
$$b(b - 9) > 0$$
$$b(b - 9) = 0$$
$$b - 9 = 0 \text{ or } b = 0$$
$$b = 9$$

Interval A: $(-\infty, 0)$ Positive

Interval B: $(0, 9)$ Negative

Interval C: $(9, \infty)$ Positive

$(-\infty, 0) \cup (9, \infty)$

17) $m^2 - 64 < 0$

$(m+8)(m-8) < 0$

$(m+8)(m-8) = 0$

$m+8 = 0$ or $m+8 = 0$

$m = -8$ $m = 8$

Interval A: $(-\infty, -8)$ Positive

Interval B: $(-8, 8)$ Negative

Interval C: $(8, \infty)$ Positive

$(-8, 8)$

19) $121 - h^2 \leq 0$

$(11+h)(11-h) \leq 0$

$(11+h)(11-h) = 0$

$11+h = 0$ or $11-h = 0$

$h = -11$ $h = 11$

Interval A: $(-\infty, -11)$ Negative

Interval B: $(-11, 11)$ Positive

Interval C: $(11, \infty)$ Negative

$(-\infty, -11] \cup [11, \infty)$

21) $(-\infty, \infty)$

23) $\varnothing$

25) $(-\infty, \infty)$

27) $(r+2)(r-5)(r-1) \leq 0$

$(r+2)(r-5)(r-1) = 0$

$r+2 = 0$ or $r-5 = 0$ or $r-1 = 0$

$r = -2$ $r = 5$ $r = 1$

Interval A: $(-\infty, -2)$ Negative

Interval B: $(-2, 1)$ Positive

Interval C: $(1, 5)$ Negative

Interval D: $(5, \infty)$ Positive

$(-\infty, -2] \cup [1, 5]$

29) $(6c+1)(c+7)(4c-3) < 0$

$(6c+1)(c+7)(4c-3) = 0$

$6c+1 = 0$ or $c+7 = 0$ or $4c-3 = 0$

$6c = -1$ $c = -7$ $4c = 3$

$c = -\dfrac{1}{6}$ $c = \dfrac{3}{4}$

Interval A: $(-\infty, -7)$ Negative

Interval B: $\left(-7, -\dfrac{1}{6}\right)$ Positive

Interval C: $\left(-\dfrac{1}{6}, \dfrac{3}{4}\right)$ Negative

Interval D: $\left(\dfrac{3}{4}, \infty\right)$ Positive

$(-\infty, -7] \cup \left[-\dfrac{1}{6}, \dfrac{3}{4}\right]$

Section 10.7: Quadratic and Rational Inequalities

31) $\dfrac{7}{p+6} > 0$

Set the numerator and denominator equal to zero and solve for p.

$7 \neq 0 \quad p+6=0$

$\qquad\qquad p=-6$

Interval A: $(-\infty, -6)$ Negative

Interval B: $(-6, \infty)$ Positive

$(-6, \infty)$

$(-\infty, 3) \cup (4, \infty)$

33) $\dfrac{5}{z+3} \leq 0$

Set the numerator and denominator equal to zero and solve for z.

$5 \neq 0 \quad z+3=0; \quad z=-3$

Interval A: $(-\infty, -3)$ Negative

Interval B: $(-3, \infty)$ Positive

$v \neq -3$ because it makes the denominator equal to zero.

$(-\infty, -3)$

37) $\dfrac{h-9}{3h+1} \leq 0$

Set the numerator and denominator equal to zero and solve for h.

$h-9=0 \quad 3h+1=0$

$\quad h=9 \qquad 3h=-1$

$\qquad\qquad\qquad h=-\dfrac{1}{3}$

Interval A: $\left(-\infty, -\dfrac{1}{3}\right)$ Positive

Interval B: $\left(-\dfrac{1}{3}, 9\right)$ Negative

Interval C: $(9, \infty)$ Positive

$h \neq -\dfrac{1}{3}$ because it makes the denominator equal to zero.

$\left(-\dfrac{1}{3}, 9\right]$

35) $\dfrac{x-4}{x-3} > 0$

Set the numerator and denominator equal to zero and solve for x.

$x-4=0 \quad x-3=0$

$\quad x=4 \qquad x=3$

Interval A: $(-\infty, 3)$ Positive

Interval B: $(3, 4)$ Negative

Interval C: $(4, \infty)$ Positive

39) $\dfrac{k}{k+3} \leq 0$

Set the numerator and denominator equal to zero and solve for k.

$k=0 \qquad k+3=0$

$\qquad\qquad k=-3$

Interval A: $(-\infty, -3)$ Positive

Interval B: $(-3, 0)$ Negative

Interval C: $(0, \infty)$ Positive

$k \neq -3$ because it makes the denominator equal to zero.

$(-3, 0]$

41)
$$\frac{7}{t+6} < 3$$

$$\frac{7}{t+6} - 3 < 0$$

$$\frac{7}{t+6} - \frac{3(t+6)}{t+6} < 0$$

$$\frac{7 - 3(t+6)}{t+6} < 0$$

$$\frac{7 - 3t - 18}{t+6} < 0$$

$$\frac{-3t - 11}{t+6} < 0$$

Set the numerator and denominator equal to zero and solve for t.

$$-3t - 11 = 0 \qquad t + 6 = 0$$
$$-3t = 11 \qquad\qquad t = -6$$
$$t = -\frac{11}{3}$$

Interval A: $(-\infty, -6)$ Negative

Interval B: $\left(-6, -\frac{11}{3}\right)$ Positive

Interval C: $\left(-\frac{11}{3}, \infty\right)$ Negative

$$(-\infty, -6) \cup \left(-\frac{11}{3}, \infty\right)$$

43)
$$\frac{3}{a+7} \geq 1$$

$$\frac{3}{a+7} - 1 \geq 0$$

$$\frac{3}{a+7} - \frac{(a+7)}{a+7} \geq 0$$

$$\frac{3 - (a+7)}{a+7} \geq 0$$

$$\frac{-a - 4}{a+7} \geq 0$$

Set the numerator and denominator equal to zero and solve for a.

$$-a - 4 = 0 \qquad\qquad a + 7 = 0$$
$$-a = 4 \qquad\qquad\qquad a = -7$$
$$a = -4$$

Interval A: $(-\infty, -7)$ Negative

Interval B: $(-7, -4)$ Positive

Interval C: $(-4, \infty)$ Negative

$a \neq -7$ because it makes the denominator equal to zero.

$(-7, -4]$

45)
$$\frac{2y}{y-6} \le -3$$

$$\frac{2y}{y-6} + 3 \le 0$$

$$\frac{2y}{y-6} + \frac{3(y-6)}{y-6} \le 0$$

$$\frac{2y+3(y-6)}{y-6} \le 0$$

$$\frac{2y+3y-18}{y-6} \le 0$$

$$\frac{5y-18}{y-6} \le 0$$

Set the numerator and denominator equal to zero and solve for y.

$$5y-18 = 0 \qquad y-6 = 0$$
$$5y = 18 \qquad\quad y = 6$$
$$y = \frac{18}{5}$$

Interval A: $\left(-\infty, \dfrac{18}{5}\right)$ Positive

Interval B: $\left(\dfrac{18}{5}, 6\right)$ Negative

Interval C: $(6, \infty)$ Positive

$y \ne 6$ because it makes the denominator equal to zero. $\left[\dfrac{18}{5}, 6\right)$

18/5

0 1 2 3 4 5 6 7 8

47)
$$\frac{3w}{w+2} > -4$$

$$\frac{3w}{w+2} + 4 > 0$$

$$\frac{3w}{w+2} + \frac{4(w+2)}{w+2} > 0$$

$$\frac{3w+4(w+2)}{w+2} > 0$$

$$\frac{3w+4w+8}{w+2} > 0$$

$$\frac{7w+8}{w+2} > 0$$

Set the numerator and denominator equal to zero and solve for w.

$$7w+8 = 0 \qquad w+2 = 0$$
$$7w = -8 \qquad\quad w = -2$$
$$w = -\frac{8}{7}$$

Interval A: $(-\infty, -2)$ Positive

Interval B: $\left(-2, -\dfrac{8}{7}\right)$ Negative

Interval C: $\left(-\dfrac{8}{7}, \infty\right)$ Positive

$$(-\infty, -2) \cup \left(-\dfrac{8}{7}, \infty\right)$$

$-\frac{8}{7}$

$-5 -4 -3 -2 -1 \ 0 \ 1 \ 2 \ 3 \ 4 \ 5$

49) $\dfrac{(4t-3)^2}{t-5} > 0$

Numerator will always be positive.
Set the denominator equal to zero
and solve for t.

$t-5=0$

$t=5$

Interval A: $(-\infty, 5)$ Negative

Interval B: $(5, \infty)$ Positive

$(5, \infty)$

51) $\dfrac{m+1}{m^2+3} \geq 0$

Denominator will always be
positive. Set the numerator equal
to zero and solve for m.

$m+1=0$

$m=-1$

Interval A: $(-\infty, -1)$ Negative

Interval B: $(-1, \infty)$ Positive

$[-1, \infty)$

53) $\dfrac{s^2+2}{s-4} \leq 0$

Numerator will always be positive.
Set the denominator equal to zero
and solve for s.

$s-4=0$

$s=4$

Interval A: $(-\infty, 4)$ Negative

Interval B: $(4, \infty)$ Positive

$s \neq 4$ because it makes the
denominator equal to zero.

$(-\infty, 4)$

55) a) $P(x) = -2x^2 + 32x - 96 > 0$

$-2(x^2 - 16x + 48) > 0$

$-2(x-4)(x-12) > 0$

$(x-4)(x-12) < 0$

$(x-4)(x-12) = 0$

$x-4=0$ or $x-12=0$

$x=4$ $x=12$

Interval A: $(-\infty, 4)$ Positive

Interval B: $(4, 12)$ Negative

Interval C: $(12, \infty)$ Positive

$(4, 12)$

Between 4000 and 12,000 units.

b) $(-\infty, 4) \cup (12, \infty)$

When they produce less than 4000
units or more than 12000 units.

57) $\overline{C}(x)=\dfrac{10x+100000}{x}\le 20$

$10x+100000\le 20x$

$-10x+100000\le 0$

$-10x\le -100000$

$x\ge 10000$

$10,000\,\text{or more}$

7) $27k^2-30=0$

$27k^2=30$

$k^2=\dfrac{30}{27}$

$k^2=\dfrac{10}{9}$

$k=\pm\sqrt{\dfrac{10}{9}}\,k=\pm\dfrac{\sqrt{10}}{3}$

$\left\{-\dfrac{\sqrt{10}}{3},\dfrac{\sqrt{10}}{3}\right\}$

Chapter 10 Review

1) $d^2=144$

$d=\pm\sqrt{144}$

$d=\pm12\qquad\{-12,12\}$

9) $d=\sqrt{(x_2-x_1)^2+(y_2-y_1)^2}$

$=\sqrt{[-12-(-8)]^2+(5-3)^2}$

$=\sqrt{(-4)^2+2^2}$

$=\sqrt{20}=2\sqrt5$

3) $v^2+4=0$

$v^2=-4$

$v=\pm\sqrt{-4}$

$v=\pm2i\qquad\{-2i,2i\}$

11) 1) $\dfrac12(10)=5$ 2) $5^2=25$

$r^2+10r+25;\qquad (r+5)^2$

5) $(b-3)^2=49$

$b-3=\pm\sqrt{49}$

$b-3=\pm7$

$b-3=7\ \text{or}\ b-3=-7$

$b=10\qquad b=-4\qquad\{-4,10\}$

13) 1) $\dfrac12(-5)=-\dfrac52$ 2) $\left(-\dfrac52\right)^2=\dfrac{25}{4}$

$c^2-5c+\dfrac{25}{4};\qquad \left(c-\dfrac52\right)^2$

15) 1) $\dfrac12\left(\dfrac23\right)=\dfrac13$ 2) $\left(\dfrac13\right)^2=\dfrac19$

$a^2+\dfrac23a+\dfrac19;\qquad \left(a+\dfrac13\right)^2$

17) $p^2 - 6p - 16 = 0$

$p^2 - 6p = 16$

$p^2 - 6p + 9 = 16 + 9$

$(p-3)^2 = 25$

$p - 3 = \pm\sqrt{25}$

$p - 3 = \pm 5$

$p - 3 = 5$ or $p - 3 = -5$

$p = 8 \qquad p = -2 \qquad \{-2, 8\}$

19) $n^2 + 10n = 6$

$n^2 + 10n + 25 = 6 + 25$

$(n+5)^2 = 31$

$n + 5 = \pm\sqrt{31}$

$n = -5 \pm \sqrt{31}$

$\{-5 - \sqrt{31}, -5 + \sqrt{31}\}$

21) $f^2 + 3f + 1 = 0$

$f^2 + 3f = -1$

$f^2 + 3f + \dfrac{9}{4} = -1 + \dfrac{9}{4}$

$\left(f + \dfrac{3}{2}\right)^2 = \dfrac{5}{4}$

$f + \dfrac{3}{2} = \pm\sqrt{\dfrac{5}{4}}$

$f + \dfrac{3}{2} = \pm\dfrac{\sqrt{5}}{2}$

$f = -\dfrac{3}{2} \pm \dfrac{\sqrt{5}}{2}$

$\left\{-\dfrac{3}{2} - \dfrac{\sqrt{5}}{2}, -\dfrac{3}{2} + \dfrac{\sqrt{5}}{2}\right\}$

23) $-3q^2 + 7q = 12$

$q^2 - \dfrac{7}{3}q = -4$

$q^2 - \dfrac{7}{3}q + \dfrac{49}{36} = -4 + \dfrac{49}{36}$

$\left(q - \dfrac{7}{6}\right)^2 = -\dfrac{95}{36}$

$q - \dfrac{7}{6} = \pm\sqrt{-\dfrac{95}{36}}$

$q - \dfrac{7}{6} = \pm\dfrac{\sqrt{95}}{6}i$

$q = \dfrac{7}{6} \pm \dfrac{\sqrt{95}}{6}i$

$\left\{\dfrac{7}{6} - \dfrac{\sqrt{95}}{6}i, \dfrac{7}{6} + \dfrac{\sqrt{95}}{6}i\right\}$

25) $m^2 + 4m - 12 = 0$

$a = 1,\ b = 4$ and $c = -12$

$m = \dfrac{-4 \pm \sqrt{4^2 - 4(1)(-12)}}{2(1)}$

$= \dfrac{-4 \pm \sqrt{16 + 48}}{2}$

$= \dfrac{-4 \pm \sqrt{64}}{2} = \dfrac{-4 \pm 8}{2}$

$\dfrac{-4 + 8}{2} = \dfrac{4}{2} = 2,$

$\dfrac{-4 - 8}{2} = \dfrac{-12}{2} = -6 \qquad \{-6, 2\}$

27) $10g - 5 = 2g^2$

$$0 = 2g^2 - 10g + 5$$

$$a = 2,\ b = -10 \text{ and } c = 5$$

$$g = \frac{-(-10) \pm \sqrt{(-10)^2 - 4(2)(5)}}{2(2)}$$

$$= \frac{10 \pm \sqrt{100 - 40}}{4}$$

$$= \frac{10 \pm \sqrt{60}}{4} = \frac{10 \pm 2\sqrt{15}}{4} = \frac{5 \pm \sqrt{15}}{2}$$

$$\left\{ \frac{5 - \sqrt{15}}{2}, \frac{5 + \sqrt{15}}{2} \right\}$$

$$a = 6,\ b = 1 \text{ and } c = -2$$

$$r = \frac{-1 \pm \sqrt{1^2 - 4(6)(-2)}}{2(6)}$$

$$= \frac{-1 \pm \sqrt{1 + 48}}{12} = \frac{-1 \pm \sqrt{49}}{12}$$

$$= \frac{-1 \pm 7}{12}$$

$$\frac{-1 + 7}{12} = \frac{6}{12} = \frac{1}{2},$$

$$\frac{-1 - 7}{12} = \frac{-8}{12} = -\frac{2}{3} \qquad \left\{ -\frac{2}{3}, \frac{1}{2} \right\}$$

33) 64; two irrational solutions

29) $\quad \frac{1}{6}t^2 - \frac{1}{3}t + \frac{2}{3} = 0$

$$6\left(\frac{1}{6}t^2 - \frac{1}{3}t + \frac{2}{3} \right) = 6(0)$$

$$t^2 - 2t + 4 = 0$$

$$a = 1,\ b = -2 \text{ and } c = 4$$

$$t = \frac{-(-2) \pm \sqrt{(-2)^2 - 4(1)(4)}}{2(1)}$$

$$= \frac{2 \pm \sqrt{4 - 16}}{2} = \frac{2 \pm \sqrt{-12}}{2}$$

$$= \frac{2 \pm 2i\sqrt{3}}{2} = 1 \pm i\sqrt{3}$$

$$\left\{ 1 - i\sqrt{3}, 1 + i\sqrt{3} \right\}$$

35) $4k^2 + bk + 9 = 0$

$$a = 4 \text{ and } c = 9$$

$$b^2 - 4ac = b^2 - 4(4)(9) = 0$$

$$b^2 - 144 = 0$$

$$b^2 = 144$$

$$b = \pm 12$$

31) $(6r + 1)(r - 4) = -2(12r + 1)$

$$6r^2 - 23r - 4 = -24r - 2$$

$$6r^2 + r - 2 = 0$$

37)

$$3k^2 + 4 = 7k$$

$$3k^2 - 7k + 4 = 0$$

$$(k-1)(3k-4) = 0$$

$$k - 1 = 0 \quad \text{or} \quad 3k - 4 = 0$$

$$k = 1 \qquad \text{or} \quad k = \frac{4}{3} \qquad \left\{1, \frac{4}{3}\right\}$$

39) $15 = 3 + (y+8)^2$

Let $u = y + 8$.

$$15 = 3 + u^2$$

$$u^2 - 12 = 0$$

$$u = \pm\sqrt{12} = \pm 2\sqrt{3}$$

$$y + 8 = \pm 2\sqrt{3}$$

$$y = -8 \pm 2\sqrt{3} \quad \left\{-8 - 2\sqrt{3}, -8 + 2\sqrt{3}\right\}$$

41) $\quad \dfrac{1}{3}w^2 + w = -\dfrac{5}{6}$

$$\frac{2w^2 + 6w}{6} = -\frac{5}{6}$$

$$2w^2 + 6w + 5 = 0$$

Solve using the Quadratic Equation,

$$w = \frac{-6 \pm \sqrt{(-6)^2 - 4(2)(5)}}{2(2)}$$

$$= \frac{-6 \pm \sqrt{36 - 40}}{4} = \frac{-6 \pm \sqrt{-4}}{4}$$

$$= -\frac{3}{2} \pm i\frac{1}{2} \quad \left\{-\frac{3}{2} - i\frac{1}{2}, -\frac{3}{2} + i\frac{1}{2}\right\}$$

43) $6 + p(p-10) = 2(4p - 15)$

$$6 + p^2 - 10p = 8p - 30$$

$$p^2 - 18p + 36 = 0$$

Solve using the Quadratic Equation,

$$p = \frac{-(-18) \pm \sqrt{(-18)^2 - 4(1)(36)}}{2(1)}$$

$$= \frac{18 \pm \sqrt{324 - 180}}{2} = \frac{18 \pm \sqrt{180}}{2}$$

$$= 9 \pm 3\sqrt{5} \qquad \left\{9 - 3\sqrt{5}, 9 + 3\sqrt{5}\right\}$$

45)

$$x^3 = x$$

$$x^3 - x = 0$$

$$x(x^2 - 1) = 0$$

$$x(x-1)(x+1) = 0$$

$$x = -1, 0, 1 \qquad \{-1, 0, 1\}$$

47) $(2x-1)^2 = 25$

$$2x - 1 = \pm\sqrt{25}$$

$$2x = 1 + 5 \quad \text{or} \quad 2x = 1 - 5$$

$$2x = 6 \qquad\qquad 2x = -4$$

$$x = 3 \qquad\qquad x = -2$$

49)
$$\frac{5}{k+1} = 3k-4$$

$$(k+1)\frac{5k}{k+1} = (k+1)(3k-4)$$

$$5k = 3k^2 - k - 4$$

$$0 = 3k^2 - 6k - 4$$

$$k = \frac{-(-6)\pm\sqrt{(-6)^2 - 4(3)(-4)}}{2(3)}$$

$$= \frac{6\pm\sqrt{36+48}}{6} = \frac{6\pm\sqrt{84}}{6}$$

$$= \frac{6\pm 2\sqrt{21}}{6} = \frac{3\pm\sqrt{21}}{3}$$

$$\left\{\frac{3-\sqrt{21}}{3}, \frac{3+\sqrt{21}}{3}\right\}$$

51)
$$f = \sqrt{7f-12}$$
$$f^2 = 7f-12$$
$$f^2 - 7f + 12 = 0$$
$$(f-3)(f-4) = 0$$
$$f-3 = 0 \text{ or } f-4 = 0$$
$$f = 3 \qquad f = 4 \qquad \{3,4\}$$

53)
$$n^4 - 17n^2 + 16 = 0$$
$$(n^2-16)(n^2-1) = 0$$
$$n^2 - 16 = 0 \text{ or } n^2 - 1 = 0$$
$$n^2 = 16 \qquad n^2 = 1$$
$$n = \pm 4 \qquad n = \pm 1$$
$$\{-4,-1,1,4\}$$

55) $q^{2/3} + 2q^{1/3} - 3 = 0; \quad u^2 = q^{2/3} \quad u = q^{1/3}$

$$u^2 + 2u - 3 = 0$$
$$(u+3)(u-1) = 0$$
$$u = -3, 1$$
$$q^{1/3} = -3 \qquad q^{1/3} = 1$$
$$\left(\sqrt[3]{q}\right)^3 = (-3)^3 \quad \left(\sqrt[3]{q}\right)^3 = 1^3$$
$$q = -27 \qquad q = 1$$
$$\{-27, 1\}$$

57) $2r^4 = 7r^2 - 2 \qquad$ Let $u = r^2$ and $u^2 = r^4$
$$2r^4 - 7r^2 + 2 = 0$$
$$2u^2 - 7u + 2 = 0$$
$$u = \frac{-(-7)\pm\sqrt{(-7)^2 - 4(2)(2)}}{2(2)}$$
$$= \frac{7\pm\sqrt{33}}{4}$$
$$u = \frac{7+\sqrt{33}}{4} \quad \text{or} \quad u = \frac{7-\sqrt{33}}{4}$$
$$u = r^2 \qquad\qquad u = r^2$$
$$r^2 = \frac{7+\sqrt{33}}{4} \qquad r^2 = \frac{7-\sqrt{33}}{4}$$
$$u = \pm\frac{\sqrt{7+\sqrt{33}}}{2} \quad u = \pm\frac{\sqrt{7-\sqrt{33}}}{2}$$
$$\left\{-\frac{\sqrt{7+\sqrt{33}}}{2}, \frac{\sqrt{7+\sqrt{33}}}{2},\right.$$
$$\left.-\frac{\sqrt{7-\sqrt{33}}}{2}, \frac{\sqrt{7-\sqrt{33}}}{2}\right\}$$

59) $(2k-5)^2 - 5(2k-5) - 6 = 0$

Let $u = 2k - 5$.

$u^2 - 5u - 6 = 0$

$(u-6)(u+1) = 0$

$u - 6 = 0$ or $u + 1 = 0$

$\quad u = 6 \qquad\qquad u = -1$

Solve for k using $u = 2k - 5$.

$\quad 6 = 2k - 5 \qquad\quad -1 = 2k - 5$

$11 = 2k \qquad\qquad 4 = 2k$

$\dfrac{11}{2} = k \qquad\qquad 2 = k$

$\left\{ 2, \dfrac{11}{2} \right\}$

61) $\qquad F = \dfrac{mv^2}{r}$

$\qquad Fr = mv^2$

$\qquad \dfrac{Fr}{m} = v^2$

$\qquad \pm\sqrt{\dfrac{Fr}{m}} = v$

$\qquad \pm\dfrac{\sqrt{Fr}}{\sqrt{m}} \cdot \dfrac{\sqrt{m}}{\sqrt{m}} = v$

$\qquad \pm\dfrac{\sqrt{Frm}}{m} = v$

63) $\quad r = \sqrt{\dfrac{A}{\pi}}$

$\quad r^2 = \dfrac{A}{\pi}$

$\quad \pi r^2 = A$

65) $kn^2 - ln - m = 0$

solve using the quadratic equation

$a = k; b = -l; c = -m$

$n = \dfrac{-(-l) \pm \sqrt{(-l)^2 - 4(k)(-m)}}{2k}$

67) $\qquad x = $ width of border

$\qquad 18 + 2x = $ length of case + borders

$\qquad 27 + 2x = $ width of case + borders

Area of case plus border $= 792$

$\qquad (18 + 2x)(27 + 2x) \qquad = 792$

length $= 18$ in.

width $= 18 - 4 = 14$ in.

$\qquad 486 + 90x + 4x^2 = 792$

$\qquad 4x^2 + 90x - 306 = 0$

$\qquad (4x + 102)(x - 3) = 0$

$\qquad 4x + 102 = 0$ or $\quad x - 3 = 0$

$\qquad\qquad 4x = -102 \qquad \boxed{x = 3}$

$\qquad\qquad x = -\dfrac{51}{2}$

The width of the border is 3 in.

69) $D = \dfrac{240}{P}; S = 4P - 2$

$\qquad D = S$

$\qquad \dfrac{240}{P} = 4P - 2$

$\qquad 240 = P(4P - 2)$

$\qquad 240 = 4P^2 - 2P$

$\qquad 0 = 4P^2 - 2P - 240$

$\qquad 0 = 2P^2 - P - 120$

$\qquad 0 = (2P + 15)(P - 8)$

$2P+15=0$ or $P-8=0$

$2P=-15$ $\boxed{P=\$8.00}$

$P=-\dfrac{15}{2}$

71) a) (h,k) b) $x=h$

c) If a is positive, the parabola opens upward. If a is negative, the parabola opens downward.

73) $f(x)=x^2-4$

$a=1;$ $b=0;$ $c=-4$

$x=\dfrac{-b}{2a}=\dfrac{0}{2}=0$

$f(0)=-4;$ $\qquad V(0,-4)$

axis of symmetry: $x=0$

y-intercept: $(0,-4)$

$\qquad 0=(x+2)(x-2)$

$\qquad 2,-2=x$

x-intercept: $(-2,0)$ and $(2,0)$

Domain: $(-\infty,\infty)$ Range:$[-4,\infty)$

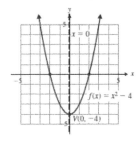

75) $f(x)=(x+2)^2-1$

$V(-2,-1)$

axis of symmetry: $x=-2$

$f(0)=(0+2)^2=4$

$y-$intercept: $(0,4)$

$\qquad 0=(x+2)^2-1$

$\qquad 0=x^2+4x+4-1$

$\qquad 0=x^2+4x+3$

$\qquad 0=(x+3)(x+1)$

$-1,3=x$

$x-$intercepts: $(-1,0)$ and $(3,0)$

Domain: $(-\infty,\infty)$ Range: $[-1,\infty)$

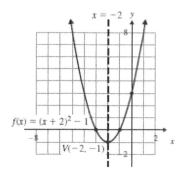

77) $y=(x-4)^2+2$

$V(4,2)$

axis of symmetry: $x=4$

$\qquad y=(0-4)^2+2=16+2=18$

$y-$intercept: $(0,18)$

$x-$intercepts: none

Domain: $(-\infty,\infty)$ Range: $[2,\infty)$

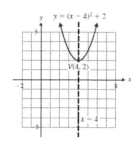

79) $g(x) = (x-6)^2$

81) $f(x) = x^2 - 2x + 3$

$$f(x) = (x^2 - 2x + 1) + 3 - 1$$
$$f(x) = (x-1)^2 + 2$$
$$(x-1)^2 + 2 = 0$$
$$(x-1)^2 = -2$$
$$x - 1 = \pm\sqrt{2}i$$
$$x = 1 \pm \sqrt{2}i$$
$$1 - \sqrt{2}i, \ 1 + \sqrt{2}i$$
$$y-\text{int}: (0,3)$$

Domain: $(-\infty, \infty)$ Range: $[2, \infty)$

There are no x-intercepts.

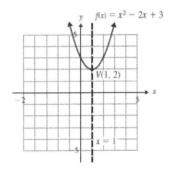

83) $y = \dfrac{1}{2}x^2 - 4x + 9$

$$y = \frac{1}{2}(x^2 - 8x) + 9$$
$$y = \frac{1}{2}(x^2 - 8x + 16) + 9 - 8$$
$$y = \frac{1}{2}(x-4)^2 + 1$$

x-int: $0 = \dfrac{1}{2}(x-4)^2 + 1$

$$-1 = \frac{1}{2}(x-4)^2$$
$$-2 = (x-4)^2$$
$$\pm i\sqrt{2} = x - 4$$
$$4 \pm i\sqrt{2} = x \qquad\qquad \text{none}$$

y-int: $y = \dfrac{1}{2}(0)^2 - 4(0) + 9 = 9$

y-int: $(0,9)$

Domain: $(-\infty, \infty)$ Range: $[1, \infty)$

85) $y = -x^2 - 6x - 10$

$$a = -1; \quad b = -6; \quad c = -10$$
$$x = \frac{-b}{2a} = \frac{-(-6)}{-2} = -3$$
$$y = -(-3)^2 - 6(-3) - 10 = -1$$
$$V = (-3, -1)$$

x int : none

y int : $0, -10$

Domain: $(-\infty, \infty)$ Range: $(-\infty, -1]$

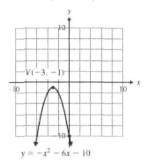

$y = -x^2 - 6x - 10$

87) a) The max height occurs at the
t-coordinate of the vertex.

$$t = -\frac{32}{2(-16)} = \frac{32}{32} = 1$$

1 sec

b) Since the object reaches max
height when $t = 1$, find $h(1)$.

$$h(1) = -16(1)^2 + 32(1) + 240$$
$$= -16 + 32 + 240 = 256$$

256 ft

c) The object hits ground when $h(t) = 0$.

$$0 = -16t^2 + 32t + 240$$
$$0 = t^2 - 2t - 15$$
$$0 = (t-5)(t+3)$$
$$t - 5 = 0 \text{ or } t + 3 = 0$$
$$\boxed{t = 5} \qquad t = -3$$

The object will hit the ground after
5 sec.

89) $V(11,3)$; $y = 3$; x-int : $(2,0)$;

y-ints : $\left(0, 3 - \sqrt{11}\right), \left(0, 3 - \sqrt{11}\right)$;

domain : $(-\infty, 11]$; range : $(-\infty, \infty)$

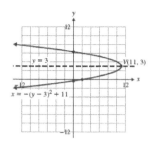

$y = 3$

$V(11, 3)$

$x = -(y-3)^2 + 11$

91) $x = (y+4)^2 - 9$; x-int : $(7,0)$;

y-ints : $\left(0, -1 - \sqrt{5}\right), \left(0, -1 + \sqrt{5}\right)$;

domain : $[-9, \infty)$; range : $(-\infty, \infty)$

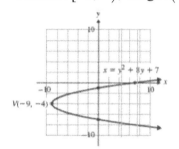

$x = y^2 + 8y + 7$

$V(-9, -4)$

93) $V(2, -3)$; x-int : $\left(-\frac{5}{2}, 0\right)$;

y-ints : $(0, -5), (0, -1)$;

domain : $(-\infty, 2]$; range : $(-\infty, \infty)$

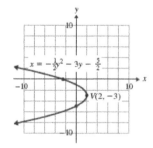

$x = -\frac{1}{2}y^2 - 3y - \frac{5}{2}$

$V(2, -3)$

95) $a^2 + 2a - 3 < 0$

$(a+3)(a-1) < 0$

$(a+3)(a-1) = 0$

$a + 3 = 0 \text{ or } a - 1 = 0$

$a = -3 \qquad a = 1$

Interval A: $(-\infty, -3)$ Positive

Interval B: $(-3, 1)$ Negative

Interval C: $(1, \infty)$ Positive

$(-3, 1)$

$$\left(-\infty, -\frac{2}{5}\right) \cup \left(\frac{1}{3}, 4\right)$$

101)

$$\frac{t+7}{2t-3} > 0$$

Set the numerator & denominator equal to zero and solve for t.

$$t + 7 = 0 \qquad 2t - 3 = 0$$

$$t = -7 \qquad t = \frac{3}{2}$$

Interval A: $(-\infty, -7)$ Positive

Interval B: $\left(-7, \frac{3}{2}\right)$ Negative

Interval C: $\left(\frac{3}{2}, \infty\right)$ Positive

$$(-\infty, -7) \cup \left(\frac{3}{2}, \infty\right)$$

97)

$$64v^2 \geq 25$$

$$64v^2 - 25 \geq 0$$

$$(8v + 5)(8v - 5) \geq 0$$

$$(8v + 5)(8v - 5) = 0; \quad v = \pm\frac{5}{8}$$

Interval A: $\left(-\infty, -\frac{5}{8}\right)$ Positive

Interval B: $\left(-\frac{5}{8}, \frac{5}{8}\right)$ Negative

Interval C: $\left(\frac{5}{8}, \infty\right)$ Positive

$$\left(-\infty, -\frac{5}{8}\right] \cup \left[\frac{5}{8}, \infty\right)$$

103)

$$\frac{z}{z-2} \leq 3$$

$$\frac{z}{z-2} - 3 \leq 0$$

$$\frac{z}{z-2} - \frac{3(z-2)}{z-2} \leq 0$$

$$\frac{z - 3(z-2)}{z-2} \leq 0$$

$$\frac{z - 3z + 6}{z-2} \leq 0$$

$$\frac{-2z + 6}{z-2} \leq 0$$

99) $(5c + 2)(c - 4)(3c - 1) < 0$

$$(5c + 2)(c - 4)(3c - 1) = 0$$

$$5c + 2 = 0 \quad \text{or} \quad c - 4 = 0 \quad \text{or} \quad 3c - 1 = 0$$

$$c = -\frac{2}{5} \qquad c = 4 \qquad c = \frac{1}{3}$$

Interval A: $\left(-\infty, -\frac{2}{5}\right)$ Negative

Interval B: $\left(-\frac{2}{5}, \frac{1}{3}\right)$ Positive

Interval C: $\left(\frac{1}{3}, 4\right)$ Negative

Interval D: $(4, \infty)$ Positive

Set the numerator and denominator equal to zero and solve for z.

$-2z+6=0 \qquad z-2=0$

$\quad -2z=-6 \qquad z=2$

$\qquad z=3$

Interval A: $(-\infty, 2)$ Negative

Interval B: $(2, 3)$ Positive

Interval C: $(3, \infty)$ Negative

$z \neq 2$ because it makes the

denominator equal to zero.

$(-\infty, 2) \cup [3, \infty)$

105) $\dfrac{r^2+4}{r-7} \geq 0$

Numerator will always be positive.

Set the denominator equal to zero

and solve for r.

$r-7=0$

$\quad r=7$

Interval A: $(-\infty, 7)$ Negative

Interval B: $(7, \infty)$ Positive

$r \neq 7$ because it makes the

denominator equal to zero.

$(7, \infty)$

Chapter 10 Test

1) $b^2+4b-7=0$

$\qquad b^2+4b=7$

$\qquad b^2+4b+4=7+4$

$\qquad (b+2)^2=11$

$\qquad b+2=\pm\sqrt{11}$

$\qquad b=-2\pm\sqrt{11}$

$\left\{-2-\sqrt{11}, -2+\sqrt{11}\right\}$

3) $(c+5)^2+8=2$

$\qquad (c+5)^2=-6$

$\qquad c+5=\pm\sqrt{-6}$

$\qquad c+5=\pm i\sqrt{6}$

$\qquad\qquad c=-5\pm i\sqrt{6}$

$\left\{-5-i\sqrt{6}, -5+i\sqrt{6}\right\}$

5) $(4n+1)^2+9(4n+1)+18=0$

Let $u=4n+1$

$\qquad u^2+9u+18=0$

$\qquad (u+6)(u+3)=0$

$u+6=0$ or $u+3=0$

$\quad u=-6 \qquad\quad u=-3$

Solve for n using $u=4n+1$.

$-6=4n+1 \qquad -3=4n+1$

$-7=4n \qquad\qquad -4=4n$

$-\dfrac{7}{4}=n \qquad\qquad -1=n$

$\left\{-\dfrac{7}{4}, -1\right\}$

7) $\quad p^4 + p^2 - 72 = 0$

$\left(p^2 + 9\right)\left(p^2 - 8\right) = 0$

$p^2 + 9 = 0 \;\text{ or }\; p^2 - 8 = 0$

$\quad p^2 = -9 \qquad p^2 = 8$

$\quad p = \pm 3i \qquad p = \pm 2\sqrt{2}$

$\quad \left\{-2\sqrt{2}, 2\sqrt{2}, -3i, 3i\right\}$

9) $\;5z^2 - 6z - 1 = 0$

$\quad a = 5,\; b = -6 \text{ and } c = -1$

$b^2 - 4ac = (-6)^2 - 4(5)(-1)$

$\qquad = 36 + 20 = 56$

two irrational solutions

11) $P(x) = 5x^2 \quad Q(x) = 2x$

$\qquad 5x^2 = 2x$

$\quad 5x^2 - 2x = 0$

$\;x(5x - 2) = 0$

$x = 0 \text{ or } 5x - 2 = 0$

$\qquad\qquad 5x = 2$

$\qquad\qquad x = \dfrac{2}{5}$

13) $\qquad x = \text{width of sheet metal}$

$\quad x + 6 = \text{length of sheet metal}$

length of box $= x + 6 - 3 - 3$

$\qquad = x$

width of box $= x - 3 - 3$

$\qquad = x - 6$

height of box $= 3$

Volume $= (\text{length})(\text{width})(\text{height})$

$\quad 273 = x(x - 6)(3)$

$\quad 273 = 3x^2 - 18x$

$\qquad 0 = 3x^2 - 18x - 273$

$\qquad 0 = x^2 - 6x - 91$

$\qquad 0 = (x + 7)(x - 13)$

$\quad x + 7 = 0 \;\text{ or }\; x - 13 = 0$

$\qquad x = -7 \qquad \boxed{x = 13}$

width $= 13$ in.

length $= 13 + 6 = 19$ in.

15) $\qquad rt^2 - st = 6$

$\quad rt^2 - st - 6 = 0$

$t = \dfrac{-(-s) \pm \sqrt{(-s)^2 - 4(r)(-6)}}{2(r)}$

$\;\; = \dfrac{s \pm \sqrt{s^2 + 24r}}{2r}$

17) $g(x) = (x + 3)^2$

19) $V(-3, 0)$;

axis of symmetry: $y = 0$

$x = -0^2 - 3$

$x = 3$

$x-$intercept: $(-3, 0)$

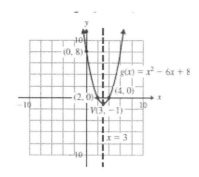

y-int: $0 = -y^2 - 3$

$3 = -y^2$

$-3 = y^2$

$\pm i\sqrt{3} = y$

$y-$intercepts: none

Domain: $[-3, \infty)$ Range: $(-\infty, \infty)$

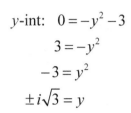

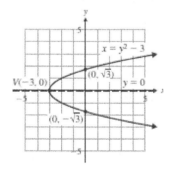

23) $y^2 + 4y - 45 \geq 0$

$(y+9)(y-5) \geq 0$

$(y+9)(y-5) = 0$

$y+9 = 0$ or $y-5 = 0$

$y = -9$ $y = 5$

Interval A: $(-\infty, -9)$ Positive

Interval B: $(-9, 5)$ Negative

Interval C: $(5, \infty)$ Positive

$(-\infty, -9] \cup [5, \infty)$

21) $g(x) = x^2 - 6x + 8$

$g(x) = (x^2 - 6x + 9) + 8 - 9$

$g(x) = (x-3)^2 - 1$

$V(3, -1)$

axis of symmetry: $x = 3$

$g(0) = 8$

$y-$int: $(0, 8)$

$0 = x^2 - 6x + 8$

$0 = (x-4)(x-2)$

$2, 4 = x$

$x-$int: $(2, 0)$ or $(4, 0)$

Domain: $(-\infty, \infty)$ Range: $[-1, \infty)$

25) $C(x) = \dfrac{5x + 80,000}{x}$

$15 = \dfrac{5x + 80,000}{x}$

$15x = 5x + 80,000$

$10x = 80,000$

$x = 8,000$

$8,000$ or more purses

Cumulative Review: Chapters 1-10

1) $\dfrac{\dfrac{12}{35}}{\dfrac{24}{49}} = \dfrac{245\left(\dfrac{12}{35}\right)}{245\left(\dfrac{24}{49}\right)} = \dfrac{84}{120} = \dfrac{7}{10}$

$$2x + 3y = 3.85$$
$$x + 2y = 2.30$$
$$\overline{2x + 3y = 3.85}$$
$$-2(\ x + 2y = 2.30)$$
$$\overline{2\!\!\!/x + 3y = 3.85}$$
$$\overline{-2\!\!\!/x - 4y = -4.60}$$
$$-y = -0.75$$
$$y = 0.75$$

3) Let x be the number of digital cameras

$$108 - x = 0.2x$$
$$108 = 1.2x$$
$$90 = x$$

$$x + 2y = 2.30$$
$$x + (2)(.75) = 2.30$$
$$x + 1.50 = 2.30$$
$$x = 0.80$$

chips: $0.80   soda: $0.75

5) a) Domain: $\{0, 3, 4\}$

   b) Range: $\{-1, 0, 1, 2\}$

   c) no

9) $3(r-5)^2 = 3(r^2 - 10r + 25) = 3r^2 - 30r + 75$

11) $a^3 + 125 = (a+5)(a^2 - 5a + 25)$

7) $x = $ chips

   $y = $ soda

13) $\dfrac{2 + \dfrac{6}{c}}{\dfrac{2}{c^2} - \dfrac{8}{c}} = \dfrac{c^2\left(2 + \dfrac{6}{c}\right)}{c^2\left(\dfrac{2}{c^2} - \dfrac{8}{c}\right)} = \dfrac{2c^2 + 6c}{2 - 8c}$

$$= \dfrac{2c(c+3)}{2(1-4c)} = \dfrac{c(c+3)}{(1-4c)}$$

15) $\sqrt{75} = \sqrt{25 \cdot 3} = 5\sqrt{3}$

17) $\sqrt{63x^7 y^4} = \sqrt{9 \cdot 7x^6 xy^4} = 3x^2 y^2 \sqrt{7x}$

19) $\dfrac{5}{2+\sqrt{3}} \cdot \dfrac{2-\sqrt{3}}{2-\sqrt{3}} = \dfrac{10-5\sqrt{3}}{4-3} = 10-5\sqrt{3}$

21) $1 - \dfrac{1}{3h-2} = \dfrac{20}{(3h-2)^2}$

$(3h-2)^2 \left[1 - \dfrac{1}{3h-2} = \dfrac{20}{(3h-2)^2} \right]$

$(3h-2)^2 - (3h-2) = 20$

$9h^2 - 12h + 4 - 3h + 2 = 20$

$9h^2 - 15h + 6 - 20 = 0$

$9h^2 - 15h - 14 = 0$

$(3h+2)(3h-7) = 0$

$$h = -\dfrac{2}{3}, \dfrac{7}{3}$$

$\left\{ -\dfrac{2}{3}, \dfrac{7}{3} \right\}$

23) $\quad r = \sqrt{\dfrac{V}{\pi h}}$

$r^2 = \dfrac{V}{\pi h}$

$r^2 \pi h = V$

25) $\qquad\qquad 25p^2 \le 144$

$25p^2 - 144 \le 0$

$(5p-12)(5p+12) \le 0$

$$p = -\dfrac{12}{5}, \dfrac{12}{5}$$

Interval A: $\left(-\infty, -\dfrac{12}{5} \right)$ Positive

Interval B: $\left(-\dfrac{12}{5}, \dfrac{12}{5} \right)$ Negative

Interval C: $\left(\dfrac{12}{5}, \infty \right)$ Positive

$p = \pm \dfrac{12}{5}$ satisfies the inequality.

$\left[-\dfrac{12}{5}, \dfrac{12}{5} \right]$

Chapter 11: Exponential and Logarithmic Functions

Section 11.1

1) $(f \circ g)(x) = f(g(x))$ so substitute the function $g(x)$ into the function $f(x)$ and simplify.

3) a) $g(x) = 2x - 9$
$g(4) = 2(4) - 9 = -1$
   b) $(f \circ g)(4) = f(g(4))$
$= f(-1)$
$= 3(-1) + 1$
$= -2$

   c) $(f \circ g)(x) = f(g(x))$
$= f(2x - 9)$
$= 3(2x - 9) + 1$
$= 6x - 26$

   d) $(f \circ g)(4) = f(g(4))$
$= 6(4) - 26$
$= -2$

5)
   a) $g(x) = x + 3$
$g(4) = 4 + 3 = 7$

   b) $(f \circ g)(4) = f(g(4))$
$= f(7)$
$= (7)^2 - 5 = 44$

   c) $(f \circ g)(x) = f(g(x))$
$= f(x + 3)$
$= (x + 3)^2 - 5$
$= x^2 + 6x + 9 - 5$
$= x^2 + 6x + 4$

d) $(f \circ g)(4) = f(g(4))$
$= (4)^2 + 6(4) + 4$
$= 16 + 24 + 4$
$= 44$

7) a) $(f \circ g)(x) = f(g(x))$
$= 5(x + 7) - 4$
$= 5x + 35 - 4$
$= 5x + 31$

   b) $(g \circ f)(x) = g(f(x))$
$= (5x - 4) + 7$
$= 5x + 3$

   c) $(f \circ g)(3) = 5(3) + 31$
$= 15 + 31$
$= 46$

9) a) $(k \circ h)(x) = k(h(x))$
$= 3(-2x + 9) - 1$
$= -6x + 27 - 1$
$= -6x + 26$

   b) $(h \circ k)(x) = h(k(x))$
$= -2(3x - 1) + 9$
$= -6x + 2 + 9$
$= -6x + 11$

   c) $(k \circ h)(-1) = -6(-1) + 26$
$= 6 + 26 = 32$

Section 11.1: Composite and Inverse Functions

11) a) $(h \circ g)(x) = h(g(x))$
$$= (x^2 - 6x + 11) - 4$$
$$= x^2 - 6x + 7$$

b) $(g \circ h)(x)$
$$= g(h(x))$$
$$= (x-4)^2 - 6(x-4) + 11$$
$$= x^2 - 8x + 16 - 6x + 24 + 11$$
$$= x^2 - 14x + 51$$

c) $(g \circ h)(4) = (4)^2 - 14(4) + 51$
$$= 16 - 56 + 51 = 11$$

13) a) $(n \circ m)(x)$
$$= n(m(x))$$
$$= -(x+8)^2 + 3(x+8) - 8$$
$$= -(x^2 + 16x + 64) + 3x + 24 - 8$$
$$= -x^2 - 16x - 64 + 3x + 16$$
$$= -x^2 - 13x - 48$$

b) $(m \circ n)(x) = m(n(x))$
$$= (-x^2 + 3x - 8) + 8$$
$$= -x^2 + 3x$$

c) $(m \circ n)(0) = -(0)^2 + 3(0) = 0$

15) a) $(f \circ g)(x) = f(g(x))$
$$= \sqrt{(x^2 - 6) + 10}$$
$$= \sqrt{x^2 + 4}$$

b) $(g \circ f)(x) = g(f(x))$
$$= (\sqrt{x+10})^2 - 6$$
$$= x + 10 - 6$$
$$= x + 4$$

c) $(f \circ g)(-3) = f(g(-3))$
$$= \sqrt{(-3)^2 + 4}$$
$$= \sqrt{13}$$

17) a) $(P \circ Q)(t) = P(Q(t))$
$$= \frac{1}{t^2 + 8}$$

b) $(Q \circ P)(t) = Q(P(t))$
$$= \left(\frac{1}{t+8}\right)^2$$
$$= \frac{1}{(t+8)^2}$$

c) $(Q \circ P)(-5) = Q(P(-5))$
$$= \frac{1}{(-5+8)^2}$$
$$= \frac{1}{9}$$

19) $f(x) = \sqrt{x} \quad g(x) = x^2 + 13$

   $(f \circ g)(x) = f(g(x))$

   $\qquad = f(x^2 + 13)$

   $\qquad = \sqrt{x^2 + 13}$

   $h(x) = \sqrt{x^2 + 13}$ answers may vary

21) $f(x) = \dfrac{1}{x} \quad g(x) = 6x + 5$

   $(f \circ g)(x) = f(g(x))$

   $\qquad = f(6x + 5)$

   $\qquad = \dfrac{1}{6x + 5}$

   $h(x) = \dfrac{1}{6x + 5}$ answers may vary

23) a) $r(5) = 4(5) = 20$

   The radius of the spill 5 min.
   after the ship started leaking
   was 20 ft.

   b) $A(20) = \pi(20)^2 = 400\pi$

   The area of the oil slick is
   400π ft^2 when its radius is 20 ft.

   c) $A(r(t)) = \pi(4t)^2 = 16\pi t^2$

   This is the area of the oil slick in
   terms of t, the number of minutes
   after the leak began.

   d) $A(r(5)) = 16\pi(5)^2$

   $\qquad = 16\pi(25) = 400\pi$

   The area of the oil slick 5
   minutes after the ship began
   leaking was 400π ft^2.

25) no

27) yes; $h^{-1} = \{(-16, -5), (-4, -1), (8, 3)\}$

29) yes

31) No; only one-to-one functions
   have inverses.

33) False; it is read "f inverse of x."

35) True

37) False; they are symmetric with
   respect to $y = x$.

39) a) yes

   b)

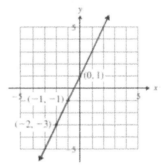

41) no

43) a) yes

   b)

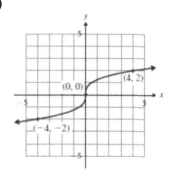

45) $f(x) = 2x - 10$

$$y = 2x - 10 \qquad \underline{\text{Replace } f(x) \text{ with } y.}$$

$$x = 2y - 10 \qquad \underline{\text{Interchange } x \text{ and } y.}$$

   Solve for y.

$$x + 10 = 2y \qquad \underline{\text{Add 10.}}$$

$$\frac{1}{2}x + 5 = y \qquad \text{Divide by 2 and simplify.}$$

$$f^{-1}(x) = \frac{1}{2}x + 5 \quad \underline{\text{Replace } y \text{ with } f^{-1}(x).}$$

47) $\quad y = x - 6$

$$x = y - 6$$

$$x + 6 = y$$

$$g^{-1}(x) = x + 6$$

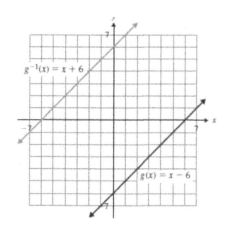

49) $\qquad y = -2x + 5$

$$x = -2y + 5$$

$$x - 5 = -2y$$

$$-\frac{1}{2}x + \frac{5}{2} = y$$

$$f^{-1}(x) = -\frac{1}{2}x + \frac{5}{2}$$

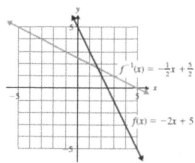

51) $\qquad y = \frac{1}{2}x$

$$x = \frac{1}{2}y$$

$$2x = y$$

$$g^{-1}(x) = 2x$$

372

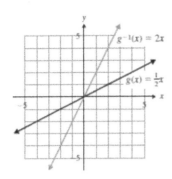

57)
$$y = -\frac{3}{2}x + 4$$
$$x = -\frac{3}{2}y + 4$$
$$x - 4 = -\frac{3}{2}y$$
$$-\frac{2}{3}x + \frac{8}{3} = y$$
$$h^{-1}(x) = -\frac{2}{3}x + \frac{8}{3}$$

53)
$$y = x^3$$
$$x = y^3$$
$$\sqrt[3]{x} = y$$
$$f^{-1}(x) = \sqrt[3]{x}$$

59)
$$y = \sqrt[3]{x+2}$$
$$x = \sqrt[3]{y+2}$$
$$x^3 = y + 2$$
$$x^3 - 2 = y$$
$$g^{-1}(x) = x^3 - 2$$

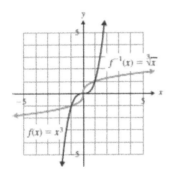

61)
$$y = \sqrt{x}$$
$$x = \sqrt{y}$$
$$x^2 = y$$
$$f^{-1}(x) = x^2, \ x \geq 0$$

55)
$$y = 2x - 6$$
$$x = 2y - 6$$
$$x + 6 = 2y$$
$$\frac{1}{2}x + 3 = y$$
$$f^{-1}(x) = \frac{1}{2}x + 3$$

Section 11.2

1) Choose values for the variable that will give positive numbers, negative numbers, and zero in the exponent.

3) Domain: $(-\infty, \infty)$; Range: $(0, \infty)$

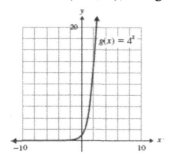

5) Domain: $(-\infty, \infty)$; Range: $(0, \infty)$

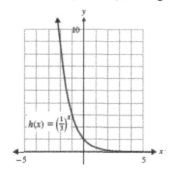

7) $(-\infty, \infty)$

9) Domain: $(-\infty, \infty)$; Range: $(0, \infty)$

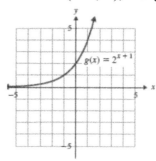

11) Domain: $(-\infty, \infty)$; Range: $(0, \infty)$

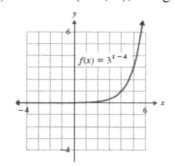

13) Domain: $(-\infty, \infty)$; Range: $(0, \infty)$

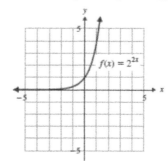

15) Domain: $(-\infty, \infty)$; Range: $(1, \infty)$

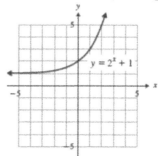

17) Domain: $(-\infty,\infty)$; Range: $(-2,\infty)$

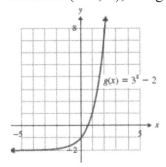

19) Domain: $(-\infty,\infty)$; Range: $(-\infty,0)$

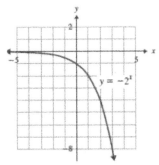

21) $g(x) = 2^x$ would grow faster because for values of $x > 2$, $2^x > 2x$.

23) Shift the graph of $f(x)$ down 2 units.

25) 2.7183

27) B

29) D

31) Domain: $(-\infty,\infty)$; Range: $(-2,\infty)$

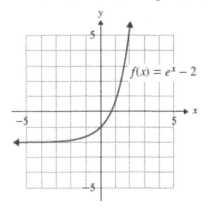

33) Domain: $(-\infty,\infty)$; Range: $(0,\infty)$

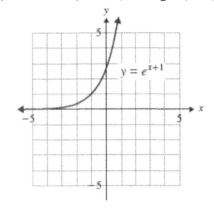

35) Domain: $(-\infty,\infty)$; Range: $(0,\infty)$

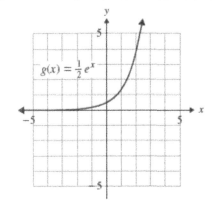

Section 11.2: Exponential Functions

37) Domain: $(-\infty, \infty)$; Range: $(-\infty, 0)$

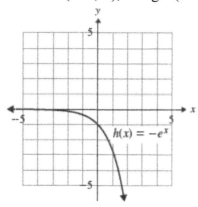

$h(x) = -e^x$

39) They are symmetric with respect to
the x-axis.

41) $6^{3n} = 36^{n-4}$

$$6^{3n} = \left(6^2\right)^{n-4}$$ Express each side with the same base.

$$6^{3n} = 6^{2(n-4)}$$ Power rule for exponents

$$6^{3n} = 6^{2n-8}$$ Distribute.

$$3n = 2n - 8$$ Set the exponents equal.

$$n = -8$$ Solve for n.

The solution set is $\{-8\}$.

43) $9^x = 81$

$9^x = 9^2$

$x = 2$ $\{2\}$

45) $4^{3a} = 64$

$4^{3a} = 4^3$

$3a = 3$

$a = 1$ $\{1\}$

47) $16^{m-2} = 2^{3m}$

$\left(2^4\right)^{m-2} = 2^{3m}$

$2^{4m-8} = 2^{3m}$

$4m - 8 = 3m$

$m = 8$ $\{8\}$

49) $7^{2k-6} = 49^{3k+1}$

$7^{2k-6} = \left(7^2\right)^{3k+1}$

$7^{2k-6} = 7^{6k+2}$

$2k - 6 = 6k + 2$

$-8 = 4k$

$-2 = k \qquad \{-2\}$

51) $6^x = \dfrac{1}{36}$

$6^x = \left(\dfrac{1}{6}\right)^2$

$6^x = 6^{-2}$

$x = -2 \qquad \{-2\}$

53) $9^r = \dfrac{1}{27}$

$\left(3^2\right)^r = \left(\dfrac{1}{3}\right)^3$

$3^{2r} = 3^{-3}$

$2r = -3$

$r = -\dfrac{3}{2} \qquad \left\{-\dfrac{3}{2}\right\}$

55) $\left(\dfrac{5}{6}\right)^{3x+7} = \left(\dfrac{36}{25}\right)^{2x}$

$\left(\dfrac{5}{6}\right)^{3x+7} = \left[\left(\dfrac{6}{5}\right)^2\right]^{2x}$

$\left(\dfrac{5}{6}\right)^{3x+7} = \left[\left(\dfrac{5}{6}\right)^{-2}\right]^{2x}$

$\left(\dfrac{5}{6}\right)^{3x+7} = \left(\dfrac{5}{6}\right)^{-4x}$

$3x + 7 = -4x$

$7 = -7x$

$-1 = x \qquad \{-1\}$

57) a) When the SUV was purchased $t = 0$. Find $V(0)$.

$V(0) = 32,700(0.812)^0$

$V(0) = 32,700$

$\$32,700$

b) Find $V(3)$.

$V(3) = 32,700(0.812)^3$

$V(3) \approx 17,507.17$

$\$17,507.17$

59) a) The value of the house in 1995 was $V(0)$.

$V(0) = 185,200(1.03)^0$

$V(0) = 185,200$

$\$185,200$

Section 11.3: Logarithmic Functions

b) In 2002, $t = 2002 - 1995 = 70$. Find $V(7)$.

$$V(7) = 185,200(1.03)^7$$

$$V(7) \approx 227,772.64$$

$227,772.64

61) $c = 2000, \ t = 18, \ r = 0.09$

$$A = 2000 \left[\frac{(1+0.09)^{18} - 1}{0.09} \right](1+0.09)$$

$$= 2000 \left[\frac{(1.09)^{18} - 1}{0.09} \right](1.09)$$

$$\approx \$90,036.92$$

63) $A(6) = 1000e^{-0.5332(6)}$

$$\approx 40.8 \text{ mg}$$

Section 11.3

1) a must be a positive real number that is not equal to 1.

3) 10

5) $7^2 = 49$

7) $9^{-2} = \frac{1}{81}$

9) $25^{1/2} = 5$

11) $13^1 = 13$

13) $\log_9 81 = 2$

15) $\log_3 \frac{1}{81} = -4$

17) $\log_{10} 1 = 0$

19) $\log_{169} 13 = \frac{1}{2}$

21) $\sqrt[3]{64} = 4$

$$64^{1/3} = 4$$

$$\log_{64} 4 = \frac{1}{3}$$

23) Write the equation in exponential form, then solve for the variable.

25) $\log_2 x = 6$

$2^6 = x$ Rewrite in exponential form.

$64 = x$ Solve for x.

The solution set is $\{64\}$.

27) $\log_{11} x = 2$

$11^2 = x$

$121 = x$ $\{121\}$

29) $\log p = 5$

$10^5 = p$

$100,000 = p$ $\{100,000\}$

31) $\log_m 49 = 2$

$\quad m^2 = 49$

$\quad m = \pm 7$

$\quad$ the base must be positive $\qquad \{7\}$

33) $\log_6 h = -2$

$\quad 6^{-2} = h$

$\quad \dfrac{1}{36} = h \qquad\qquad \left\{\dfrac{1}{36}\right\}$

35) $\log_2 (a+2) = 4$

$\quad 2^4 = a+2$

$\quad 16 = a+2$

$\quad 14 = a \qquad\qquad \{14\}$

37) $\log_{125} \sqrt{5} = c$

$\quad 125^c = \sqrt{5}$

$\quad \left(5^3\right)^c = 5^{1/2}$

$\quad 5^{3c} = 5^{1/2}$

$\quad 3c = \dfrac{1}{2}$

$\quad c = \dfrac{1}{6} \qquad\qquad \left\{\dfrac{1}{6}\right\}$

39) $\log_8 x = \dfrac{2}{3}$

$\quad 8^{2/3} = x$

$\quad \left(\sqrt[3]{8}\right)^2 = x$

$\quad 2^2 = x$

$\quad 4 = x \qquad\qquad \{4\}$

41) $\log_{(3m-1)} 25 = 2$

$\quad (3m-1)^2 = 25$

$\quad 3m-1 = \pm 5$

$\quad 3m-1 = 5 \;\text{ or }\; 3m-1 = -5$

$\quad 3m = 6 \qquad\qquad 3m = -4$

$\quad m = 2 \qquad\qquad m = -\dfrac{4}{3}$

$\quad$ the base must be positive $\qquad \{2\}$

43) $\log_2 32 = 5$ since $2^{\boxed{5}} = 32$

45) $\log 100 = 2$ since $10^{\boxed{2}} = 100$

47) Let $\log_{49} 7 = x$.

$\quad 49^x = 7$

$\quad \left(7^2\right)^x = 7^1$

$\quad 2x = 1$

$\quad x = \dfrac{1}{2}$

$\quad \log_{49} 7 = \dfrac{1}{2}$

49) Let $\log_8 \dfrac{1}{8} = x$.

$\quad 8^x = \dfrac{1}{8}$

$\quad 8^x = 8^{-1}$

$\quad x = -1$

$\quad \log_8 \dfrac{1}{8} = -1$

51) $\log_5 5 = 1$

53) Let $\log_{1/4} 16 = x.$

$$\left(\frac{1}{4}\right)^x = 16$$

$$\left(4^{-1}\right)^x = 4^2$$

$$-x = 2$$

$$x = -2$$

$$\log_{1/4} 16 = -2$$

55) Replace $f(x)$ with y, write $y = \log_a x$ in exponential form, make a table of values, then plot the points and draw the curve.

57)

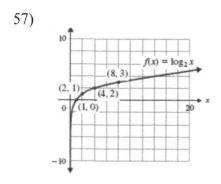

59)

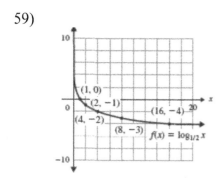

61) $y = 3^x$

$x = 3^y$

$y = \log_3 x$

$f^{-1}(x) = \log_3 x$

63) $y = \log_2 x$

$x = \log_2 y$

$y = 2^x$

$f^{-1}(x) = 2^x$

65) a) Find $L(0)$.

$$L(0) = 1800 + 68\log_3(0+3)$$

$$= 1800 + 68\log_3 3$$

$$= 1800 + 68(1) = 1868$$

b) In 2004, $t = 2004 - 1980 = 240$.

Find $L(24)$.

$$L(24) = 1800 + 68\log_3(24+3)$$

$$= 1800 + 68\log_3 27$$

$$= 1800 + 68(3)$$

$$= 1800 + 204 = 2004$$

c) Let $L(t) = 2072$ and solve for t.

$$2072 = 1800 + 68\log_3(t+3)$$

$$272 = 68\log_3(t+3)$$

$$4 = \log_3(t+3)$$

$$3^4 = t+3$$

$$81 = t+3$$

$$78 = t \qquad \text{78 years after 1980 is}$$

the year 2058.

67) a) Find $S(1)$.

$S(1) = 14\log_3\left[2(1)+1\right]$

$= 14\log_3 3$

$= 14(1) = 14 \qquad 14{,}000$

b) Find $S(4)$.

$S(4) = 14\log_3\left[2(4)+1\right]$

$= 14\log_3 9$

$= 14(2) = 28 \qquad 28{,}000$

c) Find $S(13)$ and compare this value to the actual value.

$S(13) = 14\log_3\left[2(13)+1\right]$

$= 14\log_3 27$

$= 14(3) = 42 \qquad 42{,}000$

The actual number sold was 1000 more boxes than what was predicted by the formula.

Section 11.4

1) true

3) false

5) false

7) true

9)

$\log_5 25y = \log_5 25 + \log_5 y \quad$ Product rule

$\qquad = 2 + \underline{\log_5 y} \qquad$ Evaluate $\log_5 25$

11) $\log_8(3\cdot10) = \log_8 3 + \log_8 10$

13) $\log_7 5d = \log_7 5 + \log_7 d$

15) $\log_9 \dfrac{4}{7} = \log_9 4 - \log_9 7$

17) $\log_5 2^3 = 3\log_5 2$

19) $\log p^8 = 8\log p$

21) $\log_3 \sqrt{7} = \log_3 7^{1/2} = \dfrac{1}{2}\log_3 7$

23) $\log_5 25t = \log_5 25 + \log_5 t$

$\qquad = 2 + \log_5 t$

25) $\log_2 \dfrac{8}{k} = \log_2 8 - \log_2 k = 3 - \log_2 k$

27) $\log_7 49^3 = 3\log_7 49 = 3(2) = 6$

29) $\log 1000b = \log 1000 + \log b$

$\qquad = 3 + \log b$

31) $\log_2 32^7 = 7\log_2 32 = 7(5) = 35$

33) $\log_5 \sqrt{5} = \log_5 5^{1/2} = \dfrac{1}{2}\log_5 5 = \dfrac{1}{2}$

35) $\quad \log \sqrt[3]{100} = \log 100^{1/3} = \dfrac{1}{3}\log 100$

$\qquad\qquad = \dfrac{1}{3}(2) = \dfrac{2}{3}$

37) $\log_6 w^4 z^3 = \log_6 w^4 + \log_6 z^3$

$\qquad\qquad = 4\log_6 w + 3\log_6 z$

39) $\log_7 \dfrac{a^2}{b^5} = \log_7 a^2 - \log_7 b^5$

$\qquad\qquad = 2\log_7 a - 5\log_7 b$

41) $\log \dfrac{\sqrt[5]{11}}{y^2} = \log \sqrt[5]{11} - \log y^2$

$\qquad\qquad = \log 11^{1/5} - 2\log y$

$\qquad\qquad = \dfrac{1}{5}\log 11 - 2\log y$

43) $\log_2 \dfrac{4\sqrt{n}}{m^3}$

$\quad = \log_2 4\sqrt{n} - \log_2 m^3$

$\quad = \log_2 4 + \log_2 \sqrt{n} - 3\log_2 m$

$\quad = 2 + \log_2 n^{1/2} - 3\log_2 m$

$\quad = 2 + \dfrac{1}{2}\log_2 n - 3\log_2 m$

45) $\log_4 \dfrac{x^3}{yz^2} = \log_4 x^3 - \log_4 yz^2$

$\qquad\qquad = 3\log_4 x - \left(\log_4 y + \log_4 z^2\right)$

$\qquad\qquad = 3\log_4 x - \left(\log_4 y + 2\log_4 z\right)$

$\qquad\qquad = 3\log_4 x - \log_4 y - 2\log_4 z$

47) $\log_5 \sqrt{5c} = \log_5 (5c)^{1/2} = \dfrac{1}{2}\log_5 5c$

$\qquad\qquad = \dfrac{1}{2}\left(\log_5 5 + \log_5 c\right)$

$\qquad\qquad = \dfrac{1}{2}\left(1 + \log_5 c\right)$

$\qquad\qquad = \dfrac{1}{2} + \dfrac{1}{2}\log_5 c$

49) $\log k(k-6) = \log k + \log(k-6)$

51)
$2\log_6 x + \log_6 y = \log_6 x^2 + \log_6 y$ __Power rule__

$\qquad\qquad\quad = \underline{\log_6 x^2 y}$ $\qquad$ Product rule

53) $\log_a m + \log_a n = \log_a mn$

55) $\log_7 d - \log_7 3 = \log_7 \dfrac{d}{3}$

57)

$4\log_3 f + \log_3 g = \log_3 f^4 + \log_3 g = \log_3 f^4 g$

59) $\log_8 t + 2\log_8 u - 3\log_8 v$

$= \log_8 t + \log_8 u^2 - \log_8 v^3$

$= \log_8 tu^2 - \log_8 v^3 = \log_8 \dfrac{tu^2}{v^3}$

61) $\log(r^2 + 3) - 2\log(r^2 - 3)$

$= \log(r^2 + 3) - \log(r^2 - 3)^2$

$= \log \dfrac{r^2 + 3}{(r^2 - 3)^2}$

63) $3\log_n 2 + \dfrac{1}{2}\log_n k$

$= \log_n 2^3 + \log_n k^{1/2}$

$= \log_n 8 + \log_n \sqrt{k} = \log_n 8\sqrt{k}$

65) $\dfrac{1}{3}\log_d 5 - 2\log_d z$

$= \log_d 5^{1/3} - \log_d z^2$

$= \log_d \sqrt[3]{5} - \log_d z^2 = \log_d \dfrac{\sqrt[3]{5}}{z^2}$

67) $\log_6 y - \log_6 3 - 3\log_6 z$

$= \log_6 y - \log_6 3 - \log_6 z^3$

$= \log_6 y - (\log_6 3 + \log_6 z^3)$

$= \log_6 y - \log_6 3z^3 = \log_6 \dfrac{y}{3z^3}$

69) $4\log_3 t - 2\log_3 6 - 2\log_3 u$

$= \log_3 t^4 - \log_3 6^2 - \log_3 u^2$

$= \log_3 t^4 - (\log_3 36 + \log_3 u^2)$

$= \log_3 t^4 - \log_3 36u^2 = \log_3 \dfrac{t^4}{36u^2}$

71) $\dfrac{1}{2}\log_b (c+4) - 2\log_b (c+3)$

$= \log_b (c+4)^{1/2} - \log_b (c+3)^2$

$= \log_b \sqrt{c+4} - \log_b (c+3)^2$

$= \log_b \dfrac{\sqrt{c+4}}{(c+3)^2}$

73) $\log(a^2 + b^2) - \log(a^4 - b^4)$

$= \log\left(\dfrac{a^2 + b^2}{a^4 - b^4}\right)$

$= \log\left[\dfrac{a^2 + b^2}{(a^2 - b^2)(a^2 + b^2)}\right]$

$= \log \dfrac{1}{a^2 - b^2}$

$= -\log(a^2 - b^2)$

75) $\log 45 = \log(5 \cdot 9) = \log 5 + \log 9$

$= 0.6990 + 0.9542 = 1.6532$

77) $\log 81 = \log 9^2 = 2\log 9$

$= 2(0.9542) = 1.9084$

79) $\log \dfrac{5}{9} = \log 5 - \log 9$

$= 0.6990 - 0.9542 = -0.2552$

81) $\log 3 = \log \sqrt{9} = \log 9^{1/2} = \dfrac{1}{2}\log 9$

$\qquad = \dfrac{1}{2}(0.9542) = 0.4771$

83) $\log \dfrac{1}{5} = \log 5^{-1} = -\log 5$

$\qquad = -(0.6990) = -0.6990$

85) $\log \dfrac{1}{81} = \log 1 - \log 81$

$\qquad = 0 - \log 9^2 = -2\log 9$

$\qquad = -2(0.9542) = -1.9084$

87) $\log 50 = \log(10 \cdot 5) = \log 10 + \log 5$

$\qquad = 1 + 0.6990 = 1.6990$

89) No. $\log_a xy$ is defined only if x and y are positive.

Section 11.5

1) e

3) $\log 100 = 2$ since $10^{\boxed{2}} = 100$

5) $\log \dfrac{1}{1000} = \log 10^{-3}$

$\qquad = -3\log 10 = -3 \cdot 1 = -3$

7) $\log 0.1 = \log \dfrac{1}{10} = \log 10^{-1}$

$\qquad = -1\log 10 = -1 \cdot 1 = -1$

9) $\log 10^9 = 9\log 10 = 9 \cdot 1 = 9$

11) $\log \sqrt[4]{10} = \log 10^{1/4}$

$\qquad = \dfrac{1}{4}\log 10 = \dfrac{1}{4} \cdot 1 = \dfrac{1}{4}$

13) $\ln e^6 = 6\ln e = 6 \cdot 1 = 6$

15) $\ln \sqrt{e} = \ln e^{1/2} = \dfrac{1}{2}\ln e = \dfrac{1}{2} \cdot 1 = \dfrac{1}{2}$

17) $\ln \dfrac{1}{e^5} = \ln e^{-5} = -5\ln e = -5 \cdot 1 = -5$

19) $\ln 1 = 0$

21) $\log 16 \approx 1.2041$

23) $\log 0.5 = -0.3010$

25) $\ln 3 \approx 1.0986$

27) $\ln 1.31 \approx 0.2700$

29) $\log x = 3$

$\qquad 10^3 = x$

$\qquad 1000 = x \qquad \{1000\}$

31) $\log k = -1$

$\qquad 10^{-1} = k$

$\qquad \dfrac{1}{10} = k \qquad \left\{\dfrac{1}{10}\right\}$

33) $\log(4a) = 2$

$\qquad 10^2 = 4a$

$\qquad 100 = 4a$

$\qquad 25 = a \qquad \{25\}$

35) $\log(3t+4) = 1$

$\qquad 10^1 = 3t+4$

$\qquad 10 = 3t+4$

$\qquad 6 = 3t$

$\qquad 2 = t \qquad \{2\}$

37) $\log a = 1.5$

$\qquad 10^{1.5} = a \qquad \{10^{1.5}\}; \{31.6228\}$

39) $\log r = 0.8$

$\qquad 10^{0.8} = r \qquad \{10^{0.8}\}; \{6.3096\}$

41) $\ln x = 1.6$

$\qquad e^{1.6} = x \qquad \{e^{1.6}\}; \{4.9530\}$

43) $\ln t = -2$

$\qquad e^{-2} = t$

$\qquad \dfrac{1}{e^2} = t \qquad \left\{\dfrac{1}{e^2}\right\}; \{0.1353\}$

45) $\ln(3q) = 2.1$

$\qquad e^{2.1} = 3q$

$\qquad \dfrac{e^{2.1}}{3} = q \qquad \left\{\dfrac{e^{2.1}}{3}\right\}; \{2.7221\}$

47) $\log\left(\dfrac{1}{2}c\right) = 0.47$

$\qquad 10^{0.47} = \dfrac{1}{2}c$

$\qquad 2(10)^{0.47} = c$

$\qquad \{2(10)^{0.47}\}; \{5.9024\}$

49) $\log(5y-3) = 3.8$

$\qquad 10^{3.8} = 5y-3$

$\qquad 3 + 10^{3.8} = 5y$

$\qquad \dfrac{3 + 10^{3.8}}{5} = y$

$\qquad \left\{\dfrac{3 + 10^{3.8}}{5}\right\}; \{1262.5147\}$

51) $\ln(10w+19) = 1.85$

$\qquad e^{1.85} = 10w + 19$

$\qquad e^{1.85} - 19 = 10w$

$\qquad \dfrac{e^{1.85} - 19}{10} = w$

$\qquad \left\{\dfrac{e^{1.85} - 19}{10}\right\}; \{-1.2640\}$

53) $\ln(2d-5)=0$

$$e^0 = 2d-5$$
$$5+1 = 2d$$
$$6 = 2d$$
$$3 = d \qquad\qquad \{3\}$$

55) domain: $(0,\infty)$; range: $(-\infty,\infty)$

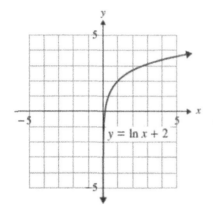

57) domain: $(2,\infty)$; range: $(-\infty,\infty)$

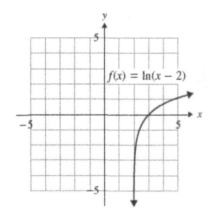

59) domain: $(0,\infty)$; range: $(-\infty,\infty)$

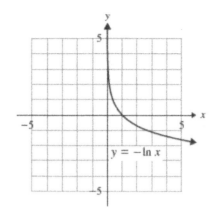

61) domain: $(0,\infty)$; range: $(-\infty,\infty)$

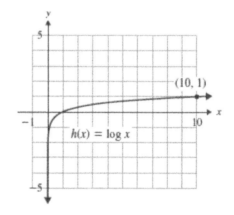

63) Shift the graph of $f(x)$ left 5 units.

65) $\log_2 13 = \dfrac{\ln 13}{\ln 2} \approx 3.7004$

67) $\log_{1/3} 16 = \dfrac{\ln 16}{\ln \dfrac{1}{3}} \approx -2.5237$

69) $L(0.000001) = 10 \log \dfrac{0.000001}{10^{-12}}$

$= 10 \log \dfrac{10^{-6}}{10^{-12}}$

$= 10 \log 10^6 = 60 \log 10$

$= 60 \text{ dB}$

71) $A = 3000\left(1 + \dfrac{0.05}{12}\right)^{12(3)}$

$\approx 3000(1.004)^{36} \approx \$3484.42$

73) $A = 3000e^{0.05(3)} = 3000e^{0.15} \approx \$3485.50$

75) a) $N(0) = 5000e^{0.0617(0)}$

$= 5000e^0$

$= 5000(1) = 5000 \text{ bacteria}$

b) $N(8) = 5000e^{0.0617(8)}$

$= 5000e^{0.4936}$

$\approx 8191 \text{ bacteria}$

77) $N(24) = 10,000e^{0.0492(24)}$

$= 10,000e^{1.1808}$

$= 32,570 \text{ bacteria}$

79) $\text{pH} = -\log\left[2 \times 10^{-3}\right]$

$= -\log(0.002) \approx 2.7; \text{ acidic}$

81) $\text{pH} = -\log\left[6 \times 10^{-12}\right]$

$= -\log(0.000000000006)$

$\approx 11.2; \text{ basic}$

83) $y = \ln x$

$x = \ln y$

$y = e^x$

Section 11.6

1) $7^x = 49$

$7^x = 7^2$

$x = 2$ $\qquad \{2\}$

3) $7^n = 15$

$\ln 7^n = \ln 15$

$n \ln 7 = \ln 15$

$n = \dfrac{\ln 15}{\ln 7}$

$\left\{\dfrac{\ln 15}{\ln 7}\right\}; \{1.3917\}$

5) $8^z = 3$

$\ln 8^z = \ln 3$

$z \ln 8 = \ln 3$

$z = \dfrac{\ln 3}{\ln 8}$

$\left\{\dfrac{\ln 3}{\ln 8}\right\}; \{0.5283\}$

Section 11.6: Solving Exponential and Logarithmic Equations

7) $6^{5p} = 36$

$6^{5p} = 6^2$

$5p = 2$

$p = \dfrac{2}{5}$ $\left\{\dfrac{2}{5}\right\}$

9) $4^{6k} = 2.7$

$\ln 4^{6k} = \ln 2.7$

$6k \ln 4 = \ln 2.7$

$k = \dfrac{\ln 2.7}{6 \ln 4}$

$\left\{\dfrac{\ln 2.7}{6 \ln 4}\right\}; \{0.1194\}$

11) $2^{4n+1} = 5$

$\ln 2^{4n+1} = \ln 5$

$(4n+1)\ln 2 = \ln 5$

$4n \ln 2 + \ln 2 = \ln 5$

$4n \ln 2 = \ln 5 - \ln 2$

$n = \dfrac{\ln 5 - \ln 2}{4 \ln 2}$

$\left\{\dfrac{\ln 5 - \ln 2}{4 \ln 2}\right\}; \{0.3305\}$

13) $5^{3a-2} = 8$

$\ln 5^{3a-2} = \ln 8$

$(3a-2)\ln 5 = \ln 8$

$3a \ln 5 - 2 \ln 5 = \ln 8$

$3a \ln 5 = \ln 8 + 2 \ln 5$

$a = \dfrac{\ln 8 + 2 \ln 5}{3 \ln 5}$

$\left\{\dfrac{\ln 8 + 2 \ln 5}{3 \ln 5}\right\}; \{1.0973\}$

15) $4^{2c+7} = 64^{3c-1}$

$4^{2c+7} = \left(4^3\right)^{3c-1}$

$4^{2c+7} = 4^{9c-3}$

$10 = 7c$

$\dfrac{10}{7} = c$ $\left\{\dfrac{10}{7}\right\}$

17) $9^{5d-2} = 4^{3d}$

$\ln 9^{5d-2} = \ln 4^{3d}$

$(5d-2)\ln 9 = 3d \ln 4$

$5d \ln 9 - 2 \ln 9 = 3d \ln 4$

$5d \ln 9 - 3d \ln 4 = 2 \ln 9$

$d(5 \ln 9 - 3 \ln 4) = 2 \ln 9$

$d = \dfrac{2 \ln 9}{5 \ln 9 - 3 \ln 4}$

$\left\{\dfrac{2 \ln 9}{5 \ln 9 - 3 \ln 4}\right\}; \{0.6437\}$

19) $e^y = 12.5$

$\ln e^y = \ln 12.5$

$y \ln e = \ln 12.5$

$y(1) = \ln 12.5$

$y = \ln 12.5$

$\{\ln 12.5\}; \{2.5257\}$

21) $e^{-4x} = 9$

$\ln e^{-4x} = \ln 9$

$-4x \ln e = \ln 9$

$-4x(1) = \ln 9$

$x = -\dfrac{\ln 9}{4}$

$\left\{-\dfrac{\ln 9}{4}\right\}; \{-1.0217\}$

23) $e^{0.01r} = 2$

$\ln e^{0.01r} = \ln 2$

$0.01r \ln e = \ln 2$

$0.01r(1) = \ln 2$

$r = \dfrac{\ln 2}{0.01}$

$\left\{\dfrac{\ln 2}{0.01}\right\}; \{69.3147\}$

25) $e^{0.006t} = 3$

$\ln e^{0.006t} = \ln 3$

$0.006t \ln e = \ln 3$

$0.006t(1) = \ln 3$

$t = \dfrac{\ln 3}{0.006}$

$\left\{\dfrac{\ln 3}{0.006}\right\}; \{183.1021\}$

27) $e^{-0.4y} = 5$

$\ln e^{-0.4y} = \ln 5$

$-0.4y \ln e = \ln 5$

$-0.4y(1) = \ln 5$

$y = -\dfrac{\ln 5}{0.4}$

$\left\{-\dfrac{\ln 5}{0.4}\right\}; \{-4.0236\}$

29) $\log_6 (k+9) = \log_6 11$

$k + 9 = 11$

$k = 2 \qquad \{2\}$

31) $\log_7 (3p-1) = \log_7 9$

$3p - 1 = 9$

$3p = 10$

$p = \dfrac{10}{3} \qquad \left\{\dfrac{10}{3}\right\}$

33) $\log x + \log (x-2) = \log 15$

$\log x(x-2) = \log 15$

$x(x-2) = 15$

$x^2 - 2x = 15$

$x^2 - 2x - 15 = 0$

$(x-5)(x+3) = 0$

$x - 5 = 0 \ \text{ or } \ x + 3 = 0$

$x = 5 \qquad\qquad x = -3$

Only one solution satisfies the original equation. $\{5\}$

Section 11.6: Solving Exponential and Logarithmic Equations

35) $\log_3 n + \log_3 (12-n) = \log_3 20$

$\qquad \log_3 n(12-n) = \log_3 20$

$\qquad\qquad n(12-n) = 20$

$\qquad\qquad 12n - n^2 = 20$

$\qquad\qquad n^2 - 12n + 20 = 0$

$\qquad\qquad (n-10)(n-2) = 0$

$\qquad\quad n - 10 = 0$ or $n - 2 = 0$

$\qquad\qquad n = 10 \qquad n = 2 \qquad \{2, 10\}$

37) $\log_2 (-z) + \log_2 (z-8) = \log_2 15$

$\qquad \log_2 \left[-z(z-8) \right] = \log_2 15$

$\qquad\qquad -z(z-8) = 15$

$\qquad\qquad -z^2 + 8z = 15$

$\qquad\qquad z^2 - 8z + 15 = 0$

$\qquad\qquad (z-5)(z-3) = 0$

$\qquad z - 5 = 0$ or $z - 3 = 0$

$\qquad\qquad z = 5 \qquad z = 3$

Neither solution satisfies the
original equation. $\varnothing$

39) $\log_6 (5b-4) = 2$

$\qquad\qquad 6^2 = 5b - 4$

$\qquad\qquad 36 = 5b - 4$

$\qquad\qquad 40 = 5b$

$\qquad\qquad 8 = b \qquad\qquad \{8\}$

41) $\log(3p+4) = 1$

$\qquad\qquad 10^1 = 3p + 4$

$\qquad\qquad 6 = 3p$

$\qquad\qquad 2 = p \qquad\qquad \{2\}$

43) $\log_3 y + \log_3 (y-8) = 2$

$\qquad\qquad \log_3 y(y-8) = 2$

$\qquad\qquad 3^2 = y(y-8)$

$\qquad\qquad 9 = y^2 - 8y$

$\qquad\qquad 0 = y^2 - 8y - 9$

$\qquad\qquad 0 = (y-9)(y+1)$

$\qquad y - 9 = 0$ or $y + 1 = 0$

$\qquad\qquad y = 9 \qquad y = -1$

Only one solution satisfies the
original equation. $\{9\}$

45) $\log_2 r + \log_2 (r+2) = 3$

$\qquad\qquad \log_2 r(r+2) = 3$

$\qquad\qquad 2^3 = r(r+2)$

$\qquad\qquad 8 = r^2 + 2r$

$\qquad\qquad 0 = r^2 + 2r - 8$

$\qquad\qquad 0 = (r+4)(r-2)$

$\qquad r + 4 = 0$ or $r - 2 = 0$

$\qquad\qquad r = -4 \qquad r = 2$

Only one solution satisfies the
original equation. $\{2\}$

47) $\log_4 20c - \log_4 (c+1) = 2$

$\qquad\qquad \log_4 \dfrac{20c}{c+1} = 2$

$\qquad\qquad 4^2 = \dfrac{20c}{c+1}$

$\qquad\qquad 16 = \dfrac{20c}{c+1}$

$\qquad\qquad 16(c+1) = 20c$

$\qquad\qquad 16c + 16 = 20c$

$\qquad\qquad 16 = 4c$

$\qquad\qquad 4 = c \qquad \{4\}$

49) $\log_2 8d - \log_2 (2d-1) = 4$

$$\log_2 \frac{8d}{2d-1} = 4$$

$$2^4 = \frac{8d}{2d-1}$$

$$16 = \frac{8d}{2d-1}$$

$$16(2d-1) = 8d$$

$$32d - 16 = 8d$$

$$-16 = -24d$$

$$\frac{-16}{-24} = d$$

$$\frac{2}{3} = d \qquad \left\{\frac{2}{3}\right\}$$

51) a) $2500 = 2000e^{0.06t}$

$$\frac{5}{4} = e^{0.06t}$$

$$\ln \frac{5}{4} = \ln e^{0.06t}$$

$$\ln \frac{5}{4} = 0.06t \ln e$$

$$\ln \frac{5}{4} = 0.06t(1)$$

$$\frac{\ln \frac{5}{4}}{0.06} = t$$

$$t \approx 3.72 \text{ yr}$$

b) $4000 = 2000e^{0.06t}$

$$2 = e^{0.06t}$$

$$\ln 2 = \ln e^{0.06t}$$

$$\ln 2 = 0.06t \ln e$$

$$\ln 2 = 0.06t(1)$$

$$\frac{\ln 2}{0.06} = t$$

$$t \approx 11.55 \text{ yr}$$

53) $7800 = 7000e^{0.075t}$

$$\frac{7800}{7000} = e^{0.075t}$$

$$\ln \frac{39}{35} = \ln e^{0.075t}$$

$$\ln \frac{39}{35} = 0.075t \ln e$$

$$\ln \frac{39}{35} = 0.075t(1)$$

$$\frac{\ln \frac{39}{35}}{0.075} = t$$

$$t \approx 1.44 \text{ yr}$$

55) $5000 = Pe^{0.08(10)}$

$$5000 = Pe^{0.80}$$

$$\frac{5000}{e^{0.80}} = P$$

$$P \approx \$2246.64$$

57) $4000 = 3000e^{r(4)}$

$$\frac{4}{3} = e^{4r}$$

$$\ln \frac{4}{3} = \ln e^{4r}$$

$$\ln \frac{4}{3} = 4r \ln e$$

$$\ln \frac{4}{3} = 4r(1)$$

$$\frac{\ln \frac{4}{3}}{4} = r$$

$$0.072 \approx r \qquad 7.2\%$$

Section 11.6: Solving Exponential and Logarithmic Equations

59) a) $5000 = 4000e^{0.0374t}$

$\dfrac{5}{4} = e^{0.0374t}$

$\ln\dfrac{5}{4} = \ln e^{0.0374t}$

$\ln\dfrac{5}{4} = 0.0374t \ln e$

$\ln\dfrac{5}{4} = 0.0374t(1)$

$\dfrac{\ln\dfrac{5}{4}}{0.0374} = t$

$t \approx 6$ hr

b) $8000 = 4000e^{0.0374t}$

$2 = e^{0.0374t}$

$\ln 2 = \ln e^{0.0374t}$

$\ln 2 = 0.0374t \ln e$

$\ln 2 = 0.0374t(1)$

$\dfrac{\ln 2}{0.0374} = t$

$t \approx 18.5$ hr

61) Let $t = 8$, $y_0 = 21{,}000$

$y = 21{,}000e^{0.036(8)}$

$y = 21{,}000e^{0.288}$

$y \approx 28{,}009$ people

63) a) Let $t = 15$, $y_0 = 2470$

$y = 2470e^{-0.013(15)}$

$y = 2470e^{-0.195}$

$y \approx 2032$ people

b) Let $y = 1600$, $y_0 = 2470$

$1600 = 2470e^{-0.013t}$

$\dfrac{1600}{2470} = e^{-0.013t}$

$\ln\dfrac{1600}{2470} = \ln e^{-0.013t}$

$\ln\dfrac{1600}{2470} = -0.013t \ln e$

$\ln\dfrac{1600}{2470} = -0.013t(1)$

$\dfrac{\ln\dfrac{1600}{2470}}{-0.013} = t$

$33 \approx t$ \qquad 2023

65) a) Let $t = 2000$, $y_0 = 15$

$y = 15e^{-0.000121(2000)}$

$y = 15e^{-0.242}$

$y \approx 11.78$ g

b) Let $y = 10$, $y_0 = 15$

$10 = 15e^{-0.000121t}$

$\dfrac{2}{3} = e^{-0.000121t}$

$\ln\dfrac{2}{3} = \ln e^{-0.000121t}$

$\ln\dfrac{2}{3} = -0.000121t \ln e$

$\ln\dfrac{2}{3} = -0.000121t(1)$

$\dfrac{\ln\dfrac{2}{3}}{-0.000121} = t$

$t \approx 3351$ yr

c) Let $y = 7.5, y_0 = 15$

$$7.5 = 15e^{-0.000121t}$$

$$\frac{1}{2} = e^{-0.000121t}$$

$$\ln\frac{1}{2} = \ln e^{-0.000121t}$$

$$\ln\frac{1}{2} = -0.000121t \ln e$$

$$\ln\frac{1}{2} = -0.000121t(1)$$

$$\frac{\ln\frac{1}{2}}{-0.000121} = t$$

$$t \approx 5728 \text{ yr}$$

67) a) $y = 0.4e^{-0.086(0)} = 0.4e^0 = 0.4 \text{ units}$

b) $y = 0.4e^{-0.086(7)} = 0.4e^{-0.602}$
$$= 0.22 \text{ units}$$

69) $\log_2(\log_2 x) = 2$

$$\log_2 x = 2^2 = 4$$

$$x = 2^4 = 16 \qquad \{16\}$$

71) $\log_3 \sqrt{n^2 + 5} = 1$

$$\sqrt{n^2 + 5} = 3^1 = 3$$

$$n^2 + 5 = 9$$

$$n^2 = 4$$

$$n = \pm 2 \qquad \{-2, 2\}$$

73) $e^{|t|} = 13$

$$\ln e^{|t|} = \ln 13$$

$$t = \pm \ln 13$$

$$\{-\ln 13, \ln 13\};$$

$$\{-2.5649, 2.5649\}$$

75) $e^{2y} + 3e^y - 4 = 0$

$$(e^y + 4)(e^y - 1) = 0$$

$$e^y = 1$$

No negative value for e^y

$$\ln(e^y) = \ln 1 = 0$$

$$y = 0 \qquad \{0\}$$

77) $5^{2c} - 4 \cdot 5^c - 21 = 0$

$$(5^c - 7)(5^c + 3) = 0$$

$$5^c = 7 \qquad \text{or} \quad 5^c = -3$$

$$\log 5^c = \log 7 \quad \text{or} \quad \text{no solution}$$

$$c = \frac{\log 7}{\log 5}$$

$$\left\{\frac{\log 7}{\log 5}\right\}; \{1.2091\}$$

79) $(\log x)^2 = \log x^3$

$$(\log x)^2 - 3\log x = 0$$

$$\log x(\log x - 3) = 0$$

$$\log_{10} x = 0 \text{ or } \log_{10} x = 3$$

$$x = 1 \text{ or } x = 1000 \qquad \{1, 1000\}$$

Chapter 11 Review

1) a) $(g \circ f)(x) = g(f(x))$
$$= 2(x+6)-9$$
$$= 2x+12-9$$
$$= 2x+3$$

    b) $(f \circ g)(x) = f(g(x))$
$$= (2x-9)+6$$
$$= 2x-3$$

    c) $(f \circ g)(5) = 2(5)-3$
$$= 10-3$$
$$= 7$$

3) a) $(N \circ G)(h) = N(G(x))$
$$= 0.8(12h) = 9.6h$$

    This is Antoine's net pay in terms of how many hours he has worked.

    b) $(N \circ G)(30) = 9.6(30) = 288$
    When Antoine works 30 hours, his net pay is $288.

    c) $(N \circ G)(40) = 9.6(40) = \$384$

5) yes; $\{(-4,-7),(1,-2),(5,1),(11,6)\}$

7) yes

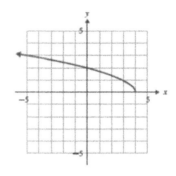

9)
$$y = x+4$$
$$x = y+4$$
$$x-4 = y$$
$$f^{-1}(x) = x-4$$

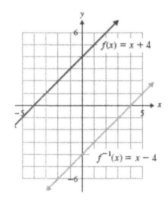

11) $y = \dfrac{1}{3}x - 1$

$x = \dfrac{1}{3}y - 1$

$x + 1 = \dfrac{1}{3}y$

$3x + 3 = y$

$h^{-1}(x) = 3x + 3$

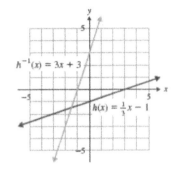

13) a) $f(2) = 6(2) - 1 = 12 - 1 = 11$

   b) $f^{-1}(11) = 2$

15) domain $(-\infty, \infty)$; range $(0, \infty)$

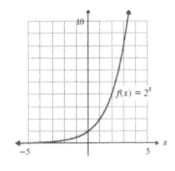

17) domain $(-\infty, \infty)$; range $(-4, \infty)$

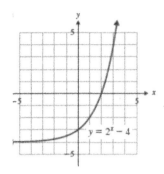

19) domain $(-\infty, \infty)$; range $(0, \infty)$

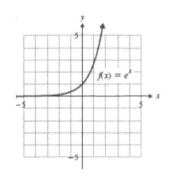

21) $2^c = 64$

   $2^c = 2^6$

   $c = 6 \qquad \{6\}$

23) $16^{3z} = 32^{2z-1}$

   $\left(2^4\right)^{3z} = \left(2^5\right)^{2z-1}$

   $2^{12z} = 2^{10z-5}$

   $12z = 10z - 5$

   $2z = -5$

   $z = -\dfrac{5}{2} \qquad \left\{-\dfrac{5}{2}\right\}$

25) $\left(\dfrac{3}{2}\right)^{x+4}=\left(\dfrac{4}{9}\right)^{x-3}$

$\left(\dfrac{3}{2}\right)^{x+4}=\left[\left(\dfrac{3}{2}\right)^{-2}\right]^{x-3}$

$\left(\dfrac{3}{2}\right)^{x+4}=\left(\dfrac{3}{2}\right)^{-2x+6}$

$x+4=-2x+6$

$3x=2$

$x=\dfrac{2}{3}$

27) x must be a positive number.
$(0,\infty)$

29) $5^3=125$

31) $10^2=100$

33) $\log_3 81=4$

35) $\log 1000=3$

37) $\log_2 x=3$

$2^3=x$

$8=x\quad\{8\}$

39) $\log_{32}16=x$

$32^x=16$

$\left(2^5\right)^x=2^4$

$2^{5x}=2^4$

$5x=4$

$x=\dfrac{4}{5}\quad\left\{\dfrac{4}{5}\right\}$

41) Let $\log_8 64=x.$

$8^x=64$

$8^x=8^2$

$x=2$

$\log_8 64=2$

43) Let $\log 1000=x.$

$10^x=1000$

$10^x=10^3$

$x=3$

$\log 1000=3$

45) Let $\log_{1/2}16=x.$

$\left(\dfrac{1}{2}\right)^x=16$

$\left(\dfrac{1}{2}\right)^x=\left(\dfrac{1}{2}\right)^{-4}$

$x=-4$

$\log_{1/2}16=-4$

47)

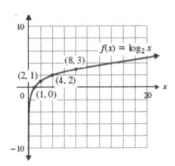

49) $y = 5^x$

$x = 5^y$

$\log_5 x = y$

$f^{-1}(x) = \log_5 x$

51) a) $S(1) = 10\log_3(2(1)+1)$

$= 10\log_3(3)$

$= 10(1) = 10 \qquad 10{,}000$

b) $S(4) = 10\log_3(2(4)+1)$

$= 10\log_3(9)$

$= 10(2) = 20 \qquad 20{,}000$

53) true

55) $\log_7 \dfrac{49}{t} = \log_7 49 - \log_7 t = 2 - \log_7 t$

57) $\log \dfrac{1}{100} = \log 10^{-2} = -2$

59) $\log_4 m\sqrt{n} = \log_4 m + \log_4 \sqrt{n}$

$= \log_4 m + \dfrac{1}{2}\log_4 n$

61) $\log_4 \dfrac{a^2}{bc^4}$

$= \log_4 a^2 - \log_4 bc^4$

$= 2\log_4 a - \left(\log_4 b + \log_4 c^4\right)$

$= 2\log_4 a - \log_4 b - 4\log_4 c$

63) $\log_6 \dfrac{r^3}{r^2-5} = \log_6 r^3 - \log_6\left(r^2-5\right)$

$= 3\log_6 r - \log_6\left(r^2-5\right)$

65) $9\log_2 a + 3\log_2 b = \log_2 a^9 + \log_2 b^3$

$= \log_2 a^9 b^3$

67) $\log_3 5 + 4\log_3 m - 2\log_3 n$

$= \log_3 5 + \log_3 m^4 - \log_3 n^2$

$= \log_3 5m^4 - \log_3 n^2 = \log_3 \dfrac{5m^4}{n^2}$

69) $\log 49 = \log 7^2 = 2\log 7$

$\approx 2(0.8451) \approx 1.6902$

71) e

73) $\log 100 = 2$ since $10^{\boxed{2}} = 100$

75) $\log \dfrac{1}{100} = \log 10^{-2}$
$= -2 \log 10 = -2 \cdot 1 = -2$

77) $\ln 1 = 0$

79) $\log 8 \approx 0.9031$

81) $\ln 1.75 \approx 0.5596$

83) $\log p = 2$
$10^2 = p$
$100 = p \quad \{100\}$

85) $\log \left(\dfrac{1}{2} c \right) = -1$
$10^{-1} = \dfrac{1}{2} c$
$\dfrac{1}{10} = \dfrac{1}{2} c$
$\dfrac{1}{5} = c \quad \left\{ \dfrac{1}{5} \right\}$

87) $\log x = 2.1$
$10^{2.1} = x \qquad \{10^{2.1}\};\{125.8925\}$

89) $\ln y = 2$
$e^2 = y \qquad \{e^2\};\{7.3891\}$

91)
$$\log(4t) = 1.75$$
$$10^{1.75} = 4t$$
$$\dfrac{10^{1.75}}{4} = t \qquad \left\{ \dfrac{10^{1.75}}{4} \right\};\{14.0585\}$$

93)   domain $(3,\infty)$; range $(-\infty,\infty)$

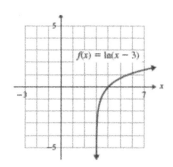

$f(x) = \ln(x - 3)$

95) $\log_4 19 = \dfrac{\log 19}{\log 4} \approx 2.1240$

97) $\log_{1/2} 38 = \dfrac{\log 38}{\log \dfrac{1}{2}} \approx -5.2479$

99) $L(0.1) = 10 \log \dfrac{0.1}{10^{-12}} = 10 \log \dfrac{10^{-1}}{10^{-12}}$
$= 10 \log 10^{11} = 110 \log 10$
$= 110(1) = 110$ dB

101) $A = 2500\left(1 + \dfrac{0.06}{4}\right)^{4(5)}$

$= 2500(1.015)^{20} \approx \$3367.14$

103) a) $N(0) = 6000e^{0.0514(0)}$

$= 6000(1) = 6000$ bacteria

b) $N(12) = 6000e^{0.0514(12)}$

$= 6000e^{0.6168}$

$\approx 11{,}118$ bacteria

105) $2^y = 16$

$2^y = 2^4$

$y = 4$ $\qquad\qquad$ $\{4\}$

107) $125^{m-4} = 25^{1-m}$

$\left(5^3\right)^{m-4} = \left(5^2\right)^{1-m}$

$5^{3m-12} = 5^{2-2m}$

$3m - 12 = 2 - 2m$

$5m = 14$

$m = \dfrac{14}{5}$ $\quad \left\{\dfrac{14}{5}\right\}$

109) $e^{0.03t} = 19$

$\ln e^{0.03t} = \ln 19$

$0.03t \ln e = \ln 19$

$0.03t = \ln 19$

$t = \dfrac{\ln 19}{0.03}$

$\left\{\dfrac{\ln 19}{0.03}\right\}; \{98.1480\}$

111) $\log_2 x + \log_2 (x+2) = \log_2 24$

$\log_2 x(x+2) = \log_2 24$

$x(x+2) = 24$

$x^2 + 2x = 24$

$x^2 + 2x - 24 = 0$

$(x+6)(x-4) = 0$

$x + 6 = 0$ or $x - 4 = 0$

$x = -6 \qquad x = 4$

Only one solution satisfies the

original equation. $\{4\}$

113)

$\log_4 k + \log_4 (k-12) = 3$

$\log_4 k(k-12) = 3$

$4^3 = k(k-12)$

$64 = k^2 - 12k$

$0 = k^2 - 12k - 64$

$0 = (k-16)(k+4)$

$k - 16 = 0$ or $k + 4 = 0$

$k = 16 \qquad k = -4$

Only one solution satisfies the

original equation. $\{16\}$

115) $10{,}000 = Pe^{0.065(6)}$

$10{,}000 = Pe^{0.39}$

$\dfrac{10{,}000}{e^{0.39}} = P$

$P \approx \$6770.57$

Chapter 11: Test

117)   a) $y = 16,410e^{0.016(5)}$

$\quad\quad = 16,410e^{0.08} \approx 17,777$ people

  b) $23,000 = 16,410e^{0.016t}$

$$\frac{23,000}{16,410} = e^{0.016t}$$

$$\ln\frac{23,000}{16,410} = \ln e^{0.016t}$$

$$\ln\frac{23,000}{16,410} = 0.016t \ln e$$

$$\frac{\ln\frac{23,000}{16,410}}{0.016} = t$$

$$21 \approx t$$

The year 2011

Chapter 11 Test

1)   a) $(h \circ k)(x) = h(k(x))$

$\quad\quad = 2(x^2 + 5x - 3) + 7$

$\quad\quad = 2x^2 + 10x - 6 + 7$

$\quad\quad = 2x^2 + 10x + 1$

  b) $(k \circ h)(x) = k(h(x))$

$\quad\quad = (2x + 7)^2 + 5(2x + 7) - 3$

$\quad\quad = 4x^2 + 28x + 49 + 10x + 35 - 3$

$\quad\quad = 4x^2 + 38x + 81$

  c) $(k \circ h)(-3) = 4(-3)^2 + 38(-3) + 82$

$\quad\quad = 36 - 114 + 81$

$\quad\quad = 3$

3)   no

5)   yes

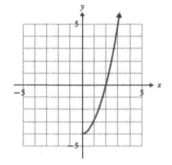

7)

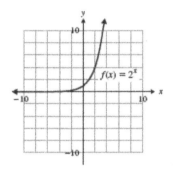

9)   a) $(0, \infty)$

  b) $(-\infty, \infty)$

11) $\log_3 \dfrac{1}{9} = -2$

13) $125^{2c} = 25^{c-4}$

$\quad\quad \left(5^3\right)^{2c} = \left(5^2\right)^{c-4}$

$\quad\quad 5^{6c} = 5^{2c-8}$

$\quad\quad 6c = 2c - 8$

$\quad\quad 4c = -8$

$\quad\quad c = -2 \qquad\qquad \{-2\}$

15) $\log(3r+13)=2$

$$10^2 = 3r+13$$
$$100 = 3r+13$$
$$87 = 3r$$
$$29 = r \qquad \{29\}$$

17) a) Let $\log_2 16 = x$.

$$2^x = 16$$
$$2^x = 2^4$$
$$x = 4$$

b) Let $\log_7 \sqrt{7} = x$.

$$7^x = \sqrt{7}$$
$$7^x = 7^{1/2}$$
$$x = \frac{1}{2}$$

19) $\log_8(5n) = \log_8 5 + \log_8 n$

21) $2\log x - 3\log(x+1)$

$$= \log x^2 - \log(x+1)^3 = \log \frac{x^2}{(x+1)^3}$$

23) $e^{0.3t} = 5$

$$\ln e^{0.3t} = \ln 5$$
$$0.3t \ln e = \ln 5$$
$$t = \frac{\ln 5}{0.3} \qquad \left\{\frac{\ln 5}{0.3}\right\}; \{5.3648\}$$

25) $\qquad 4^{4a+3} = 9$

$$\ln 4^{4a+3} = \ln 9$$
$$(4a+3)\ln 4 = \ln 9$$
$$4a \ln 4 + 3\ln 4 = \ln 9$$
$$4a \ln 4 = \ln 9 - 3\ln 4$$
$$a = \frac{\ln 9 - 3\ln 4}{4\ln 4}$$
$$\left\{\frac{\ln 9 - 3\ln 4}{4\ln 4}\right\}; \{-0.3538\}$$

27) $\qquad$ domain $(-1,\infty)$; range $(-\infty,\infty)$

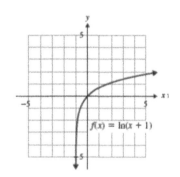

$f(x) = \ln(x + 1)$

29) $A = 6000e^{0.074(5)} = 6000e^{0.37}$

$$\approx \$8686.41$$

Cumulative Review: Chapters 1-11

1) $\left(-5a^2\right)\left(3a^4\right) = -15a^{2+4} = -15a^6$

3) $0.00009231 = 9.231 \times 10^{-5}$

11)

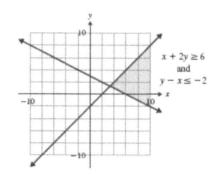

5) $-4x + 7 < 13$

$\quad -4x < 6$

$\quad\quad x > -\dfrac{6}{4}$

$\quad\quad x > -\dfrac{3}{2} \qquad \left(-\dfrac{3}{2}, \infty\right)$

13) $\sqrt{\dfrac{36a^5}{a^3}} = \sqrt{36a^2} = 6a$

7) $2c - 5 \overline{\smash{\big)}\, 6c^3 - 7c^2 - 22c + 5}$ with quotient $3c^2 + 4c - 1$

$\quad \underline{-(6c^3 - 15c^2)}$

$\quad\quad 8c^2 - 22c$

$\quad\quad \underline{-(8c^2 - 20c)}$

$\quad\quad\quad -2c + 5$

$\quad\quad\quad \underline{-(-2c + 5)}$

$\quad\quad\quad\quad 0$

15) $k^2 - 8k + 4 = 0$

$\quad k^2 - 8k = -4$

$\quad k^2 - 8k + 16 = -4 + 16$

$\quad (k - 4)^2 = 12$

$\quad k - 4 = \pm\sqrt{12}$

$\quad k = 4 \pm 2\sqrt{3}$

$\quad \left\{ 4 - 2\sqrt{3}, 4 + 2\sqrt{3} \right\}$

9) $\quad x^2 + 14x = -48$

$\quad x^2 + 14x + 48 = 0$

$\quad (x + 6)(x + 8) = 0$

$\quad x + 6 = 0 \;\; \text{or} \;\; x + 8 = 0$

$\quad x = -6 \qquad x = -8 \qquad \{-8, -6\}$

17) $\quad t^2 = 10t - 41$

$\quad t^2 - 10t = -41$

$\quad t^2 - 10t + 25 = -41 + 25$

$\quad (t - 5)^2 = -16$

$\quad t - 5 = \pm 4i$

$\quad t = 5 \pm 4i$

$\quad \{5 - 4i, 5 + 4i\}$

19) $\left(\dfrac{81}{16}\right)^{-5/4} = \left(\dfrac{16}{81}\right)^{5/4} = \left[\left(\dfrac{2}{3}\right)^4\right]^{5/4}$

$= \left(\dfrac{2}{3}\right)^{4(5/4)} = \left(\dfrac{2}{3}\right)^5$

$= \dfrac{32}{243}$

21) domain $(-\infty, \infty)$; range $(-3, \infty)$

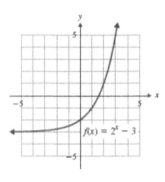

$f(x) = 2^x - 3$

23) $\log a + 2\log b - 5\log c$

$= \log a + \log b^2 - \log c^5$

$= \log ab^2 - \log c^5 = \log \dfrac{ab^2}{c^5}$

25) $e^{-0.04t} = 6$

$\ln e^{-0.04t} = \ln 6$

$-0.04t \ln e = \ln 6$

$t = -\dfrac{\ln 6}{0.04}$

$\left\{-\dfrac{\ln 6}{0.04}\right\}; \{-44.7940\}$

Section 12.1

1)

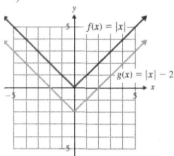

3)

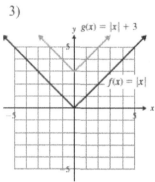

5)

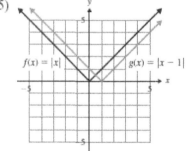

7)

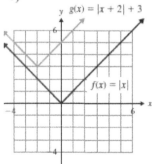

9)

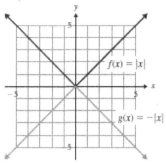

11)

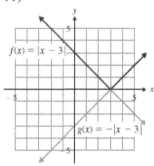

13) a) $g(x)$ b) $k(x)$

c) $h(x)$ d) $f(x)$

15) $g(x) = |x-1| + 4$

17) $f(x) = \begin{cases} -x-3, & x \le -1 \\ 2x+2, & x > -1 \end{cases}$

x	$f(x) = -x-3$
-1	$-(-1)-3 = 1-3 = -2$
-2	$-(-2)-3 = 2-3 = -1$
-3	$-(-3)-3 = 3-3 = 0$

x	$f(x) = 2x+2$
-1	$2(-1)+2 = -2+2 = 0$
0	$2(0)+2 = 0+2 = 2$
1	$2(1)+2 = 2+2 = 4$

x	$h(x) = \dfrac{1}{2}x+1$
2	$\dfrac{1}{2}(2)+1 = 1+1 = 2$
1	$\dfrac{1}{2}(1)+1 = \dfrac{1}{2}+1 = 1\dfrac{1}{2}$
0	$\dfrac{1}{2}(0)+1 = 0+1 = 1$

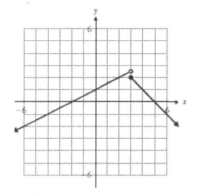

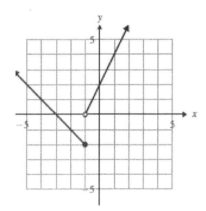

19) $h(x) = \begin{cases} -x+5, & x \ge 3 \\ \dfrac{1}{2}x+1, & x < 3 \end{cases}$

X	$h(x) = -x+5$
3	$-3+5 = 2$
4	$-4+5 = 1$
5	$-5+5 = 0$

21) $g(x) = \begin{cases} -\dfrac{3}{2}x-3, & x < 0 \\ 1, & x \ge 0 \end{cases}$

x	$g(x) = -\dfrac{3}{2}x-3$
-2	$-\dfrac{3}{2}(-2)-3 = 3-3 = 0$
-4	$-\dfrac{3}{2}(-4)-3 = 6-3 = 3$
-6	$-\dfrac{3}{2}(-6)-3 = 9-3 = 6$

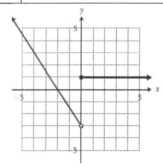

23) $k(x) = \begin{cases} x+1, & x \geq -2 \\ 2x+8, & x < -2 \end{cases}$

x	$k(x) = x+1$
-2	$-2+1 = -1$
0	$0+1 = 1$
1	$1+1 = 2$

x	$k(x) = 2x+8$
-3	$2(-3)+8 = 2$
-4	$2(-4)+8 = 0$
-5	$2(-5)+8 = -2$

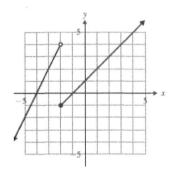

25) $f(x) = \begin{cases} 2x-4, & x > 1 \\ -\dfrac{1}{3}x - \dfrac{5}{3}, & x \leq 1 \end{cases}$

x	$f(x) = 2x-4$
2	$2(2)-4 = 0$
3	$2(3)-4 = 2$
4	$2(4)-4 = 4$

x	$f(x) = -\dfrac{1}{3}x - \dfrac{5}{3}$
1	$-\dfrac{1}{3}(1) - \dfrac{5}{3} = -\dfrac{1}{3} - \dfrac{5}{3} = -\dfrac{6}{3} = -2$
0	$-\dfrac{1}{3}(0) - \dfrac{5}{3} = -\dfrac{5}{3}$
-1	$-\dfrac{1}{3}(-1) - \dfrac{5}{3} = \dfrac{1}{3} - \dfrac{5}{3} = -\dfrac{4}{3}$

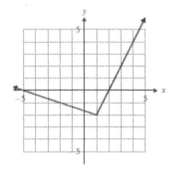

27) $f\left(3\dfrac{1}{4}\right) = \left[\!\left[3\dfrac{1}{4}\right]\!\right] = 3$

29) $f(9.2) = [\![9.2]\!] = 9$

31) $f(8) = [\![8]\!] = 8$

33) $f\left(-6\dfrac{2}{5}\right) = \left[\!\left[-6\dfrac{2}{5}\right]\!\right] = -7$

35) $f(-8.1) = [\![-8.1]\!] = -9$

37) $f(x) = [\![x]\!] + 1$

x	$f(x) = [\![x]\!] + 1$
0	$[\![0]\!] + 1 = 1$
1	$[\![1]\!] + 1 = 2$
2	$[\![2]\!] + 1 = 3$

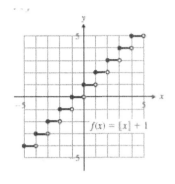

41) $g(x) = [\![x+2]\!]$

x	$g(x) = [\![x+2]\!]$
0	$[\![0+2]\!] = 2$
1	$[\![1+2]\!] = [\![3]\!] = 3$
2	$[\![2+2]\!] = [\![4]\!] = 4$

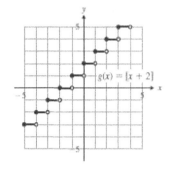

39) $h(x) = [\![x]\!] - 4$

x	$h(x) = [\![x]\!] - 4$
0	$[\![0]\!] - 4 = 0 - 4 = -4$
1	$[\![1]\!] - 4 = 1 - 4 = -3$
2	$[\![2]\!] - 4 = 2 - 4 = -2$

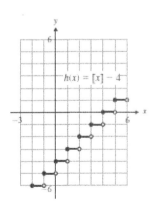

43) $k(x) = \left[\!\left[\dfrac{1}{2}x\right]\!\right]$

x	$k(x) = \left[\!\left[\dfrac{1}{2}x\right]\!\right]$
0	$\left[\!\left[\dfrac{1}{2}(0)\right]\!\right] = 0$
1	$\left[\!\left[\dfrac{1}{2}(1)\right]\!\right] = \left[\!\left[\dfrac{1}{2}\right]\!\right] = 0$
2	$\left[\!\left[\dfrac{1}{2}(2)\right]\!\right] = [\![1]\!] = 1$

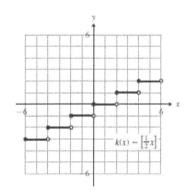

45) $C(x) = 3.75 + 1.10[\![x]\!]$

x	$C(x) = 3.75 + 1.10[\![x]\!]$
0	$C(x) = 3.75 + 1.10[\![0]\!] = 3.75$
1	$C(x) = 3.75 + 1.10[\![1]\!] = 4.85$
2	$C(x) = 3.75 + 1.10[\![2]\!] = 5.95$
3	$C(x) = 3.75 + 1.10[\![3]\!] = 7.05$
4	$C(x) = 3.75 + 1.10[\![4]\!] = 8.15$
5	$C(x) = 3.75 + 1.10[\![5]\!] = 9.25$

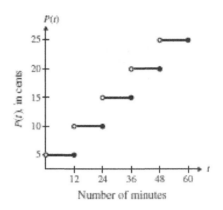

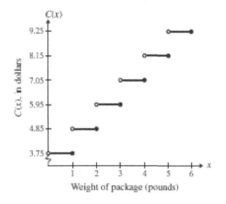

47) $P(t) = 5 + 5\left[\!\!\left[\dfrac{t}{12}\right]\!\!\right]$

t	$P(t) = 5 + 5\left[\!\!\left[\dfrac{t}{12}\right]\!\!\right]$
0	$5 + 5[\![0]\!] = 5$
12	$5 + 5[\![1]\!] = 10$
24	$P(t) = 5 + 5[\![2]\!] = 15$
36	$P(t) = 5 + 5[\![3]\!] = 20$
48	$P(t) = 5 + 5[\![4]\!] = 25$
60	$P(t) = 5 + 5[\![5]\!] = 30$

49) $f(x) = x^3$

a) Domain: $(-\infty, \infty)$; Range:

$(-\infty, \infty)$

x	$f(x) = x^3$
0	0
1	1
−1	−1
2	8
−2	−8

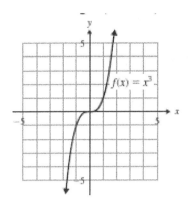

8	2
−8	−2

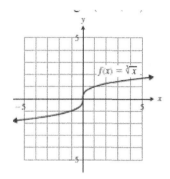

b) Domain: $[0,\infty)$; Range: $[0,\infty)$

x	$f(x)=\sqrt{x}$
0	0
1	1
4	2

51) $y=\sqrt{x-1}$

Domain: $[1,\infty)$; Range: $(0,\infty)$

x	$y=\sqrt{x-1}$
1	0
2	1
5	2

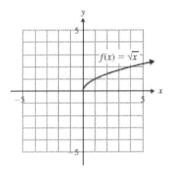

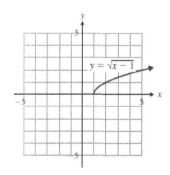

c) Domain: $(-\infty,\infty)$; Range: $(-\infty,\infty)$

x	$f(x)=\sqrt[3]{x}$
0	0
1	1
−1	−1

53) $h(x) = x^3 - 3$

Domain: $(-\infty, \infty)$; Range: $(-\infty, \infty)$

x	$f(x) = x^3 - 3$
0	-3
1	$(1)^3 - 3 = 1 - 3 = -2$
-1	$(-1)^3 - 3 = -1 - 3 = -4$
2	$(2)^3 - 3 = 8 - 3 = 5$

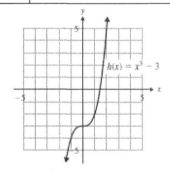

55)

$k(x) = \sqrt[3]{x-2}$; graph of $\sqrt[3]{x}$ shifted right 2.

Domain: $(-\infty, \infty)$; Range: $(-\infty, \infty)$

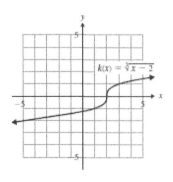

57) $g(x) = |x+2| + 3$

Graph of $|x|$ shifted left 2 and up 3.

Domain: $(-\infty, \infty)$; Range: $[3, \infty)$

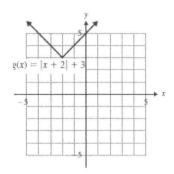

59) $k(x) = \frac{1}{2}|x|$

Graph of $|x|$ but wider.

Domain: $(-\infty, \infty)$; Range: $[0, \infty)$

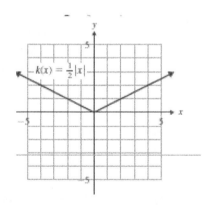

61) $f(x) = \sqrt{x+4} - 2$

Graph of $\sqrt{x}$ shifted left 4 and down 2.

Domain: $[-4, \infty)$; Range: $[-2, \infty)$

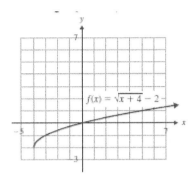

63) $h(x) = -x^3$

Reflection of the graph of x^3.

Domain: $(-\infty, \infty)$; Range: $(-\infty, \infty)$

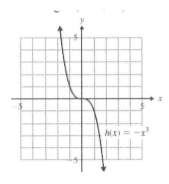

65) $g(x) = \sqrt{x+5}$

67) $g(x) = (x+2)^3 - 1$

69) $g(x) = -\sqrt[3]{x}$

Section 12.2

1) No; there are values in the domain that give more than one value in the range. The graph fails the vertical line test.

3) Center: $(-2, 4)$; $r = \sqrt{9} = 3$

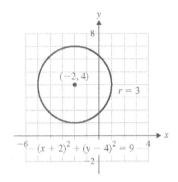

5) Center: $(5, 3)$; $r = \sqrt{1} = 1$

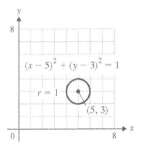

7) Center: $(-3,0)$; $r = \sqrt{4} = 2$

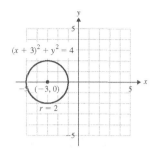

9) Center: $(6,-3)$; $r = \sqrt{16} = 4$

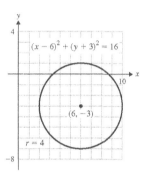

11) Center: $(0,0)$; $r = \sqrt{36} = 6$

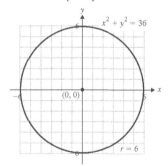

13) Center: $(0,0)$; $r = \sqrt{9} = 3$

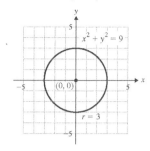

15) Center: $(0,1)$; $r = \sqrt{25} = 5$

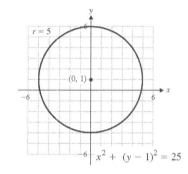

17) $(x-4)^2 + (y-1)^2 = 5^2$

$(x-4)^2 + (y-1)^2 = 25$

19) $(x-(-3))^2 + (y-2)^2 = 1^2$

$(x+3)^2 + (y-2)^2 = 1$

21) $(x-(-1))^2 + (y-(-5))^2 = (\sqrt{3})^2$

$(x+1)^2 + (y+5)^2 = 3$

23) $(x-0)^2 + (y-0)^2 = (\sqrt{10})^2$

$x^2 + y^2 = 10$

25) $(x-6)^2 + (y-0)^2 = (4)^2$

$(x-6)^2 + y^2 = 16$

27) $(x-0)^2 + (y-(-4))^2 = (2\sqrt{2})^2$

$x^2 + (y+4)^2 = 8$

29) $x^2 + y^2 - 8x + 2y + 8 = 0$

$(x^2 - 8x) + (y^2 + 2y) = -8$

$\boxed{\text{Group } x \text{ and } y \text{ terms separately.}}$

$\boxed{(x^2 - 8x + 16) + (y^2 + 2y + 1) = -8 + 16 + 1}$

Complete the square.

$\boxed{(x-4)^2 + (y+1)^2 = 9}$

Factor.

31)

$x^2 + y^2 + 2x + 10y + 17 = 0$

$(x^2 + 2x) + (y^2 + 10y) = -17$

$(x^2 + 2x + 1) + (y^2 + 10y + 25) = -17 + 1 + 25$

$(x+1)^2 + (y+5)^2 = 9$

Center: $(-1, -5)$; $r = \sqrt{9} = 3$

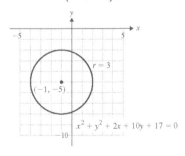

33) $x^2 + y^2 + 8x - 2y - 8 = 0$

$(x^2 + 8x) + (y^2 - 2y) = 8$

$(x^2 + 8x + 16) + (y^2 - 2y + 1) = 8 + 16 + 1$

$(x+4)^2 + (y-1)^2 = 25$

Center: $(-4,1)$; $r = \sqrt{25} = 5$

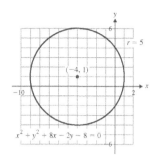

35)

$x^2 + y^2 - 10x - 14y + 73 = 0$

$(x^2 - 10x) + (y^2 - 14y) = -73$

$(x^2 - 10x + 25) + (y^2 - 14y + 49) = -73 + 25 + 49$

$(x-5)^2 + (y-7)^2 = 1$

Center: $(5,7)$; $r = \sqrt{1} = 1$

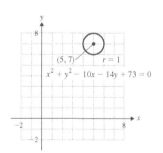

37) $x^2 + y^2 + 6y + 5 = 0$

$x^2 + (y^2 + 6y) = -5$

$x^2 + (y^2 + 6y + 9) = -5 + 9$

$x^2 + (y + 3)^2 = 4$

Center: $(0, -3)$; $r = \sqrt{4} = 2$

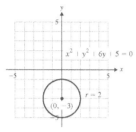

39) $x^2 + y^2 - 4x - 1 = 0$

$(x^2 - 4x) + y^2 = 1$

$(x^2 - 4x + 4) + y^2 = 1 + 4$

$(x - 2)^2 + y^2 = 5$

Center: $(2, 0)$; $r = \sqrt{5}$

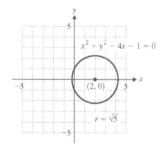

41) $x^2 + y^2 - 8x + 8y - 4 = 0$

$(x^2 - 8x) + (y^2 + 8y) = 4$

$(x - 8x + 16) + (y^2 + 8y + 16) = 4 + 16 + 16$

$(x - 4)^2 + (y + 4)^2 = 36$

Center: $(4, -4)$; $r = \sqrt{36} = 6$

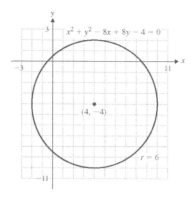

43) $4x^2 + 4y^2 - 12x - 4y - 6 = 0$

$x^2 + y^2 - 3x - y - \dfrac{3}{2} = 0$

$(x^2 - 3x) + (y^2 - y) = \dfrac{3}{2}$

$\left(x^2 - 3x + \dfrac{9}{4}\right) + \left(y^2 - y + \dfrac{1}{4}\right) = \dfrac{3}{2} + \dfrac{9}{4} + \dfrac{1}{4}$

$\left(x - \dfrac{3}{2}\right)^2 + \left(y - \dfrac{1}{2}\right)^2 = 4$

Center: $\left(\dfrac{3}{2}, \dfrac{1}{2}\right)$; $r = \sqrt{4} = 2$

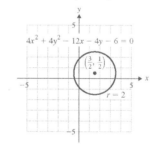

45) a) $128\,\text{m}$

b) $64\,\text{m}$

c) $(0,71)$

d) $x^2+(y-71)^2=4096$

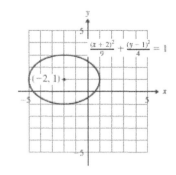

47) $A=\pi\left(r_2^2-r_1^2\right)$
$=3.14(3600-56.25)$
$=3.14(3543.75)$
$\approx 11,127\,\text{mm}^2$

11) $\dfrac{(x-3)^2}{9}+\dfrac{(y+2)^2}{16}=1$

$h=3 \quad k=-2$

center: $(3,-2)$

$a=\sqrt{9}=3 \quad b=\sqrt{16}=4$

Section 12.3

1) false

3) true

5) true, because the coefficients of x^2 and y^2 are different.

7) false, because both terms on the left must be positive.

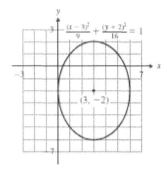

13) $\dfrac{x^2}{36}+\dfrac{y^2}{16}=1$

$h=0 \quad k=0$

center: $(0,0)$

$a=\sqrt{36}=6 \quad b=\sqrt{16}=4$

9) $\dfrac{(x+2)^2}{9}+\dfrac{(y-1)^2}{4}=1$

$h=-2 \quad k=1$

center: $(-2,1)$

$a=\sqrt{9}=3 \quad b=\sqrt{4}=2$

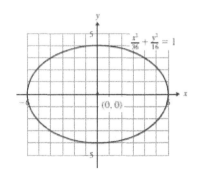

15) $x^2 + \dfrac{y^2}{4} = 1$

$h = 0 \quad k = 0$

center: $(0,0)$

$a = \sqrt{1} = 1 \quad b = \sqrt{4} = 2$

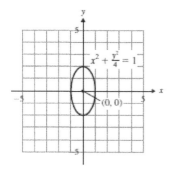

17) $\dfrac{x^2}{25} + (y+4)^2 = 1$

$h = 0 \quad k = -4$

center: $(0,-4)$

$a = \sqrt{25} = 5 \quad b = \sqrt{1} = 1$

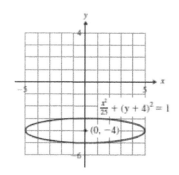

19) $\dfrac{(x+1)^2}{4} + \dfrac{(y+3)^2}{9} = 1$

$h = -1 \quad k = -3$

center: $(-1,-3)$

$a = \sqrt{4} = 2 \quad b = \sqrt{9} = 3$

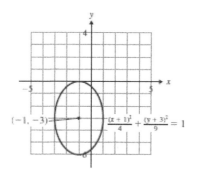

21) $4x^2 + 9y^2 = 36$

$\dfrac{4x^2}{36} + \dfrac{9y^2}{36} = \dfrac{36}{36}$

$\dfrac{x^2}{9} + \dfrac{y^2}{4} = 1$

$h = 0 \quad k = 0$

center: $(0,0)$

$a = \sqrt{9} = 3 \quad b = \sqrt{4} = 2$

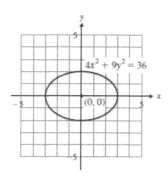

23) $25x^2 + y^2 = 25$

$\dfrac{25x^2}{25} + \dfrac{y^2}{25} = \dfrac{25}{25}$

$x^2 + \dfrac{y^2}{25} = 1$

$h = 0 \quad k = 0$

center: $(0,0)$

$a = \sqrt{1} = 1 \quad b = \sqrt{25} = 5$

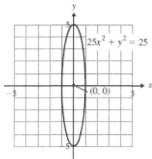

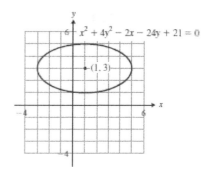

25) $3x^2 + 2y^2 - 6x + 4y - 7 = 0$

$$\left(3x^2 - 6x\right) + \left(2y^2 + 4y\right) = 7$$

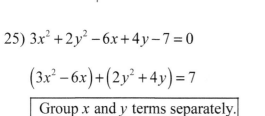

$\boxed{\text{Group } x \text{ and } y \text{ terms separately.}}$

$\boxed{3\left(x^2 - 2x\right) + 2\left(y^2 + 2y\right)}$ Complete the square.

$$3\left(x^2 - 2x + 1\right) + 2\left(y^2 + 2y + 1\right) = 7 + 3(1) + 2(1)$$
$$3\left(x^2 - 2x + 1\right) + 2\left(y^2 + 2y + 1\right) = 12$$
$$\boxed{\frac{3(x-1)^2}{12} + \frac{2(y+1)^2}{12} = \frac{12}{12}}$$
$$\boxed{\frac{(x-1)^2}{4} + \frac{(y+1)^2}{6} = 1}$$

29) $9x^2 + y^2 + 72x + 2y + 136 = 0$

$$9x^2 + y^2 + 72x + 2y = -136$$
$$\left(9x^2 + 72x\right) + \left(y^2 + 2y\right) = -136$$
$$9\left(x^2 + 8x + 16\right) + \left(y^2 + 2y + 1\right) = -136 + 144 + 1$$
$$9(x+4)^2 + (y+1)^2 = 9$$
$$\frac{9(x+4)^2}{9} + \frac{(y+1)^2}{9} = \frac{9}{9}$$
$$(x+4)^2 + \frac{(y+1)^2}{9} = 1$$

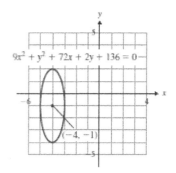

27) $x^2 + 4y^2 - 2x - 24y + 21 = 0$

$$x^2 + 4y^2 - 2x - 24y = -21$$
$$\left(x^2 - 2x\right) + \left(4y^2 - 24y\right) = -21$$
$$\left(x^2 - 2x + 1\right) + 4\left(y^2 - 6y + 9\right) = -21 + 1 + 36$$
$$(x-1)^2 + 4(y-3)^2 = 16$$
$$\frac{(x-1)^2}{16} + \frac{4(y-3)^2}{16} = \frac{16}{16}$$
$$\frac{(x-1)^2}{16} + \frac{(y-3)^2}{4} = 1$$

Section 12.3: The Ellipse

31) $4x^2 + 9y^2 - 16x - 54y + 61 = 0$

$$4x^2 + 9y^2 - 16x - 54y = -61$$

$$\left(4x^2 - 16x\right) + \left(9y^2 - 54y\right) = -61$$

$$4\left(x^2 - 4x + 4\right) + 9\left(y^2 - 6y + 9\right) =$$

$$-61 + 16 + 81$$

$$4(x-2)^2 + 9(y-3)^2 = 36$$

$$\frac{4(x-2)^2}{36} + \frac{9(y-3)^2}{36} = \frac{36}{36}$$

$$\frac{(x-2)^2}{9} + \frac{(y-3)^2}{4} = 1$$

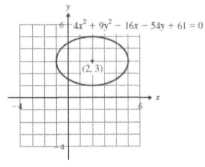

33) $25x^2 + 4y^2 + 150x + 125 = 0$

$$25x^2 + 4y^2 + 150x = -125$$

$$\left(25x^2 + 150x\right) + 4y^2 = -125$$

$$25\left(x^2 + 6x + 9\right) + 4y^2 = -125 + 225$$

$$25(x+3)^2 + 4y^2 = 100$$

$$\frac{25(x+3)^2}{100} + \frac{4y^2}{100} = \frac{100}{100}$$

$$\frac{(x+3)^2}{4} + \frac{y^2}{25} = 1$$

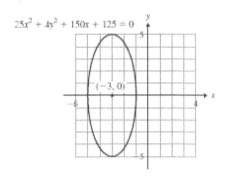

35) $a = 3;\quad a^2 = 9 \qquad b = 5;\quad b^2 = 25$

$$\frac{x^2}{9} + \frac{y^2}{25} = 1$$

37) $a = 7;\quad a^2 = 49 \qquad b = 1;\quad b^2 = 1$

$$\frac{x^2}{49} + y^2 = 1$$

39) center: $(3,1)$

$$a = 2;\quad a^2 = 4 \qquad b = 4;\quad b^2 = 16$$

$$\frac{(x-3)^2}{4} + \frac{(y-1)^2}{16} = 1$$

41) Answers may vary.

43) $\dfrac{x^2}{99^2} + \dfrac{y^2}{106.5^2} = 1$

$$\frac{x^2}{9801} + \frac{y^2}{11,342.25} = 1$$

45) $\dfrac{x^2}{36} + \dfrac{y^2}{16} = 1$

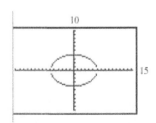

Section 12.4

1) False, it is an ellipse.

3) true

5) $\dfrac{x^2}{9} - \dfrac{y^2}{25} = 1$

Center: $(0,0)$

$a = \sqrt{9} = 3$

$b = \sqrt{25} = 5$

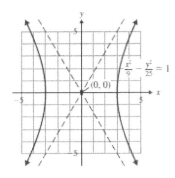

7) $\dfrac{y^2}{16} - \dfrac{x^2}{4} = 1$

Center: $(0,0)$

$a = \sqrt{4} = 2$

$b = \sqrt{16} = 4$

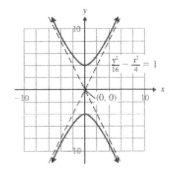

9) $\dfrac{(x-2)^2}{9} - \dfrac{(y+3)^2}{16} = 1$

Center: $(2,-3)$

$a = \sqrt{9} = 3$

$b = \sqrt{16} = 4$

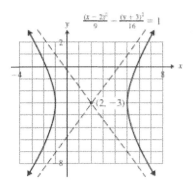

11) $\dfrac{(y+1)^2}{25} - \dfrac{(x+4)^2}{4} = 1$

Center: $(-4,-1)$

$a = \sqrt{4} = 2$

$b = \sqrt{25} = 5$

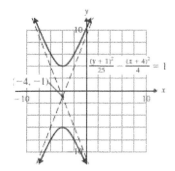

Section 12.4: The Hyperbola

13) $y^2 - \dfrac{(x-1)^2}{9} = 1$

Center: $(0,1)$

$a = \sqrt{9} = 3$

$b = \sqrt{1} = 1$

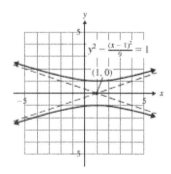

15) $\dfrac{(x-1)^2}{25} - \dfrac{(y-2)^2}{25} = 1$

Center: $(1,2)$

$a = \sqrt{25} = 5$

$b = \sqrt{25} = 5$

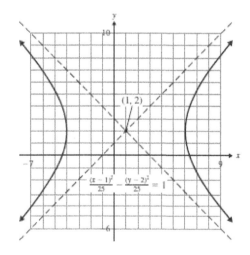

17) $9x^2 - y^2 = 36$

$\dfrac{9x^2}{36} - \dfrac{y^2}{36} = \dfrac{36}{36}$

$\dfrac{x^2}{4} - \dfrac{y^2}{36} = 1$

Center: $(0,0)$

$a = \sqrt{4} = 2$

$b = \sqrt{36} = 6$

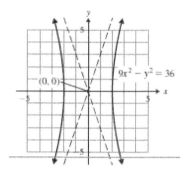

19) $y^2 - x^2 = 1$

Center: $(0,0)$

$a = \sqrt{1} = 1$

$b = \sqrt{1} = 1$

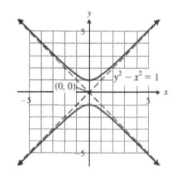

21) $\dfrac{x^2}{9} - \dfrac{y^2}{25} = 1$

$a = \sqrt{9} = 3$ and $b = \sqrt{25} = 5$

Asymptotes

$y = \dfrac{b}{a} = \dfrac{5}{3}x$ and $y = -\dfrac{b}{a} = -\dfrac{5}{3}x$

23)

$\dfrac{y^2}{16} - \dfrac{x^2}{4} = 1$

$a = \sqrt{4} = 2$ and $b = \sqrt{16} = 4$

Assymptotes

$y = \dfrac{b}{a} = \dfrac{4}{2}x = 2x$ and $y = -\dfrac{b}{a} = -\dfrac{4}{2}x = -2x$

25) $9x^2 - y^2 = 36$

$\dfrac{9x^2}{36} - \dfrac{y^2}{36} = \dfrac{36}{36}$

$\dfrac{x^2}{4} - \dfrac{y^2}{36} = 1$

$a = \sqrt{4} = 2$ and $b = \sqrt{36} = 6$

Asymptotes

$y = \dfrac{b}{a} = \dfrac{6}{2}x = 3x$ and $y = -\dfrac{b}{a} = -\dfrac{6}{2}x = -3x$

27)

$y^2 - x^2 = 1$

$a = \sqrt{1} = 1$ and $b = \sqrt{1} = 1$

Asymptotes

$y = \dfrac{b}{a} = \dfrac{1}{1}x = x$ and $y = -\dfrac{b}{a} = -\dfrac{1}{1}x = -x$

29) $xy = 1$

$y = \dfrac{1}{x}$

x	y
1	1
2	$\dfrac{1}{2}$
3	$\dfrac{1}{3}$
−1	−1
−2	$-\dfrac{1}{2}$
−3	$-\dfrac{1}{3}$

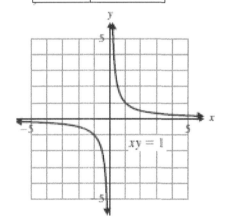

31) $xy = 10 \rightarrow y = \dfrac{10}{x}$

x	y
1	10

2	$\dfrac{10}{2}=5$
3	$\dfrac{10}{3}$
−1	−10
−2	$-\dfrac{10}{2}=-5$
−3	$-\dfrac{10}{3}$

−3	$\dfrac{-4}{-3}=\dfrac{4}{3}$

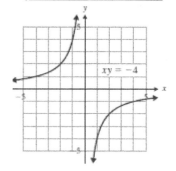

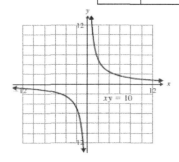

35) $xy=-6$

$$y=-\dfrac{6}{x}$$

33) $xy=-4 \rightarrow y=-\dfrac{4}{x}$

x	y
1	−4
2	$\dfrac{-4}{2}=-2$
3	$-\dfrac{4}{3}$
−1	$\dfrac{-4}{-1}=4$
−2	$\dfrac{-4}{-2}=2$

x	y
1	−6
2	$\dfrac{-6}{2}=-3$
3	$-\dfrac{6}{3}=-2$
−1	$-(-6)=6$
−2	$\dfrac{-6}{-2}=3$
−3	$\dfrac{-6}{-3}=2$

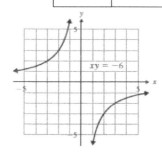

422

37)

$$f(x) = \sqrt{9 - x^2}$$

$$y = \sqrt{9 - x^2}$$

$$y^2 = 9 - x^2$$

$$x^2 + y^2 = 9$$

$$\frac{x^2}{9} + \frac{y^2}{9} = 1$$

Center: $(0,0)$; top half of circle

Domain: $[-3,3]$; Range: $[0,3]$

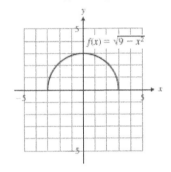

39) $h(x) = -\sqrt{1 - x^2}$

$$-y = \sqrt{1 - x^2}$$

$$y^2 = 1 - x^2 \rightarrow x^2 + y^2 = 1$$

Center: $(0,0)$; top half of circle

Domain: $[-1,1]$; Range: $[-1,0]$

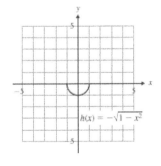

41)

$$g(x) = -2\sqrt{1 - \frac{x^2}{9}}$$

$$\frac{y}{-2} = \sqrt{1 - \frac{x^2}{9}}$$

$$\frac{y^2}{4} = 1 - \frac{x^2}{9}$$

$$\frac{x^2}{9} + \frac{y^2}{4} = 1$$

Center: $(0,0)$; bottom half of ellipse

Domain: $[-3,3]$; Range: $[-2,0]$

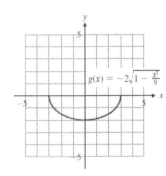

43) $g(x) = -3\sqrt{\frac{x^2}{4} - 1}$

$$\frac{y}{-3} = \sqrt{\frac{x^2}{4} - 1}$$

$$\frac{y^2}{9} = \frac{x^2}{4} - 1$$

$$-\frac{x^2}{4} + \frac{y^2}{9} = -1$$

$$\frac{x^2}{4} - \frac{y^2}{9} = 1$$

Section 12.4: The Hyperbola

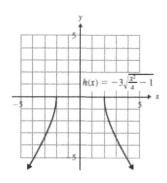

45) $x = \sqrt{16 - y^2}$

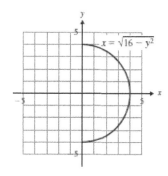

47) $x = -3\sqrt{1 - \dfrac{y^2}{4}}$

Center: $(0,0)$; bottom half of hyperbola

Domain: $(-\infty, -2] \cup [2, \infty)$; Range: $(-\infty, 0]$

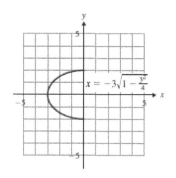

49) $4x^2 - 9y^2 - 8x - 18y - 41 = 0$

Group x and y terms separately.

Complete the square.

$4(x^2 - 2x + 1) - 9(y^2 + 2y + 1) = 41 + 4 - 9$

$4(x-1)^2 - 9(y+1)^2 = 36$

$\dfrac{4(x-1)^2}{36} - \dfrac{9(y+1)^2}{36} = \dfrac{36}{36}$

$\dfrac{(x-1)^2}{9} - \dfrac{(y+1)^2}{4} = 1$

51) $x^2 - 4y^2 - 2x - 24y - 51 = 0$

$x^2 - 2x - 4y^2 - 24y = 51$

$(x^2 - 2x + 1) - 4(y^2 + 6y + 9) = 51 + 1 - 36$

$(x-1)^2 - 4(y+3)^2 = 16$

$\dfrac{(x-1)^2}{16} - \dfrac{4(y+3)^2}{16} = \dfrac{16}{16}$

$\dfrac{(x-1)^2}{16} - \dfrac{(y+3)^2}{4} = 1$

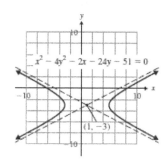

53)

$$16y^2 - 9x^2 + 18x - 64y - 89 = 0$$

$$16y^2 - 64y - 9x^2 + 18x = 89$$

$$16(y^2 - 4y + 4) - 9(x^2 - 2x + 1) = 89 + 64 - 9$$

$$\frac{16(y-2)^2}{144} - \frac{9(x-1)^2}{144} = \frac{144}{144}$$

$$\frac{(y-2)^2}{9} - \frac{(x-1)^2}{16} = 1$$

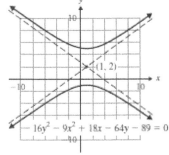

63) $y = \frac{1}{2}x$ and $y = -\frac{1}{2}x$ are asymptotes

$$\frac{b}{a} = \frac{1}{2}; \quad b^2 = 1 \quad a^2 = 4$$

opens in the $y-$direction; Center: $(0,0)$

$$y^2 - \frac{x^2}{4} = 1$$

65) $\dfrac{(x-2)^2}{9} - \dfrac{(y+3)^2}{16} = 1$

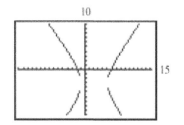

55) $y + 3 = \dfrac{4}{3}(x-2)$ and $y + 3 = -\dfrac{4}{3}(x-2)$

67) $\dfrac{(y+1)^2}{25} - \dfrac{(x+4)^2}{4} = 1$

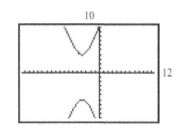

57) $y + 1 = \dfrac{5}{2}(x+4)$ and $y + 1 = -\dfrac{5}{2}(x+4)$

59) $y = \dfrac{1}{3}(x-1)$ and $y = -\dfrac{1}{3}(x-1)$

61) $y - 2 = x - 1$ and $y - 2 = -(x-1)$

$\qquad y = x + 1 \qquad\qquad y = -x + 3$

Chapter 12: Putting It All Together

1) parabola

$$V = \left[x, f(x) \right]$$

$$x = \frac{-b}{2a} = \frac{-4}{2(1)} = -2$$

$$f(-2) = (-2)^2 + 4(-2) + 8 = 4$$

$$V = (-2, 4)$$

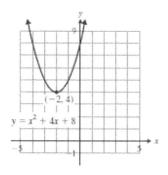

3) $\dfrac{(y+4)^2}{9} - \dfrac{(x+1)^2}{4} = 1$

Center: $(-1, -4)$

$$a = \sqrt{4} = 2 \quad b = \sqrt{9} = 3$$

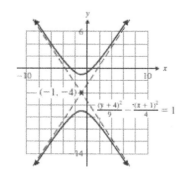

5) $16x^2 + 9y^2 = 144$

$$\frac{16x^2}{144} + \frac{9y^2}{144} = 1$$

$$\frac{x^2}{9} + \frac{y^2}{16} = 1$$

Center: $(0, 0)$

$$a = \sqrt{9} = 3 \quad b = \sqrt{16} = 4$$

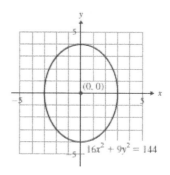

7) $\qquad x^2 + y^2 + 8x - 6y - 11 = 0$

$$(x^2 + 8x + 16) + (y^2 - 6y + 9) = 11 + 16 + 9$$

$$(x+4)^2 + (y-3)^2 = 36$$

Center: $(-4, 3)$ Radius: 6

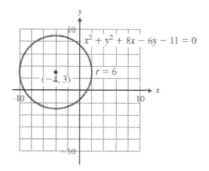

9) $(x-1)^2 + \dfrac{y^2}{16} = 1$

Center: $(1, 0)$

$$a = \sqrt{1} = 1 \quad b = \sqrt{16} = 4$$

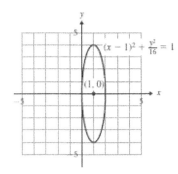

11) $x = -(y+4)^2 - 3$

$V = (-3, -4)$

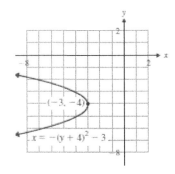

13) $25x^2 - 4y^2 = 100$

$$\frac{25x^2}{100} - \frac{4y^2}{100} = 1$$

$$\frac{x^2}{4} - \frac{y^2}{25} = 1$$

Center: $(0,0)$

$a = \sqrt{4} = 2 \quad b = \sqrt{25} = 5$

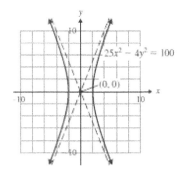

15) $(x-3)^2 + y^2 = 16$

Center: $(3,0)$ Radius: 4

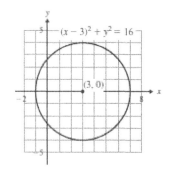

17) $x = \dfrac{1}{2}y^2 + 2y + 3$

$$y = \frac{-2}{2 \cdot \dfrac{1}{2}} = \frac{-2}{1} = -2$$

$$f(-2) = \frac{1}{2}(-2)^2 + 2(-2) + 3 = 1$$

$V = (1, -2)$

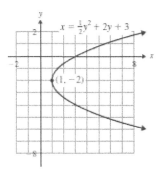

19) $(x-2)^2 - (y+1)^2 = 9$

Center: $(2,-1)$

$a = \sqrt{9} = 3 \quad b = \sqrt{9} = 3$

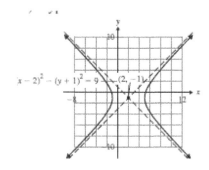

21) $xy = 5$

$y = \dfrac{5}{x}$

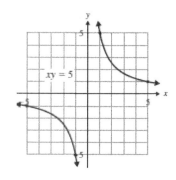

23) $9x^2 + y^2 - 54x + 4y + 76 = 0$

$$(9x^2 - 54x) + (y^2 + 4y) = -76$$

$$9(x^2 - 6x + 9) + (y^2 + 4y + 4) = -76 + 81 + 4$$

$$9(x-3)^2 + (y+2)^2 = 9$$

$$\frac{9(x-3)^2}{9} + \frac{(y+2)^2}{9} = 1$$

$$(x-3)^2 + \frac{(y+2)^2}{9} = 1$$

Center: $(3,-2)$

$a = \sqrt{1} = 1 \quad b = \sqrt{9} = 3$

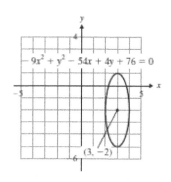

25)

$$9y^2 - 4x^2 - 18y + 16x - 43 = 0$$

$$(9y^2 - 18y) - (4x^2 - 16x) = 43$$

$$9(y^2 - 2y + 1) - 4(x^2 - 4x + 4) = 43 + 9 - 16$$

$$\frac{9(y-1)^2}{36} - \frac{4(x-2)^2}{36} = 1$$

$$\frac{(y-1)^2}{4} - \frac{(x-2)^2}{9} = 1$$

Center: $(1,2)$

$a = \sqrt{9} = 3 \quad b = \sqrt{4}2$

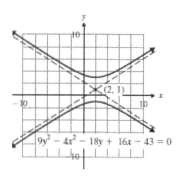

27) $x^2 + y^2 + 8x - 6y - 11 = 0$

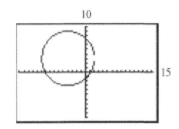

29) $x = -(y+4)^2 - 3$

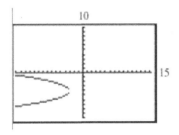

Section 12.5

1) a)

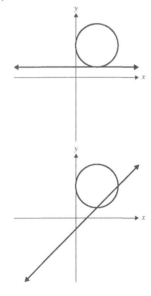

b)

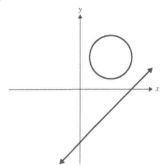

c) $0, 1,$ or 2

3) a)

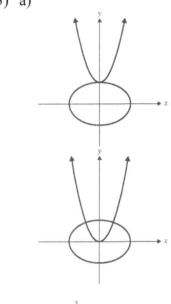

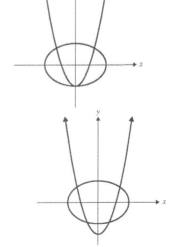

b)

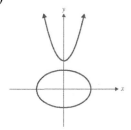

c) $0, 1, 2, 3,$ or 4

5) a)

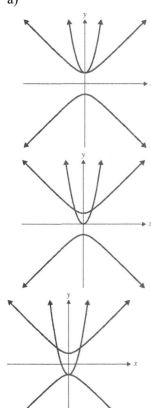

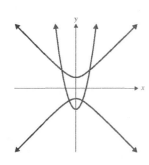

b)

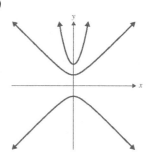

c) $0, 1, 2, 3,$ or 4

7) $x^2 + 4y = 8$ (1)

$x + 2y = -8$ (2)

Substitute $x = -2y - 8$ into (1).

Step 1:

$$(-2y-8)^2 + 4y = 8$$
$$4y^2 + 32y + 64 + 4y = 8$$
$$4y^2 + 36y + 56 = 0$$
$$y^2 + 9y + 14 = 0$$
$$(y+7)(y+2) = 0$$
$$y + 7 = 0 \text{ or } y + 2 = 0$$
$$y = -7 \qquad y = -2$$

Step 2:

$$y = -7: x + 2(-7) = -8$$
$$x - 14 = -8$$
$$x = 6$$

Step 3:

$$y = -2: x + 2(-2) = -8$$
$$x - 4 = -8$$
$$x = -4$$

Verify by substituting into (1).

$$\{(-4, -2), (6, -7)\}$$

9) $x+2y=5$ (1)

$x^2+y^2=10$ (2)

Substitute $x=5-2y$ into (2).

Step 1:

$$(5-2y)^2+y^2=10$$

$$4y^2-20y+25+y^2=10$$

$$5y^2-20y+15=0$$

$$y^2-4y+3=0$$

$$(y-3)(y-1)=0$$

$$y-3=0 \text{ or } y-1=0$$

$$y=3 \qquad\qquad y=1$$

Step 2:

$$y=3: x+2(3)=5$$

$$x+6=5$$

$$x=-1$$

Step 3:

$$y=1: x+2(1)=5$$

$$x+2=5$$

$$x=3$$

Verify by substituting into (2).

$$\{(-1,3),(3,1)\}$$

11) $y=x^2-6x+10$ (1)

$y=2x-6$ (2)

Substitute (2) into (1).

$$2x-6=x^2-6x+10$$

$$0=x^2-8x+16$$

$$0=(x-4)^2$$

$$0=x-4$$

$$4=x$$

$$x=4: y=2(4)-6=8-6=2$$

Verify by substituting into (1).

$$\{(4,2)\}$$

13) $x^2+2y^2=11$ (1)

$x^2-y^2=8$ (2)

Solve using elimination.

$$x^2+2y^2=11 \qquad (1)$$

$$+ \quad -x^2+y^2=-8 \qquad -1\cdot(2)$$

$$3y^2=3$$

$$y^2=1$$

$$y=\pm\sqrt{1}=\pm1$$

$$y=1: x^2-1^2=8$$

$$x^2-1=8$$

$$x^2=9$$

$$x=\pm3$$

$$y=-1: x^2-(-1)^2=8$$

$$x^2-1=8$$

$$x^2=9$$

$$x=\pm3$$

Verify by substituting into (1).

$$\{(3,1),(3,-1),(-3,1),(-3,-1)\}$$

15) $x^2+y^2=6$ (1)

$2x^2+5y^2=18$ (2)

Solve using elimination.

$$-2x^2-2y^2=-12 \qquad -2\cdot(1)$$

$$+ \quad 2x^2+5y^2=18 \qquad (2)$$

$$3y^2=6$$

$$y^2=2$$

$$y=\pm\sqrt{2}$$

$y = \sqrt{2} : x^2 + \left(\sqrt{2}\right)^2 = 6$

$$x^2 + 2 = 6$$

$$x^2 = 4$$

$$x = \pm 2$$

$y = -\sqrt{2} : x^2 + \left(-\sqrt{2}\right)^2 = 6$

$$x^2 + 2 = 6$$

$$x^2 = 4$$

$$x = \pm 2$$

Verify by substituing into (2).

$$\left\{\left(2, \sqrt{2}\right), \left(2, -\sqrt{2}\right), \left(-2, \sqrt{2}\right), \left(-2, -\sqrt{2}\right)\right\}$$

17) $3x^2 + 4y = -1$ (1)

     $x^2 + 3y = -12$ (2)

       $3x^2 + 4y = -1$ (1)

     $\underline{+ \quad -3x^2 - 9y = 36 \qquad -3 \cdot (2)}$

           $-5y = 35$

             $y = -7$

$y = -7 : x^2 + 3(-7) = -12$

       $x^2 - 21 = -12$

          $x^2 = 9$

           $x = \pm 3$

Verify by substituting into (2).

$$\left\{(3, -7), (-3, -7)\right\}$$

19) $y = 6x^2 - 1$ (1)

  $2x^2 + 5y = -5$ (2)

Substitute (1) into (2).

$2x^2 + 5\left(6x^2 - 1\right) = -5$

  $2x^2 + 30x^2 - 5 = -5$

         $32x^2 = 0$

          $x^2 = 0$

           $x = 0$

$x = 0 : y = 6(0)^2 - 1 = -1$

Verify by substituting into (2).

$$\left\{(0, -1)\right\}$$

21) $x^2 + y^2 = 4$ (1)

  $-2x^2 + 3y = 6$ (2)

Solve using elimination.

       $2x^2 + 2y^2 = 8$ $2 \cdot (1)$

  $\underline{+ \quad -2x^2 + 3y = 6 \qquad\qquad (2)}$

        $2y^2 + 3y = 14$

      $2y^2 + 3y - 14 = 0$

    $(2y + 7)(y - 2) = 0$

    $2y + 7 = 0$ or $y - 2 = 0$

       $2y = -7$ $y = 2$

        $y = -\dfrac{7}{2}$

$y = -\dfrac{7}{2} : x^2 + \left(-\dfrac{7}{2}\right)^2 = 4$

        $x^2 + \dfrac{49}{4} = 4$

          $x^2 = -\dfrac{33}{4}$

           $x = \pm \dfrac{\sqrt{33}}{2} i$

does not give real number solutions.

$y = 2 : x^2 + 2^2 = 4$

       $x^2 + 4 = 4$

         $x^2 = 0$

          $x = 0$

Verify by substituting into (2).

$$\left\{(0, 2)\right\}$$

432

23) $x^2 + y^2 = 3$ (1)

    $x + y = 4$ (2)

    Substitute $y = 4 - x$ into (1).

    $x^2 + (4 - x)^2 = 3$

   $x^2 + 16 - 8x + x^2 = 3$

    $2x^2 - 8x + 13 = 0$

   $x = \dfrac{-(-8) \pm \sqrt{(-8)^2 - 4(2)(13)}}{2(2)}$

    $= \dfrac{8 \pm \sqrt{64 - 104}}{4} = \dfrac{8 \pm \sqrt{-40}}{4}$

    No real number values for x. $\varnothing$

25) $x = \sqrt{y}$ (1)

   $x^2 - 9y^2 = 9$ (2)

    Substitute (1) into (2).

    $\left(\sqrt{y}\right)^2 - 9y^2 = 9$

     $y - 9y^2 = 9$

      $0 = 9y^2 - y + 9$

   $y = \dfrac{-(-1) \pm \sqrt{(-1)^2 - 4(9)(9)}}{2(9)}$

    $= \dfrac{1 \pm \sqrt{1 - 324}}{18} = \dfrac{1 \pm \sqrt{-323}}{18}$

    No real number values for y. $\varnothing$

27) $9x^2 + y^2 = 9$ (1)

    $x^2 + y^2 = 5$ (2)

    Solve using elimination.

      $9x^2 + y^2 = 9$ (1)

   $+$ $-x^2 - y^2 = -5$ $-1 \cdot (2)$

         $8x^2 = 4$

         $x^2 = \dfrac{1}{2}$

     $x = \pm\sqrt{\dfrac{1}{2}} = \pm\dfrac{\sqrt{2}}{2}$

$x = \dfrac{\sqrt{2}}{2} : \left(\dfrac{\sqrt{2}}{2}\right)^2 + y^2 = 5$

         $\dfrac{2}{4} + y^2 = 5$

         $y^2 = \dfrac{18}{4}$

         $y = \pm\dfrac{3\sqrt{2}}{2}$

$x = -\dfrac{\sqrt{2}}{2} : \left(-\dfrac{\sqrt{2}}{2}\right)^2 + y^2 = 5$

         $\dfrac{2}{4} + y^2 = 5$

         $y^2 = \dfrac{18}{4}$

         $y = \pm\dfrac{3\sqrt{2}}{2}$

    Verify by substituting into (1).

$$\left\{ \left(\dfrac{\sqrt{2}}{2}, \dfrac{3\sqrt{2}}{2}\right), \left(\dfrac{\sqrt{2}}{2}, -\dfrac{3\sqrt{2}}{2}\right), \right.$$

$$\left. \left(-\dfrac{\sqrt{2}}{2}, \dfrac{3\sqrt{2}}{2}\right), \left(-\dfrac{\sqrt{2}}{2}, -\dfrac{3\sqrt{2}}{2}\right) \right\}$$

Section 12.6: Second-Degree Inequalities and System of Inequalities

29) Substitute (1) into (2).

$$x^2 + \left(-x^2 - 2\right)^2 = 4$$
$$x^2 + x^4 + 4x^2 + 4 = 4$$
$$x^4 + 5x^2 = 0$$
$$x^2\left(x^2 + 5\right) = 0$$

$$x^2 + 5 = 0 \ \text{ or } \ x^2 = 0$$
$$x^2 = -5 \qquad \boxed{x = 0}$$
$$x = \pm i\sqrt{5}$$
$$x = \pm i\sqrt{5} \ \text{does not give}$$
$$\text{real number solutions.}$$
$$x = 0: y = -(0)^2 - 2 = -2$$
Verify by substituting into (2).

$$\left\{(0, -2)\right\}$$

31) x = one number

y = other number

$$xy = 40$$
$$x + y = 13$$

Substitute $y = 13 - x$ into (1).

$$x(13 - x) = 40$$
$$13x - x^2 = 40$$
$$0 = x^2 - 13x + 40$$
$$0 = (x - 8)(x - 5)$$
$$x - 8 = 0 \ \text{ or } \ x - 5 = 0$$
$$x = 8 \qquad x = 5$$
$$x = 8: y = 13 - 8 = 5$$
$$x = 5: y = 13 - 5 = 8$$
The numbers are 8 and 5.

33) l = length of screen

w = width of screen

$$2l + 2w = 38 \qquad (1)$$
$$lw = 88 \qquad (2)$$
Solve (1) for l.
$$2l = 38 - 2w$$
$$l = 19 - w \qquad (3)$$
Substitute (3) into (2).
$$(19 - w)w = 88$$
$$19w - w^2 = 88$$
$$0 = w^2 - 19w + 88$$
$$0 = (w - 11)(w - 8)$$
$$w - 11 = 0 \ \text{ or } \ w - 8 = 0$$
$$w = 11 \qquad w = 8$$
$$w = 11: l = 19 - 11 = 8$$
$$w = 8: l = 19 - 8 = 11$$
The dimensions of the screen are 8 in×11 in.

35)
$$15x^2 = 6x^2 + 33x + 12$$
$$9x^2 - 33x - 12 = 0$$
$$3x^2 - 11x - 4 = 0$$
$$(3x + 1)(x - 4) = 0$$
$$3x + 1 = 0 \ \text{ or } \ x - 4 = 0$$
$$3x = -1 \qquad \boxed{x = 4}$$
$$x = -\frac{1}{3}$$
$$x = 4: y = 15(4)^2 = 240$$
The break-even point is 4000 basketballs and \$240.

434

Section 12.6

1) Three points that satisfy the inequality are $(7,0),(9,0),(4,5)$. Three points that do not satisfy the inequality are $(0,0),(2,-1),(1,2)$. Answers may vary.

3) Three points that satisfy the inequality are $(0,0),(-3,1),(4,1)$. Three points that do not satisfy the inequality are $(0,-3),(2,2),(-2,-2)$. Answers may vary.

5) $\dfrac{x^2}{9}+\dfrac{y^2}{16}<1$

Center: $(0,0)$

$a=\sqrt{9}=3 \quad b=\sqrt{16}=4$

Test $(0,0)$ is true, shade inside the ellipse.

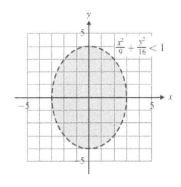

7)

$y<(x+1)^2+3$

$(h,k)=(-1,3)$

Test $(2,0)$ is true, shade outside the parabola.

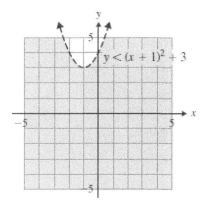

9) $25y^2-4x^2\geq100$

$\dfrac{25y^2}{100}-\dfrac{4x^2}{100}\geq1$

$\dfrac{y^2}{4}-\dfrac{x^2}{25}\geq1$

Center: $(0,0)$

$a=\sqrt{25}=5 \quad b=\sqrt{4}=2$

Test $(0,3)$ is true, shade inside the hyperbola.

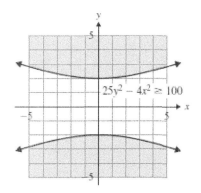

11) $x^2+y^2\geq3$

Center: $(0,0)$ Radius: 1.73

Test $(0,1)$ is not true, shade outside the circle.

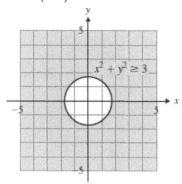

13) $x > y^2 - 2$

$$y = \frac{-b}{2a} = \frac{0}{2 \cdot 1} = 0$$

$$x = -0^2 - 2 = -2$$

$$V = (-2, 0)$$

Test $(3, 0)$ is true, shade inside the parabola.

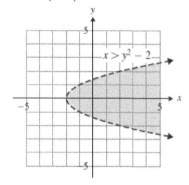

15) $\dfrac{x^2}{4} - \dfrac{y^2}{9} \leq 1$

   Center: $(0,0)$

   $a = \sqrt{4} = 2 \quad b = \sqrt{9} = 3$

Test $(0,3)$ is true, shade outside the hyperbola.

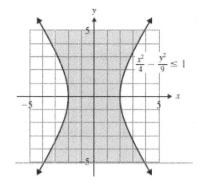

17) $x^2 + (y+4)^2 < 9$

   Center: $(0,-4)$ Radius: 3

Test $(0,-3)$ is true, shade inside the circle.

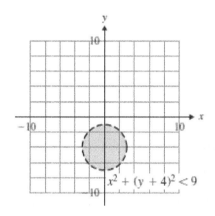

19) $x^2 + 9y^2 \geq 9$

   $$\frac{x^2}{9} + \frac{9y^2}{9} \leq 1$$

   $$\frac{x^2}{9} + y^2 \leq 1$$

   Center: $(0,0)$

   $a = \sqrt{9} = 3 \quad b = \sqrt{1} = 1$

Test $(0,3)$ is true, shade outside the ellipse.

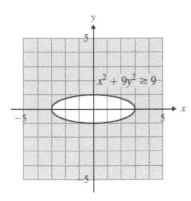

21) $y \geq -x^2 - 2x + 3$

$$x = \frac{-b}{2a} = \frac{2}{2 \cdot -1} = -1$$

$$y = -(-1)^2 - 2(-1) + 3 = 4$$

$$V = (-1, 4)$$

Test $(0, 2)$ is true, shade inside the parabola.

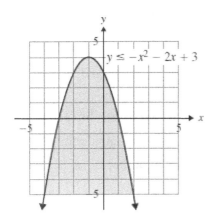

23) $y \geq x - 2$

$$m = \frac{1}{1}; \ b = -2$$

$$y \leq -x^2 + 1$$

$$x = \frac{-b}{2a} = \frac{0}{2(-1)} = 0$$

$$y = -0^2 + 1 = 1$$

$$V = (0, 1)$$

Test $(0,0)$ is true, shade inside the parabola and above the line.

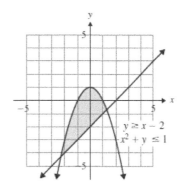

25) $2y - x > 4$

$$y > \frac{1}{2}x + 2$$

$$m = \frac{1}{2}; \ b = 2$$

$$4x^2 + 9y^2 > 36$$

$$\frac{4x^2}{36} + \frac{9y^2}{36} > 1$$

$$\frac{x^2}{9} + \frac{y^2}{4} > 1$$

Center: $(0, 0)$

$$a = \sqrt{9} = 3 \quad b = \sqrt{4} = 2$$

Test $(0, 3)$ is true, shade outside the ellipse and above the line.

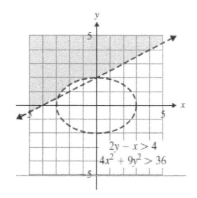

27) $x^2 + y^2 \geq 16$

Center: $(0,0)$ Radius: 4

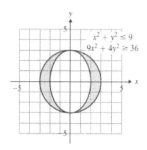

$25x^2 - 4y^2 \leq 100$

$\dfrac{25x^2}{100} - \dfrac{4y^2}{100} \leq 1$

$\dfrac{x^2}{4} - \dfrac{y^2}{25} \leq 1$

Center: $(0,0)$

$a = \sqrt{4} = 2 \quad b = \sqrt{25} = 5$

Test $(0,5)$ is true, shade outside the circle and outside the hyperbola.

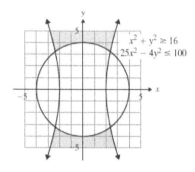

29) $x^2 + y^2 \leq 9$

Center: $(0,0)$ Radius: 3

$\dfrac{9x^2}{36} + \dfrac{4y^2}{36} \geq 1$

$\dfrac{x^2}{4} + \dfrac{y^2}{9} \geq 1$

Center: $(0,0)$

$a = \sqrt{4} = 2 \quad b = \sqrt{9} = 3$

Test $(0,0)$ is not true for the ellipse and true for the circle, shade outside the ellipse and inside the circle.

31) $y^2 - x^2 < 1$

Center: $(0,0)$

$a = \sqrt{1} = 1 \quad b = \sqrt{1} = 1$

$4x^2 + y^2 > 16$

$\dfrac{4x^2}{16} + \dfrac{y^2}{16} > 1$

$\dfrac{x^2}{4} + \dfrac{y^2}{16} > 1$

Center: $(0,0)$

$a = \sqrt{4} = 2 \quad b = \sqrt{16} = 4$

Test $(3,0)$ is true for the ellipse and hyperbola, shade outside the ellipse and outside the hyperbola.

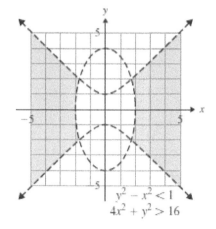

33) $x < -y^2$

$$y = \frac{-b}{2a} = \frac{0}{2 \cdot -1} = 0$$

$$x = (0)^2 = 0$$

$$V = (0,0)$$

$$x^2 + y^2 < 16$$

$$\frac{x^2}{16} + \frac{y^2}{16} < 1$$

Center: $(0,0)$ Radius: 4

Test $(-1,0)$ is true for the parabola and the circle, shade inside the parabola and circle.

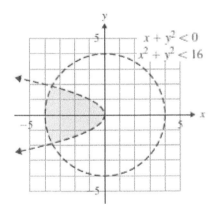

35) $\dfrac{x^2}{16} + \dfrac{y^2}{9} \leq 1$

Center: $(0,0)$

$$a = \sqrt{16} = 4 \quad b = \sqrt{9} = 3$$

$$4x^2 - y^2 \geq 16$$

$$\frac{4x^2}{16} - \frac{y^2}{16} \geq 1$$

$$\frac{x^2}{4} - \frac{y^2}{16} \geq 1$$

Center: $(0,0)$

$$a = \sqrt{4} = 2 \quad b = \sqrt{16} = 4$$

Test $(3,0)$ is true for the hyperbola and for the ellipse, shade inside the ellipse and hyperbola.

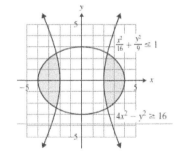

37) $y \leq -x$

$$m = -1; \quad b = 0$$

$$y \leq 2x^2 - 2$$

$$x = \frac{-b}{2a} = \frac{0}{2(2)} = 0$$

$$y = -0^2 - 2 = -2$$

$$V = (0,-2)$$

Test $(0,-3)$ is true for the parabola and the line, shade outside the parabola and below the line.

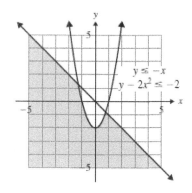

39) $x \geq 0$

$x^2 + y^2 \leq 25$

Center: $(0,0)$;   Radius: 5

Test $(1,0)$ is true for the circle and the line, shade inside the circle and right of the line.

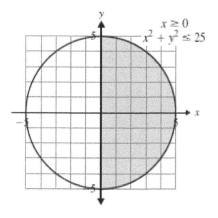

41) $y < 0$

$4x^2 - 9y^2 < 36$

$\dfrac{4x^2}{36} - \dfrac{9y^2}{36} < 1$

$\dfrac{x^2}{9} - \dfrac{y^2}{4} < 1$

Center: $(0,0)$

$a = \sqrt{9} = 3$ $b = \sqrt{4} = 2$

Test $(0,-1)$ is true for the hyperbola and true for the line, shade outside the hyperbola and below the line.

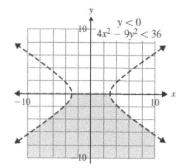

43) $x \geq 0$

$y \geq x^2 + 4$

$x = \dfrac{-b}{2a} = \dfrac{0}{2(1)} = 0$

$y = -0^2 + 4 = 4$

$V = (0,4)$

$x + 2y \leq 12$

$y \leq -\dfrac{1}{2}x + 6$

$m = -\dfrac{1}{2};$ $b = 6$

Test $(1,5)$ is true for the parabola and true for both lines, shade inside the parabola and right of the line $x \geq 0$ and below the line $y \leq -\dfrac{1}{2}x + 6.$

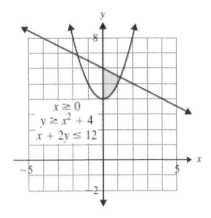

45) $y < 0$

$x^2 + y^2 < 9$

Center: $(0,0)$;   Radius: 3

$y > x + 1$

$m = 1$; $b = 1$

Test $(0, -1)$ is true for the circle and the line $y < 0$, not true for $y > x + 1$. Shade inside the circle, below the line $y < 0$ and above the line $y > x + 1$.

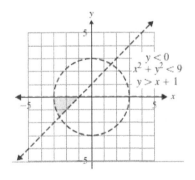

47) $y \geq 0$

$y \leq x^2$

$x = \dfrac{-b}{2a} = \dfrac{0}{2(-1)} = 0$

$y = -0^2 = 0$

$V = (0,0)$

$\dfrac{x^2}{4} + \dfrac{y^2}{9} \leq 1$

Center: $(0,0)$

$a = \sqrt{4} = 2$ $b = \sqrt{9} = 3$

Test $(0,1)$ is true for the ellipse and the line $y \geq 0$, not true for the parabola. Shade inside the ellipse, above the line $y \geq 0$, and outside the parabola.

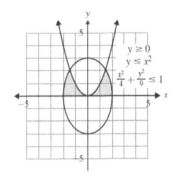

49) $y < 0$

$$4x^2 + 9y^2 < 36$$

$$\frac{4x^2}{36} + \frac{9y^2}{36} < 1$$

$$\frac{x^2}{9} + \frac{y^2}{4} < 1$$

Center: $(0,0)$

$a = \sqrt{9} = 3 \quad b = \sqrt{4} = 2$

$$x^2 - y^2 > 1$$

Center: $(0,0)$

$a = \sqrt{1} = 1 \quad b = \sqrt{1} = 1$

Test $(2,-1)$ is true for the ellipse, the line $y < 0$, and for the hyperbola. Shade inside the ellipse, below the line $y < 0$, and inside the hyperbola.

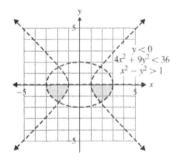

51) $x \geq 0$

$y \geq 0$

$y^2 \leq x$

$$x = \frac{-b}{2a} = \frac{0}{2(1)} = 0$$

$y = 0^2 = 0$

$V = (0,0)$

$$x^2 + 4y^2 \geq 16$$

Center: $(0,0)$; radius: 2

Test $(3,1)$ is true for the parabola, both lines, and for the circle. Shade inside the parabola, right of $x \geq 0$, above $y \geq 0$, and outside the circle.

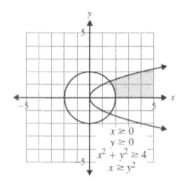

Chapter 12 Review

1) $g(x) = |x|$

Domain: $(-\infty, \infty)$ Range: $(0, \infty)$

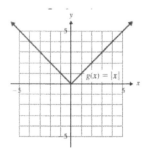

3) $h(x) = |x| - 4$

Domain: $(-\infty, \infty)$ Range: $(-4, \infty)$

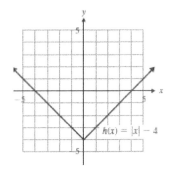

5) $k(x) = -|x| + 5$

Domain: $(-\infty, \infty)$ Range: $(-\infty, 0)$

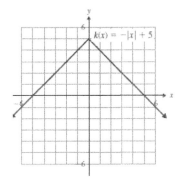

7) $f(x) = \begin{cases} -\dfrac{1}{2}x - 2, & x \le 2 \\ x - 3, & x > 2 \end{cases}$

x	$f(x) = -\dfrac{1}{2}x - 2$
0	$-\dfrac{1}{2}(0) - 2 = -2$
-2	$-\dfrac{1}{2}(-2) - 2 = -1$
-4	$-\dfrac{1}{2}(-4) - 2 = 0$

X	$f(x) = x - 3$
3	$3 - 3 = 0$
4	$4 - 3 = 1$
5	$5 - 3 = 2$

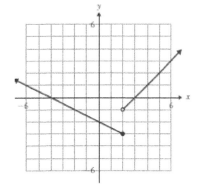

9) $f(x) = [\![x]\!] = \left[\!\left[7\dfrac{2}{3}\right]\!\right] = 7$

11) $f(x) = [\![x]\!] = \left[\!\left[-8\dfrac{1}{2}\right]\!\right] = -9$

13) $f(x) = [\![x]\!] = \left[\!\left[\dfrac{3}{8}\right]\!\right] = 0$

15) $g(x) = \left[\!\left[\dfrac{1}{2}x\right]\!\right]$

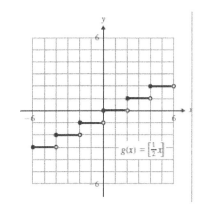

17) $(x+3)^2 + (y-5)^2 = 36$

$$\frac{(x+3)^2}{36} + \frac{(y-5)^2}{36} = 1$$

Center: $(-3,5)$ Radius: 6

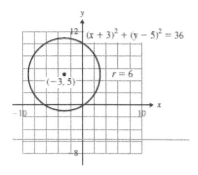

25) $\dfrac{(x+3)^2}{9} + \dfrac{(y-3)^2}{4} = 1$

Center: $(-3,3)$

$a = \sqrt{9} = 3$ $b = \sqrt{4} = 2$

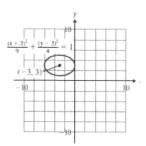

19)

$$x^2 + y^2 - 10x - 4y + 13 = 0$$

$$\left(x^2 - 10x + 25\right) + \left(y^2 - 4y + 4\right) = -13 + 25 + 4$$

$$(x-5)^2 + (y-2)^2 = 16$$

Center: $(5,2)$ Radius: 4

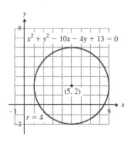

27) $25x^2 + 4y^2 = 100$

$$\frac{25x^2}{100} + \frac{4y^2}{100} = 1$$

$$\frac{x^2}{4} + \frac{y^2}{25} = 1$$

Center: $(0,0)$

$a = \sqrt{4} = 2$ $b = \sqrt{25} = 5$

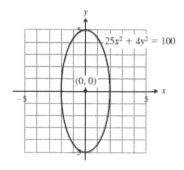

21) $(x-3)^2 + y^2 = 16$

23) When $a = b$ in $\dfrac{(x-h)^2}{a^2} + \dfrac{(y-k)^2}{b^2} = 1$

29) $25x^2 + 4y^2 - 100x = 0$

$$25x^2 - 100x + 4y^2 = 0$$
$$25(x^2 - 4x + 4) + 4y^2 = 0 + 100$$
$$25(x-2)^2 + 4y^2 = 100$$
$$\frac{25(x-2)^2}{100} + \frac{4y^2}{100} = 1$$
$$\frac{(x-2)^2}{4} + \frac{y^2}{25} = 1$$

Center: $(2,0)$

$a = \sqrt{4} = 2 \quad b = \sqrt{25} = 5$

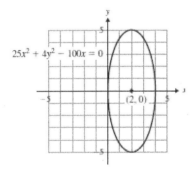

31) $\dfrac{y^2}{9} - \dfrac{x^2}{25} = 1$

Center: $(0,0)$

$a = \sqrt{9} = 3 \quad b = \sqrt{25} = 5$

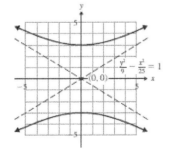

33) $\dfrac{(y+1)^2}{4} - \dfrac{(x+2)^2}{4} = 1$

Center: $(-1,-2)$

$a = \sqrt{4} = 2 \quad b = \sqrt{4} = 2$

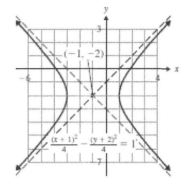

35) $xy = 6$

$$y = \frac{6}{x}$$

x	y
1	6
2	$\dfrac{6}{2} = 3$
3	$\dfrac{6}{3} = 2$
-1	-6
-2	$-\dfrac{6}{2} = -3$
-3	$-\dfrac{6}{3} = -2$

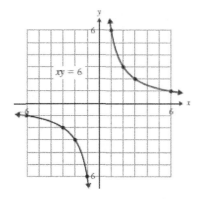

39) $h(x) = 2\sqrt{1 - \dfrac{x^2}{9}}$

Domain: $[-3, 3]$ Range: $[0, 2]$

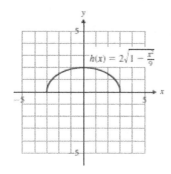

37)

$$16y^2 - x^2 + 2x + 96y = -127$$

$$\left(16y^2 + 96y\right) - \left(x^2 - 2x\right) = -127$$

$$16\left(y^2 + 6y + 9\right) - \left(x^2 - 2x + 1\right) = -127 + 144 - 1$$

$$\frac{16(y+3)^2}{16} - \frac{(x-1)^2}{16} = 1$$

$$(y+3)^2 - \frac{(x-1)^2}{16} = 1$$

41) $x^2 + 9y^2 = 9$; ellipse

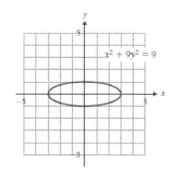

Center: $(1, -3)$

$a = \sqrt{16} = 4$ $b = \sqrt{1} = 1$

43) $x = -y^2 + 6y - 5$; parabola

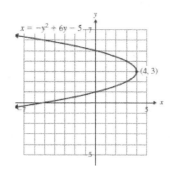

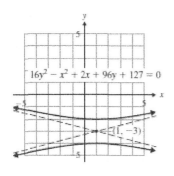

45) $\dfrac{(x-3)^2}{16} - \dfrac{(y-4)^2}{25} = 1$; hyperbola

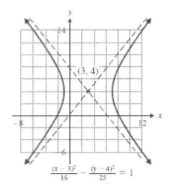

47) $x^2 + y^2 - 2x + 2y - 2 = 0$

$\left(x^2 - 2x + 1\right) + \left(y^2 + 2y + 1\right) = 2 + 1 + 1$

$(x-1)^2 + (y+1)^2 = 4$; circle

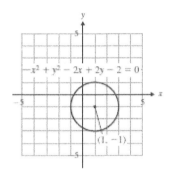

49) $y = \dfrac{1}{2}(x+2)^2 + 1$; parabola

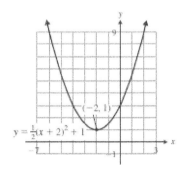

51) $0, 1, 2, 3,$ or 4

53) $-4x^2 + 3y^2 = 3$ \hfill (1)

$ 7x^2 - 5y^2 = 7$ \hfill (2)

Solve using elimination.

$5\left(-4x^2 + 3y^2 = 3\right)$

$\underline{+ \quad 3\left(7x^2 - 5y^2 = 7\right)}$

$-20x^2 + 15y^2 = 15$

$\underline{21x^2 - 15y^2 = 21}$

$x^2 = 36$

$x = \pm 6$

$x = 6: -4(6)^2 + 3y^2 = 3$

$-144 + 3y^2 = 3$

$3y^2 = 3 + 144$

$y^2 = \dfrac{147}{3}$

$y = \sqrt{49}$

$y = \pm 7$

$x = -6: -4(-6)^2 + 3y^2 = 3$

$-144 + 3y^2 = 3$

$3y^2 = 3 + 144$

$y^2 = \dfrac{147}{3}$

$y = \sqrt{49}$

$y = \pm 7$

$\left\{(6,7), (6,-7), (-6,7), (-6,-7)\right\}$

Chapter 12: Review

55)
$$x^2 + y = 3 \quad (1)$$
$$\underline{ x - y = -1 \quad (2)}$$
$$x^2 + x = 2$$
$$x^2 + x - 2 = 0$$
$$(x+2)(x-1) = 0$$
$$x = -2 \ \text{ or } \ x = 1$$

$x = -2:$
$$x - y = -1$$
$$-2 - y = -1$$
$$-y = -1 + 2$$
$$y = -1$$

$x = 1:$
$$x - y = -1$$
$$1 - y = -1$$
$$-y = -1 - 1$$
$$-y = -2$$
$$y = 2$$

$$\{(1,2), (-2,-1)\}$$

57)
$$4x^2 + 9y^2 = 36 \quad (1)$$
$$y = \frac{1}{3}x - 5 \quad (2)$$

Use Substitution

$$4x^2 + 9\left(\frac{1}{3}x - 5\right)^2 = 36$$
$$4x^2 + 9\left(\frac{1}{9}x^2 - \frac{10}{3}x + 25\right) = 36$$
$$4x^2 + x^2 - 30x + 225 = 36$$
$$5x^2 - 30x + 189 = 0$$

Discriminant $b^2 - 4ac = -2880$
is negative, so there is no solution. $\varnothing$

59)
$$xy = 36$$
$$x + y = 13$$

Use substitution.
$$y = \frac{36}{x}$$
$$x\left(x + \frac{36}{x} = 13\right)$$
$$x^2 + 36 = 13x$$
$$x^2 - 13x + 36 = 0$$
$$(x-9)(x-4) = 0$$
$$x = 9 \ \text{ or } \ x = 4$$

$x = 9:$
$$xy = 36$$
$$9y = 36$$
$$y = 4$$

9 and 4

61) $x^2 + y^2 \le 4$

Center: $(0,0)$ Radius: 2

Test $(0,0)$ is true, shade inside the circle.

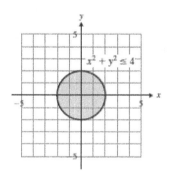

63) $\dfrac{x^2}{9}+\dfrac{y^2}{4}>1$

Center: $(0,0)$

$a=\sqrt{9}=3 \quad b=\sqrt{4}=2$

Test $(0,0)$ is not true, shade outside
the ellipse.

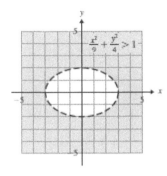

65) $x<-y^2+2y-12$

$y=\dfrac{-b}{2a}=\dfrac{-2}{2(-1)}=1$

$x=-(1)^2+2(1)-12=-11$

$V=(-11,1)$

Test $(0,0)$ is not true, shade inside
the parabola.

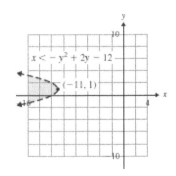

67) $y>x^2+2$

$x=\dfrac{-b}{2a}=\dfrac{0}{2(-1)}=0$

$x=(0)^2+2=2$

$V=(0,2)$

$y>-x+5$

$m=-1 \quad b=5$

Test $(0,6)$ is true, shade inside
the parabola and above the line.

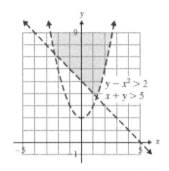

69) $\dfrac{x^2}{16}+\dfrac{y^2}{25}<1$

Center: $(0,0)$

$a=\sqrt{16}=4 \quad b=\sqrt{25}=5$

$y>x^2-3$

$x=\dfrac{-b}{2a}=\dfrac{0}{2(1)}=0$

$x=-(0)^2-3=-3$

$V=(0,-3)$

Test $(0,0)$ is true for both the circle and
the parabola, shade inside the circle and
the parabola.

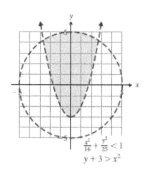

$$\frac{x^2}{16}+\frac{y^2}{25}<1$$
$$y+3>x^2$$

71) $y \geq 0$

$x^2 + y^2 \leq 36$

Center: $(0,0)$ Radius: 6

$x \leq -y^2$

$y = \dfrac{-b}{2a} = \dfrac{0}{2(-1)} = 0$

$x = -(0)^2 = 0$

$V = (0,0)$

Test $(-2,1)$ is true for the circle, the line,
and the parabola; shade inside the circle,
inside the parabola and above the line.

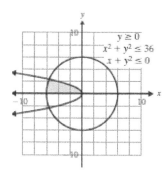

$y \geq 0$
$x^2 + y^2 \leq 36$
$x + y^2 \leq 0$

Chapter 12 Test

1) $g(x) = |x| + 2$

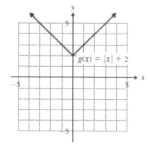

$g(x) = |x| + 2$

3) $f(x) = \begin{cases} x+3, & x > -1 \\ -2x-5, & x \leq -1 \end{cases}$

x	$f(x) = x+3$
0	$0+3 = 3$
1	$1+3 = 4$
2	$2+3 = 5$

x	$f(x) = -2x-5$
-2	$-2(-2)-5 = -1$
-3	$-2(-3)-5 = 1$
-4	$-2(-4)-5 = 3$

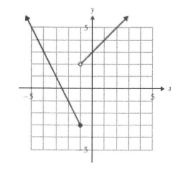

5) $y = -2x^2 + 6$; parabola

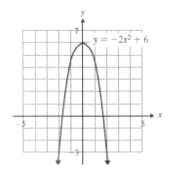

7) $x^2 + (y-1)^2 = 9$; circle

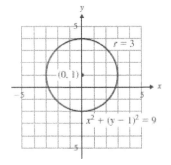

9) $\quad\quad\quad x^2 + y^2 + 2x - 6y = 6$

$$(x^2 + 2x + 1) + (y^2 - 6y + 9) = 6 + 1 + 9$$

$$(x+1)^2 + (y-3)^2 = 16$$

Center: $(-1,3)$ Radius: 4

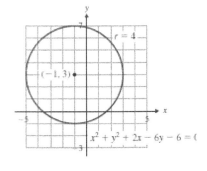

11) $\dfrac{x^2}{8836} + \dfrac{y^2}{6084} = 1$

13) Parabola and a circle.

a)

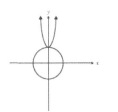 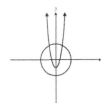

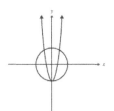 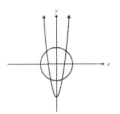

b)

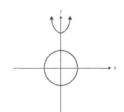

c) $0, 1, 2, 3,$ or 4

15) $2x^2 + 3y^2 = 21$

$2(-x^2 + 12y^2 = 3)$

$2x^2 + 3y^2 = 21$

$-2x^2 + 24y^2 = 6$

$27y^2 = 27$

$y^2 = 1$

$y = \pm 1$

$y = 1:\quad 2x^2 + 3y^2 = 21$

$2x^2 + 3(1)^2 = 21$

$2x^2 = 18$

$x^2 = 9$

$x = \pm 3$

$y = -1:\quad 2x^2 + 3y^2 = 21$

$2x^2 + 3(-1)^2 = 21$

$2x^2 = 18$

$x^2 = 9$

$x = \pm 3$

$\{(3,1),(3,-1),(-3,1),(-3,-1)\}$

17) $y \geq x^2 - 2$

$x = \dfrac{-b}{2a} = \dfrac{0}{2\cdot 1} = 0$

$y = 0^2 - 2 = -2$

$V = (0,-2)$

Test $(0,0)$ is true, shade inside the parabola.

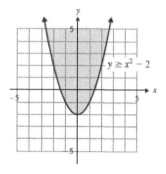

19) $x^2 + y^2 > 9$

Center: $(0,0)$; Radius: 3

$y < 2x - 1$

$m = 2 \quad b = -1$

Test $(4,0)$ is true for the circle and the line, shade outside the circle and below the line.

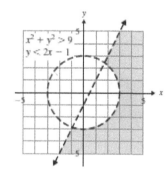

452

Cumulative Review: Chapters 1-12

1) $\dfrac{1}{6} - \dfrac{11}{12} = \dfrac{2}{12} - \dfrac{11}{12} = -\dfrac{9}{12} = -\dfrac{3}{4}$

3) $A = \dfrac{1}{2}bh = \dfrac{1}{2}(7.2)(5) = 18$ cm.

$P = 3s = 7.2 + 5 + 7 = 19.2$ cm.

5) $(-1)^5 = -1$

7) $\left(\dfrac{2a^8 b}{a^2 b^{-4}}\right)^{-3} = \left(2a^6 b b^4\right)^{-3} = \left(2a^6 b^5\right)^{-3}$

$= \dfrac{1}{\left(2a^6 b^5\right)^3} = \dfrac{1}{8a^{18} b^{15}}$

9) $an + z = c$

$an = c - z$

$n = \dfrac{c - z}{a}$

11) $x = 1^{st}$ integer

$x + 2 = 2^{nd}$ integer

$x + 3 = 3^{rd}$ integer

$x + x + 2 + x + 4 = 2(x + 4) + 13$

$3x + 6 = 2x + 8 + 13$

$3x + 6 = 2x + 21$

$3x - 2x = 21 - 6$

$x = 15$

$x = 15,\ 17,\ 19$

13) slope is 0

15) $(-4, 7)$ and $(4, 1)$

$m = \dfrac{y_2 - y_1}{x_2 - x_1} = \dfrac{1 - 7}{4 - (-4)} = \dfrac{-6}{8} = -\dfrac{3}{4}$

$y = mx + b$

$1 = -\dfrac{3}{4}(4) + b$

$1 = -3 + b$

$4 = b$

$y = -\dfrac{3}{4}x + 4$

17)

Percent	Amount	
.08	x	$.08x$
.16	y	$.16y$
.14	20	2.8

$x + y = 20$

$.08x + .16y = 2.8$

Use substitution method.

$y = -x + 20$

$.08x + .16(-x + 20) = 2.8$

$.08x - .16x + 3.2 = 2.8$

$-.08x = -.4$

$x = 5$

$x + y = 20$

$5 + y = 20$

$y = 15$

5 mL of 8% solution

15 mL of 16% solution

19) $(4w-3)(2w^2+9w-5)$

$8w^3+36w^2-20w-6w^2-27w+15$

$8w^3+30w^2-47w+15$

21) $6c^2-14c+8=2(3c^2-7c+4)$

$\qquad =2(3c-4)(c-1)$

23) $\qquad (x+1)(x+2)=2(x+7)+5x$

$\qquad x^2+3x+2=2x+14+5x$

$x^2+3x-7x+2-14=0$

$x^2-4x-12=0$

$(x-6)(x+2)=0$

$\qquad x=6 \text{ or } x=-2$

25) $\dfrac{\dfrac{t^2-9}{4}}{\dfrac{t-3}{24}}=\dfrac{24\cdot\dfrac{t^2-9}{4}}{24\cdot\dfrac{t-3}{24}}$

$\qquad =\dfrac{6\left(\cancel{t-3}\right)(t+3)}{\cancel{t-3}}$

$\qquad =6(t+3)$

27) $|5r+3|>12$

$5r+3<-12$

$5r<-15$

$r<-3$

$5r+3>12$

$5r>9$

$r>\dfrac{9}{5}$

$(-\infty,-3)\cup\left(\dfrac{9}{5},\infty\right)$

29) $\sqrt[3]{48}=\sqrt[3]{8\cdot6}=2\sqrt[3]{6}$

31) $(16)^{-\frac{3}{4}}=\dfrac{1}{(16)^{\frac{3}{4}}}=\dfrac{1}{\left(\sqrt[4]{16}\right)^3}=\dfrac{1}{8}$

33) $\dfrac{5}{\sqrt{3}+4}\cdot\dfrac{\sqrt{3}-4}{\sqrt{3}-4}=\dfrac{5\sqrt{3}-20}{3-16}$

$\qquad =-\dfrac{5\sqrt{3}-20}{13}$

$\qquad =\dfrac{20-5\sqrt{3}}{13}$

35) $y^2=7y-3$

$y^2-7y+3=0$

$a=1; b=-7; c=3$

$x=\dfrac{-(-7)\pm\sqrt{7^2-4(1)(3)}}{2}$

$x=\dfrac{7\pm\sqrt{37}}{2}$

$\left\{\dfrac{7+\sqrt{37}}{2},\dfrac{7-\sqrt{37}}{2}\right\}$

37) $\dfrac{t-3}{2t+5}>0;\quad t\neq-\dfrac{5}{2};\quad t-3=0$

$\qquad\qquad\qquad\qquad t=3$

test of 0 is not true, do not shade.

$\left(-\infty,-\dfrac{5}{2}\right)\cup(3,\infty)$

39) $f(x) = \sqrt{x}$ and $g(x) = \sqrt{x+3}$

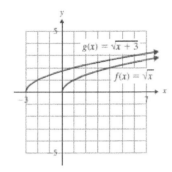

41) $f(x) = \{(-5,9),(-2,11),(3,14),(7,9)\}$

   a) no

   b) no

43) $8^{5t} = 4^{t-3}$

$$2^{3 \cdot 5t} = 4^{2(t-3)}$$

$$15t = 2t - 6$$

$$13t = -6$$

$$t = -\frac{6}{13}$$

45) $\log 100 = 2$

47) $e^{3k} = 8$

    $\ln e^{3k} = \ln 8$

    $3k = \ln 8$

    $k = \dfrac{\ln 8}{3}$

    $\left\{\dfrac{\ln 8}{3}\right\}; \{0.6931\}$

49) $x^2 + y^2 - 2x + 6y - 6 = 0$

$$(x^2 - 2x + 1) + (y^2 + 6y + 9) = 6 + 1 + 9$$

$$\frac{(x-1)^2}{16} + \frac{(y+3)^2}{16} = 16$$

Center: $(1,-3)$ Radius: 4

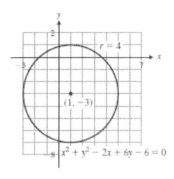

NOTES

NOTES

NOTES

NOTES

NOTES